Hillerød, maj 1997

Kære John
Ferskvandsbiologisk Laboratorium
siger tak for godt samarbejde
Venlig hilsen
hg Ole

Tak for hjælp med de sproglige rettelser!

Kaj Sand-Jensen and Ole Pedersen, editors

Freshwater Biology

Priorities and Development
in Danish Research

*On the one hundredth anniversary of the
Danish Freshwater Biological Laboratory*

G·E·C Gad, Copenhagen 1997

Freshwater Biology
Priorities and Development in Danish Research

© 1997 The Freshwater Biological Laboratory,
University of Copenhagen and
G.E.C. Gad Publishers Ltd., Vimmelskaftet 32,
DK-1161 Copenhagen Denmark

1. edition 1997

Cover photo: Bert Wiklund
Cover design: Axel Surland
PrePress and printing: Narayana Press, Gylling
Project editor Lars Boesgaard

ISBN 87-12-03135-6

Contents

Preface

This book has been written by scientists from the Freshwater Biological Laboratory, University of Copenhagen, to mark its one hundredth anniversary. In May 1897, Carl Wesenberg-Lund established a small field laboratory on the banks of Lake Fure, North of Copenhagen. Later, in 1911, Wesenberg-Lund moved the laboratory to Hillerød and it soon became an integrated part of Copenhagen University.

Although the one hundredth anniversary of the Freshwater Biological Laboratory is the stimulus for writing this book, and we have briefly described the historical and scientific perspectives provided by the 100-years of research in Chapter 1 and 2, the book's main emphasis is on the ecology of the types of lakes, streams and coastal waters which are typical of Denmark. The research priorities and themes of the Freshwater Biological Laboratory over many years provide the focus for the topics in this book. The long history of research has allowed long-term changes in lakes and brackish waters as a result of anthropogenic influences to be assessed, and this is discussed in Chapter 3, 4, 7 and 8. The ecology and ecophysiology of macrophytes has been a recurrent theme in the research of the laboratory: the key-role of macrophytes in the ecology of streams and shallow estuaries is discussed in Chapter 6 and 7 and their ecophysiology in Chapter 12. Studies on invertebrate communities at this laboratory originated in the pioneering work of Wesenberg-Lund, and the tradition is continued here with a discussion of the factors responsible for regulating their communities in lakes and streams in Chapter 4 and 13 and their ecophysiology in the profundal zone in lakes in Chapter 11. The recent interests in the dynamics of microbial food webs and resource and predator regulation of macrozooplankton in lakes are discussed in Chapter 9 and 10.

We have tried to write in a non-technical and accessible style and hope that the result will be readable and useful, not only to freshwater ecologists, but also to a wide spectrum of students, biologists and ecologists in many countries.

We are grateful to the Carlsberg Foundation for financial support to the publication of this book. We thank John Anderson, Gary Banta, Carolyn W. Burns, Carlos Duarte, Erik Jeppesen, Louis Kaplan, Stephen Maberly, Tom V. Madsen, Robert Moeller, Bo Riemann, Melbourne C. Whiteside, Michael J. Winterbourn, Johnstone O. Young, Justin Healy and Anna Garde for reviews and linguistic help. We are also grateful to Charlotte Andersen, Karen Kristensen, Anne Lise Middelboe, Jon Theil-Nielsen, Ena Poulsen, Finn Pedersen and Jacob Worm for help with illustrations and secretarial assistance.

1 The origin of the Freshwater Biological Laboratory

By Kaj Sand-Jensen

One day in the early summer of 1897 you could see a strange procession of horses, wagons and people moving down the steep hill from the small village of Virum towards the banks of Lake Fure. The procession was carrying the materials for the small wooden laboratory which was to become the first Freshwater Biological Laboratory in Denmark.

The words are Carl Wesenberg-Lund's in his description, fortythree years later, of the modest start of the Freshwater Biological Laboratory (Wesenberg-Lund 1940). In 1896, Wesenberg-Lund had participated in the Ingolf Expedition to the North Atlantic. Upon his return, it occurred to him that the small special house that had been installed upon the deck of the schooner Ingolf to serve as a laboratory for the participating scientists could be of future use as a small freshwater field laboratory (Fig. 1.1). Influential people supported his idea. Count Schulin, who owned the Frederiksdal Estate located at the outlet from Lake Fure, offered a piece of land on the banks of the lake. The Carlsberg Foundation provided the financial support for rebuilding the house, and later – together with the Danish State – granted the necessary money for a rowing boat, a motor boat and a landing stage.

A very active early period of freshwater research followed, usually on the initiative of Wesenberg-Lund, but with the engagement of many new students and the cooperation of university specialists from other disciplines

Figure 1.1. The first Freshwater Biological Laboratory in Denmark was established by C. Wesenberg-Lund on the banks of Lake Fure in 1897. The small wooden laboratory was originally built as a laboratory for the scientists on the Ingolf Expedition to the North Atlantic in 1895-1896.

such as chemistry, geology and physiology. All of them were apparently attracted by this exciting new environment that Wesenberg-Lund talked and wrote so inspiringly about. Most of the studies took place in Lake Fure and the numerous ponds in the vicinity, but other lakes distributed around the country were soon included.

Why a freshwater laboratory and why Wesenberg-Lund?

The two questions: why initiate a freshwater laboratory? and why was Wesenberg-Lund the founder? are of course inseparable (Fig. 1.2). Wesenberg-Lund was born in 1867 and had spent much of his time as a schoolboy in the woods surrounding Hillerød, where he had explored the numerous ponds and observed the life of the animals they contained. So he really had an affection for freshwater biology which had been marked in his mind from his very early days. This interest was not generated by reading other people's work; the passion was already there when he started studying at Copenhagen University.

Throughout his life he continued to build on his own perceptions by observing the organisms week after week, year after year, out in the natural environment, before he eventually sat down to write the overall story. It was the natural history of the organisms and their behaviour and adaptation to the environment that attracted him. This perspective was new in Danish university circles which were strongly dominated by studies of morphology, anatomy and taxonomy of skeletons, dried plants and organisms in formaline.

We often tend to think, with hindsight, that historical developments are predetermined to reach the final result, let it be the rise and the fall of the Roman Empire, the formulation of the mechanism for biological evolution by Charles Darwin or, in more modest terms, the foundation of the Danish Freshwater Biological Laboratory. Such a predetermined development is of course false, because numerous factors and coincidences interact and combine to allow a certain development to take place. The foundation of the Freshwater Biological Laboratory was also caused by a number of fortunate circumstances: Wesenberg-Lund was passionate about the idea, influential people and the Carlsberg Foudation supported him, and he suddenly realized the practical

Figure 1.2. C. Wesenberg-Lund, the founder of the Danish Freshwater Biological Laboratory. The photograph was taken in the 1930's outside the laboratory in Hillerød with the Frederiksborg Castle in the background.

opportunity. Moreover, many students and colleagues shared his interest in this "new" exciting natural environment.

Perhaps an additional aspect has helped the Freshwater Biological Laboratory in the first years and certainly will do so in the years to come. Wesenberg-Lund was a well-known figure, not only among biologists, but also in wider academic and public circles. He was an inspired author of scientific publications and books in French, English and German, but he was also an excellent writer in Danish about the natural environment and the biology of insects and birds. He devoted a large part of his time to writing articles about biology and nature for popular magazines and newspapers and, later in his life, to broadcasting on radio. This gave him an opportunity to repay his debt to the society which gave him the optimal opportunities to study the freshwater environments and the animals, which were his lifelong passion. I am certain that this deep commitment impressed the authorities in the funding agencies, at the university as well as in the ministries.

Biology in Denmark 100 years ago

The biological community in Denmark in the late 1800's was influenced not only by the scientists at the university. Equally important were the studies performed at the Carlsberg Laboratory and the Royal Veterinary High School. The use of chemical methods in biology and the new disciplines of genetics and plant physiology were first introduced in Denmark at these two institutions. In the case of animal physiology, the inspiration and techniques came from the Medical Faculty with August Krogh (Nobel prize in 1920), who was essential to the progress of Danish research in physiology and ecology.

The new disciplines had an immediate impact on the freshwater biological studies. Chemical and physical methods were first applied in Danish freshwater biology by the physicist J. N. Brønsted who made the first measurements of inorganic ions and dissolved oxygen in Danish lakes in 1905-1906 in cooperation with Wesenberg-Lund (Brønsted and Wesenberg-Lund 1911-1912). The increased interest in animal physiology was reflected by the very talented studies by A. Krogh on the regulatory function of the gas bladders in the phantom midge, *Corethra flavicans* (Krogh 1911) and by R. Ege on the respiratory function of the air stores of Corixidae and Dytiscidae and on the respiration of *Donacia* larvae living on the roots of swamp plants rooted in anoxic sediments (Ege 1915).

In many respects, however, zoology in Denmark was very conservative in the late 1800's and slow to assimilate the new approaches in chemistry, genetics and physiology (Wolff 1979). Moreover the biological fields lacked the new perspectives that expeditions to tropical regions might have brought back, as it happened in England with the return of Charles Darwin and Alfred Wallace. The publication of *The Origin of Species* in 1859 was barely noticed in Denmark, though Darwin sent his book as a gift to the Danish professor of zoology, J. Steenstrup (Robson 1985). The debate on Darwinism did not start in Denmark until 15 years later, introduced by young students following the translation of Darwin's books. The Danish university professors did not appear very interested in the general questions on the origin of species and the mechanisms of natural evolution.

So if the main interests of Danish biology in the 1880's and 1890's were not the fundamental question of species evolution, and if most scientists did not undertake long expeditions to remote continents to collect new species, which type of original studies did they pursue, and how were new important scientific approaches, nonetheless, initiated? One reason, at least, was that some of the scientists instead were deeply engaged in the studies of the natural Danish habitats. Wesenberg-Lund's life-project was to study the freshwater organisms in lakes and ponds and their interactions with the natural physical-chemical environment. In this approach, he differed from the prevailing zoological traditions in Copenhagen, and it was probably no coincidence that he avoided these circles and instead cooperated with other scientists in botany, geology and physiology in the laboratory at Lake Fure and later in Hillerød. His old mentor, the botanist Eugenius Warming, had a similar idea of studying the plants, the life forms and the vegetation processes in the natural terrestrial habitats of Denmark, forming the early foundation of plant ecology (Warming 1895), in between his other outstanding international contributions to plant taxonomy and morphology. Outside the university circles, the marine zoologist C. G. Joh. Petersen was important to the progress of marine ecology and fishery sciences. He made several outstanding contributions to the quantitative studies of animal communities on the sea bottom, the techniques of measuring fish growth and migration, and he initiated the study of quantitative production ecology by showing the importance of the detritus food chain in the nutrition of invertebrates and fish (Spärck 1960).

The three scientists, Wesenberg-Lund, Warming and Petersen, therefore, mark an early start of ecology with a background in natural history, but supported by repeated and careful field observations and often followed by some type of general classification. C. G. Joh. Petersen, inspired by the German plankton scientist V. Hensen, clearly advanced the ecological approach by counting and weighing the marine animals and plants and computing their biomass and production (Petersen and Boysen-Jensen 1911, Petersen 1913). Such a quantitative ecological approach was never taken by Wesenberg-Lund and did not appear in a comprehensive manner in Danish freshwater science until K. Berg's studies of macroinvertebrates in Lake Esrom (Berg 1938).

Interestingly, the focus on production ecology and energy conversion generated by C. G. Joh. Petersen and P. Boysen-Jensen had a lasting impact on Danish aquatic science with the later introduction of the ^{14}C-method for measuring oceanic primary production (Steemann-Nielsen 1952) and the comparative studies of planktonic oxygen production and respiration in lakes (Nygaard 1955). The historical heritage is probably felt more strongly in a small country like Denmark with only one historical centre for freshwater research and one for marine research.

The first 20 years at Lake Fure

Right from the start, the studies carried out at the Freshwater Laboratory reflected a strong interest in the biology of the organisms in relation to environmental conditions. These early studies also reflected an awareness of the advantages which combinations of physico-chemical, botanical and zoological approaches could bring to freshwater science. The interdisciplinary approach differed from the separate directions which the botanical and zoological sciences had taken at the main institutes in Copenhagen and other universities.

These holistic ideas were not uncommon among freshwater scientists 100 years ago. They were apparent already in Forel's first limnological work on Lake Geneva in Switzerland in the 1870's, J. Murray's studies in Scotland, and in D. Birge's and C. Juday's studies in Wisconsin, just to mention a few active limnologists around the turn of the century. So it is certainly no coincidence that systems ecology and theories of food web interactions have had such a strong foundation in limnology (Forbes 1889, Lindemann 1942).

The holistic approach is well illustrated by the programme of work undertaken by Wesenberg-Lund during the early years of the Laboratory up to the publication of the extensive monograph on Lake Fure in 1917 (Wesenberg-Lund et al. 1917). When Wesenberg-Lund started the Freshwater Laboratory in 1897, he had already worked intensively on the morphology and biology of rotifers. He considered, however, that this topic was too specialized to stimulate the work of the new laboratory. Instead, he initiated the studies of bottom deposits in the Danish lakes, with help from geologists (Wesenberg-Lund 1901), and the better known studies on the lake plankton (Wesenberg-Lund 1904, 1908). The plankton studies on Lake Fure started in 1898, only a year after the Freshwater Biological Laboratory was founded. From the beginning, Wesenberg-Lund realized that if he was to contribute substantially to the knowledge within this field, which was growing rapidly in mid-Europe, he had to take a comprehensive approach. His decided to sample the plankton in 14 ponds and 9 lakes frequently for a full year and to include as many lake types, distributed around the country, as possible. He combined the biological investigations with measurements of temperature and transparency and postponed the chemical analysis until later. The logistic difficulties of this comprehensive comparative approach were great, but he had extensive financial support from the Carlsberg Foundation, specialist assistance for studying the phytoplankton and local help in taking the samples.

Wesenberg-Lund published the plankton investigations of Danish lakes and ponds in two books in 1904 and 1908. One of the main topics of the books was the statistical analysis of the temporal variation of the size of several zooplankton species (e.g. *Daphnia cucullata* and *Bosmina coregoni*, Fig. 1.3). The temporal size variation was mainly discussed in relation to changes in temperature and viscosity and the ability of the organisms to remain suspended in the water column. The books contained many interesting observations on the biology of planktonic organisms. One observation has attracted my attention because of a personal interest in freshwater symbionts. Wesenberg-Lund (1909) observed

the ciliate *Stentor* with numerous zoochlorellae dominating in the periphyton of a small pond in summer and in the pelagic waters in autumn, while chlorellae, supposedly released from the ciliates, dominated in the pond water in winter. He proposed an interesting life cycle and the existence of enclosed and free-living zoochlorellae, which have not been further examined since then.

The monograph on Lake Fure is based on field studies undertaken from 1911 to 1914 and describes the bathymetry and the spatial distribution of macrophytes and molluscs (Wesenberg-Lund et al. 1917). The botanical work by J. Boye Petersen and A. S. Raunkiær is particularly noteworthy because it is a semiquantitative study of the depth distribution of submerged macrophytes along several

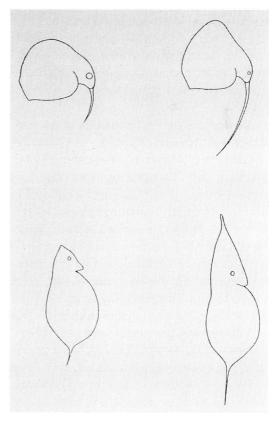

Figure 1.3. Seasonal variations in the size and form of *Bosmina coregoni* and *Daphnia cucullata* during winter and summer. From Wesenberg-Lund (1908).

transects, and it offers detailed maps of the distribution of emergent vegetation along protected and exposed coasts (reviewed by Hutchinson 1975). The semiquantitative approach is important because it allows reconstruction of the historical development of the vegetation in Lake Fure by comparison with more recent studies (Sand-Jensen 1996).

The monograph on Lake Fure was the first ecosystem-oriented study in Danish limnology, but it marked the final retreat of Wesenberg-Lund from Lake Fure and the transfer of the Freshwater Laboratory to Hillerød. The main reason for abandoning Lake Fure was the spread of private villas along the banks and the loss of the tranquillity (Wesenberg-Lund 1940). For a very extended period, until his retirement in 1940 and his death in 1955, Wesenberg-Lund concentrated his studies at the Freshwater Biological Laboratory in Hillerød on the biology of freshwater animals. These studies resulted in numerous publications and comprehensive overviews in several books on the biology of the freshwater fauna (Wesenberg-Lund 1937, 1939).

Contributions of early limnology to contemporary science

There is no reason to overstate nor neglect the inspiration that the early limnological studies have had on contemporary science. Most of the early limnological studies are outdated, use a verbose form and lack clearly formulated hypotheses. Wesenberg-Lund's publications are, nonetheless, enjoyable reading because his drawings are artistic (Fig. 1.4), and his language is elegant and full of opinions; though some of them reveal misunderstandings due to an insufficient foundation in physics. The old observations of the biology and behaviour of freshwater organisms can also form a platform for future more specific investigations. There is, therefore, every reason to read, not only Wesenberg-Lund, but also other old literature, to pay tribute to those people who originally

formulated the ideas instead of following the unfortunate practice of citing some later minor modifications to the original findings. It is also possible to come across a few splendid ideas for future work. From the recent literature I recall two such examples of highly original studies, first presented by letters in Science, which clearly acknowledge a background in very old studies (Dacey 1980, Bird and Kalff 1986).

Another obvious reason for consulting the old literature is the possibility of reconstructing the historical development of environmental and biological conditions in lakes and streams. This comparative possibility has been used in the case of Lake Fure and Lake Esrom, and the approach is applicable to several lakes, streams and coastal waters examined around the turn of the century in Denmark as well as in other countries. The old scientists were usually careful to note their observations, and the notebooks are sometimes kept in museums, available for closer inspection to supplement the data presented in the original publications.

Scientists 100 years ago had a different scientific approach than is common today. Their studies commonly lacked clear testable hypotheses and would have a very difficult time gaining acceptance with the present

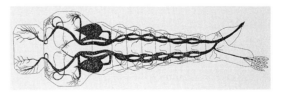

Figure 1.4. The highly specialized mosquito larva *Mansonia ricardii* lives on the surface of submerged higher plants. A stiletto on the abdomen is deeply buried into the stem tissue connecting the tracheae of the insect to the lacunal air system of the plant to ensure larval respiration. Wesenberg-Lund (1918) first discovered the rare and, then, unidentified larva in a small pond. He revisited the pond on numerous occasions for several years to attract the adults to sting him until he finally caught one and identified the species.

funding agencies and scientific journals. On the other hand, there were perhaps valuable elements in the old, mainly descriptive, approaches which are worthwhile considering? In every time period, scientists tend to believe that they have reached the optimal approach. I guess the earlier workers, if they had the opportunity to review contemporary freshwater science, would consider some modern approaches unfit to assimilate sudden unexpected and perhaps highly original findings. Certainly, some of the recent experimental approaches and field manipulations, covering limited spatial and temporal scales, reach conclusions which are unlikely to encompass the general ecological regulations. In that respect, wide-scale comparisons of ecosystems, based on descriptive work as it used to be practised, are likely to reach safer general conclusions. It is perhaps appropriate to point out that the main reasons for the success of freshwater sciences are the holistic studies on ecosystems, food webs, and anthropogenic influences; coupled with an interest among freshwater scientists in incorporating their studies of finer details in a general body of theory. This final remark is important, considering the increasing fragmentation within ecological sciences which restricts the use of ecological knowledge in environmental planning in the attempt to reach a sustainable future development.

Literature cited

Bird, D. F., and J. Kalff. 1986. Bacterial grazing by planktonic lake algae. Science 231: 493-495.

Dacey, J. W. H. 1980. Internal winds in water lillies: an adaptation for life in anaerobic sediments. Science 210: 1017-1019.

Forbes, S. A. 1887. The lake as a microcosm. Bulletin of the Peoria Scientific Association. Reprinted in Bulletin of the Illinois State Natural History Survey 15 (1925): 537-550.

Hutchinson, G. E. 1975. A treatise on limnology. Volume III. Limnological Botany. Wiley, New York, USA.

Krogh, A. 1911. On the hydrostatic pressure of the *Corethra* larva with an account of methods of microscopical gas analyses. Scandinavische Archiv für Physiologie 25: 183-203.

Lindeman, R. A. 1942. The trophic dynamic aspect of ecology. Ecology 23: 399-418.

Petersen, C. G. J. 1913. Havets bonitering II. Om havbundens dyresamfund og om disses betydning for den marine zoogeografi. Report Danish Biological Station 21: 1-42.

Petersen, C. G. J., and P. Boysen-Jensen. 1911. Valuation of the sea. I. Animal life of the sea bottom, its food and quantity. Report Danish Biological Station 20: 3-81.

Robson, M. 1985. Darwinismens modtagelse i Norden. Pages 365-282 (Kapitel 14) *in* N. Bonde and H. Stangerup, editors. Naturens historiefortællere I. Gad, Copenhagen, Denmark.

Sand-Jensen, K. 1997. Eutrophication and plant communities in Lake Fure during 100 years. Chapter 3 *in* K. Sand-Jensen and O. Pedersen, editors. Freshwater Biology. Priorities and Development in Danish Research. Gad, Copenhagen, Denmark.

Spärck, R. 1960. Ved 100-Aarsdagen for C. G. Joh. Petersens fødsel. Naturens Verden 321-328.

Steemann Nielsen, E. 1952. The use of radioactive carbon (C^{14}) for measuring organic production in the sea. Journal du Conseil International pour l'exploration de la Mers 18: 117-140.

Warming, E. 1895. Plantesamfund. Copenhagen, Denmark. German translation in 1896.

Wolff, T. 1979. Zoologi. Universitet 1479-1979, bind XIII, Pages 1-162 *in* T. Wolff, editor. Københavns Universitet 1479-1979, bind XIII, Det matematisk-naturvidenskabelige Fakultet, 2. del., Copenhagen, Denmark.

A complete list of publications from the Freshwater Biological Laboratory is found in Chapter 14 of this book.

2 The Freshwater Biological Laboratory in Hillerød – History and scientific priorities

By Kaj Sand-Jensen and Peter C. Dall

The research at the laboratory in Hillerød from 1911 to 1965 focused on the biology and ecology of animals in lakes. After 1965, the research staff expanded from a few to ten permanent scientists whose combined expertise covered chemistry, botany, microbiology and zoology. Consequently, more recent research has comprised a variety of ecological studies of entire ecosystems, selected biological communities as well as eco-physiological studies of important species. There was a gradual expansion of the research to include lakes, streams and brackish waters and a growing interest in environmental and biological interactions. The common ecological background provided by studies of basic properties of aquatic ecosystems has ensured a close coupling of research and teaching, and has permitted an informed scientific contribution to the public debate on the management of aquatic environments.

The Freshwater Laboratory from 1911 to 1965

In 1911, C. Wesenberg-Lund moved the Freshwater Biological Laboratory from Lake Fure to a 3-room flat in Hillerød, close to Frederiksborg Castle Lake. He soon expanded and received funds for a permanent research assistant. A few years later, the laboratory acquired an outdoor facility in Hillerød with artificial ponds for biological experiments, and a field station at Lake Tystrup in South Zealand.

Until his death in 1955, Wesenberg-Lund studied numerous aspects of the biology of small freshwater organisms in lakes and ponds. He was among the earliest workers who studied the cyclomorphosis of plankton organisms in relation to seasonal changes in temperature and viscosity. He also studied life cycles and morphological adaptations of rotifers and aquatic insects (Wesenberg-Lund 1915, 1923, 1930). His early work had a practical application as well. He studied the distribution and behaviour of mosquitoes in Denmark, including *Anopheles maculipennis,* which had caused serious outbreaks of malaria in the late 1800's (Wesenberg-Lund 1920, 1921). He also made detailed studies on the life cycles of parasitic trematodes which have aquatic molluscs as hosts for some of their infectious stages (Wesenberg-Lund 1931, 1934). Later these studies inspired extensive Danish work on the bilharzia disease. Many of Wesenberg-Lund's investigations were presented in voluminous books (Wesenberg-Lund 1937, 1939, 1944). They are still a pleasure to read because of the elegant writing, the wealth of information and the artistic illustrations.

In 1925, K. Berg came to the laboratory as

a research assistant. He managed to develop an independent research programme, and he later succeeded Wesenberg-Lund as professor from 1939 to 1969. Berg's early interest was in the distribution and biology of Danish cladocerans in ponds and lakes (Berg 1929, 1931) and the cyclic reproduction and changes between parthenogenetic and gammogenetic generations (Berg 1934, 1937). Among his contributions in this field was the confirmation that planktonic cladocerans can be induced to reproduce sexually and to produce resting egg stages (ephippia) even in populations that solely reproduce asexually in nature. Berg's next major project was the study of environmental conditions and the quantitative composition of macroinvertebrate communities in Lake Esrom (Berg 1938). This study demonstrated the marked differences between the species-rich invertebrate communities in the littoral and the species-poor community in the profundal mud. The investigation formed a baseline for later studies and has allowed us to evaluate possible changes in the lake benthos from 1933 to the present (Lindegaard et al. 1997).

The facilities for freshwater biological studies were greatly improved in 1930 when the Carlsberg Foundation purchased a villa on the shore of Fredriksborg Castle Lake and equipped it for quantitative field studies and experimental work. Berg used the improved facilities as a platform for several interdisciplinary studies. The River Susaa project (Berg 1943, 1948) was the first major investigation of a lowland stream. The study of the humic Lake Gribsø (Berg and Clemens-Petersen 1956) is noteworthy for the clarification of the special flora and fauna associated with this lake type. Lake Gribsø was one of the first lakes in which the postglacial development was based on fossil diatoms in the sediment (Nygaard and Wolthers 1956). The study of sediment diatoms later inspired international evaluations of changes in lake pH following anthropogenic acidification.

Berg (1958) also edited the second major investigation of Lake Fure which examined the changes in environmental conditions and communities of plants and animals following eutrophication since the turn of the century. This study documented the changes in the composition of phytoplankton (Nygaard 1958) and the distribution of submerged macrophytes (Christensen and Andersen 1958). The international audience has particularly noted the first studies of sediment phosphorus kinetics with the application of radioactive ^{32}P (Olsen 1958).

Berg resumed the interst on respiratory physiology of aquatic animals in the original Danish tradition of August Krogh. Between 1951 to 1965, he and several colleagues examined the respiration of freshwater invertebrates in relation to season, habitat and temperature conditions (e.g. Berg 1951, Berg et al. 1958, 1962, Berg and Jónasson 1965). The respiratory kinetics were linked to essential adaptive features and environmental conditions in different stream and lake habitats. The investigations also evaluated the influence of feeding status and body size on respiratory rates (Hemmingsen 1960). An important result was the demonstration that respiratory rates of species from the littoral of lakes decline with reduced oxygen concentrations, whereas species from the profundal zone are able to regulate respiration and maintain unaltered rates down to the very low oxygen levels which they regularly experience during summer stratification (Berg et al. 1962, Berg and Jónasson 1965).

Until the late 1960's, the main freshwater research was performed by relatively few persons, most of them with a zoological background (Fig. 2.1). A. Nielsen (1950, 1951) was particularly interested in the zoogeography and speciation of the fauna in the Danish springs, and he evaluated the morphological adaptations of the torrential invertebrate fauna. J. Thorup (1963, 1966) studied the life cycles and habitat requirements of

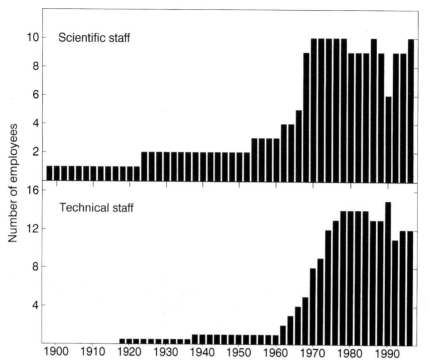

Figure 2.1. The size of the permanent scientific (above) and technical staff (below) during the 100 years existence of the Freshwater Biological Laboratory.

stream invertebrates. By 1953, P. M. Jónasson initiated the long-term studies on the population dynamics of the profundal fauna in Lake Esrom, showing the strong influence of temperature, oxygen and sedimenting phytoplankton for the growth and life cycle of the abundant chironomids (Jónasson 1961, 1964).

However, a combined chemical-botanical approach was also evident in the freshwater biological studies by G. Nygaard, E. Steemann Nielsen and S. Olsen. Nygaard developed a well-known numerical index (Compound quotient) to characterize the trophic status of lakes based on the number of phytoplankton species within taxonomic groups typical of eutrophic versus oligotrophic waters (Nygaard 1949). Later studies showed an increasing concern for the quality of lake ecosystems. Evaluations of the changes in phytoplankton biomass and productivity with the continued eutrophication were apparent in several studies (Nygaard 1955, Johnsen et al. 1962, Kristiansen and

Mathiesen 1965, Jónasson and Kristiansen 1967). While these studies focused on phytoplankton communities, others looked at aquatic macrophytes. The distinct differences in the distribution of submerged elodeid angiosperms between softwater and hardwater localities, already described by Iversen (1929), finally got a more firm explanation by Steemann Nielsen's experimental studies (1944, 1947) of carbon dioxide and bicarbonate use by different macrophyte species and growth forms. S. Olsen (1950, 1964) expanded the relationship between water chemistry and the distribution of macrophytic plants to include streams and lakes.

The Freshwater Biological Laboratory from 1965 to 1997

The permanent research staff at the laboratory increased markedly to about ten positions during the late 1960's and early 1970's along with an expansion of the technical staff (Fig. 2.1). At this time the available space for laboratories and offices increased to four

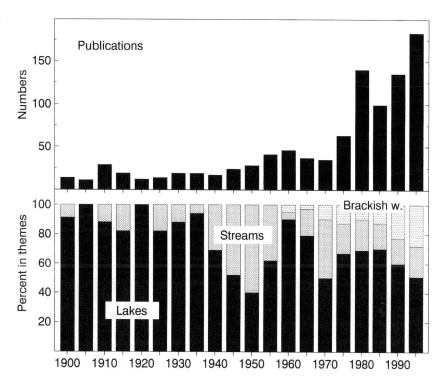

Figure 2.2. Number of publications (above) and their percentage distribution between studies in lakes, streams and brackish waters (below) for 5 years periods during the 100 years' existence of the Freshwater Biological Laboratory.

large villas in Hillerød. Finally, a well-equipped laboratory, located in the Danish lake district in mid-Jutland, became available for research and field courses.

The new research positions were designed to ensure a combination of chemical, micro-biological, botanical and zoological expertise in the scientific staff. A broader representation allowed for a more comprehensive coverage of the ecological disciplines both in teaching and science. Subsequent develop-ment has shown that the interdisciplinary influence has been particularly prominent in the university courses offered in freshwater and aquatic ecology as well as in the scien-tific attitude of the research staff. The com-bined expertise was reflected in several interdisciplinary studies of plant-animal interactions, analyses of stream ecosystems, and evaluations of resource and predator re-gulation in lakes. However, a combined effort of all researchers directed at just one fresh-water locality, or one broad theme, has never taken place. Instead, much of the cooperative

effort has been directed towards other na-tional and international aquatic ecologists with similar interests. Thus, to characterize the research interests and priorities during the last 40 years, it is fair to say that many research themes have concurrently been re-presented, including ecosystem studies of lakes, streams and brackish waters as well as more detailed experimental studies of selected communities and specific organisms (Fig. 2.2).

Lake studies – Studies on individual lakes have included detailed analysis of alkaline, mesotrophic Lake Esrom and softwater, oligo-trophic Lake Kalgaard and Grane Langsø as well as analysis of the subarctic Lake Mývatn and Lake Thingvallavatn in Iceland.

P. M. Jónasson was professor at the labora-tory from 1977 to 1990. He continued studies of the coupling between phytoplankton production and population dynamics of the profundal zoobenthos in Lake Esrom (Jónas-son 1972). These investigations showed how the temporal nutrient limitation of the phyto-plankton, causing a bimodal annual production and sedimentation, and the temporal changes in oxygen availability influence the life cycles of the profundal benthos. The change in growth and voltinism of the chironomid *Chironomus anthracinus* in various depth zones emphasized the critical conditions in the hypolimnion of the eutrophic lake (Jónasson et al. 1974, Lastein 1976, Jónasson 1996). Other permanent inhabitants in the profundal zone, such as the oligochaete *Potamothrix hammoniensis* and the bivalve *Pisidium* spp., can also sustain the critical life conditions in the hypolimnion (Thorhauge 1976, Holopainen and Jónasson 1983, 1989), but apply different survival strategies including dormancy. The populations of *C. anthracinus* have been maintained from 1932 to 1997, despite the increased oxygen depletion in the hypolimnion, while populations of *P. hammoniensis* have increased (Lindegaard et al. 1997). Evaluation of the adaptation and survival strategies of chironomids and oligochaetes has demonstrated the importance of glycogen as the main carbon reserve for anoxic survival during summer (Hamburger et al. 1997). Critical conditions for summer survival have been related to body size and short-time intrusion of oxygen by downward mixing.

Another major research theme in Lake Esrom has been the population dynamics of zooplankton (Bosselmann 1974, 1975, 1979) and the regulation of life history, production and respiration of littoral macroinvertebrates (Dall et al. 1984). The high species richness and the variable composition of macroinvertebrates in the littoral zone can, in part, be explained by differences in physical exposure and growth of periphyton (Dall et al. 1984, Brodersen 1995). The littoral leech fauna includes ten species, of which six are quantitatively important (Dall 1987). Coexistence among species is facilitated by spatial seperation and partitioning of food resources (Dall

1983). Quantitative studies of most of the faunal groups have permitted an overall evaluation of the role of macroinvertebrates in the energy budget of the lake (Lindegaard and Dall 1988, Lindegaard et al. 1990, 1994, Dall and Hamburger 1996).

While phytoplankton is the main primary producer in the relatively large and deep Lake Esrom, submerged macrophytes and phytoplankton contribute about equally to primary production in the oligotrophic, soft-water Lake Kalgaard and Lake Grane Langsø (Søndergaard and Sand-Jensen 1978, Nygaard and Sand-Jensen 1981). One of the main features of these *Lobelia*-lakes is the carpet-like cover of small, perennial, and slow-growing isoetid species in shallow water (Sand-Jensen and Søndergaard 1978, 1979). The deepest parts of the lake bottom in Grane Langsø are covered by characeans and mosses (Nygaard 1958). The submerged macrophytes have, therefore, a key-role in the carbon, nutrient and oxygen dynamics of the lakes (Sand-Jensen and Søndergaard, 1997).

Eco-physiological studies have revealed remarkable plant-sediment interactions for the isoetid species covering the nutrient-poor sandy sediments. The use of sediment-CO_2 for photosynthesis was first shown in the plant kingdom in Wium-Andersen's studies (1971) of *Lobelia dortmanna*. Later laboratory and field experiments confirmed that the majority of CO_2 was derived from the sediment rather than the lake water in *L. dortmanna* as well as in *Littorella uniflora* and *Isoetes lacustris* (Søndergaard and Sand-Jensen 1979, Sand-Jensen 1987). *L. dortmanna* and *L. uniflora* are amphibious species which retain the sediment as the main CO_2 source on land, though *L. uniflora* forms new and thinner leaves with numerous stomata to facilitate use of atmospheric CO_2, whereas *L. dortmanna* maintains the same leaf form with few stomata (Nielsen et al. 1991, Pedersen and Sand-Jensen 1992). *L. dortmanna* also releases the majority of oxygen produced during photosynthesis via the roots (Sand-Jensen and Prahl 1982). Oxygen and CO_2 pools, therefore, follow complementary rhythmic pulses during day and night on sandy sediments inhabited by *L. dortmanna* (Pedersen et al. 1995). Lake sediments are normally anoxic just a few mm below the sediment surface, but sandy sediments covered by isoetids can have extensive oxic zones influencing redox potentials, nitrification-denitrification, distribution of reduced/oxidized iron and manganese, and availability of phosphorus (Wium-Andersen and Andersen 1972, Christensen et al. 1995, Sand-Jensen and Søndergaard 1997). The isoetid vegetation, therefore, influences the biogeochemical processes in the entire lake.

Inter-Nordic cooperation has permitted ecosystem analyses of Lake Mývatn and Lake Thingvallavatn in Iceland (Jónasson 1979, 1992). The Danish studies of the zoobenthos have shown that the shallow, productive Lake Mývatn had a high abundance (max. 230,000 m⁻²) and production of chironomids, especially *Tanytarsus gracilentus* and *Chironomus islandicus,* which were the basic food source for vast populations of ducks (Lindegaard and Jónasson 1979). The deep, oligotrophic Lake Thingvallavatn housed a diverse bottom fauna (about 60 taxa) of low abundances (3,500-21,000 m⁻², Lindegaard 1992). Thingvallavatn is the only known lake with four different morphs of arctic charr demonstrating a unique case of resource polymorphism due to different feeding modes and habitats (Sandlund et al. 1992). The subarctic Icelandic lakes can, like the Danish lakes, have a significant contribution of benthic plants to the primary production and biological structure of the lake ecosystem (Hunding 1979, Kairesalo et al. 1992)

Detailed lake studies provided comprehensive knowledge on individual lakes and lake types. However, there were numerous lakes whose ecology and environmental quality needed consideration. Consequently, models evaluating water quality as a function of external nutrient loading in lakes with different morphometry, hydraulic retention time and sediment composition, were developed at the laboratory. The first steps in the research were to establish the mechanisms of nitrogen and phosphorus kinetics in the sediments and subsequently use the information to improve model predictions. Andersen (1974, 1977) analyzed nitrification and denitrification processes in oxic and anoxic sediments from different lake types and demonstrated the substantial denitrification losses in a series of shallow lakes of short retention time. Kamp-Nielsen (1975, 1978) and Jacobsen (1975, 1978) studied the exchange and adsorption processes of phosphorus in lake sediments, the immobilization of phosphorus in deeper sediments, and defined the mobile sediment phosphorus pool, included as an important state variable in phosphorus models. Subsequent work attempted to improve the dynamic models to predict the response of lakes to different scenarios of altered nutrient inputs (Kamp-Nielsen 1980, 1983). The simple empirical models have been used extensively by local officials to set policies for water purification and lake management.

The strong emphasis on nutrient regulation of lake biota in the 1960's and 1970's subsequently stimulated increased interest in biotic interactions and cascading trophic effects; initially in opposition to the main emphasis on resource regulation. Riemann and Søndergaard (1986) studied the details of energy flow in different lake ecosystems, emphasizing the importance of bacteria and microbial food webs. Riemann (1985, and later) also advanced the studies of possible cascading influence of fish in whole lake and enclosure manipulations. These early initiatives were continued in recent studies on grazing impacts on bacteria and microzooplankton (Christoffersen et al. 1990, 1995) and evaluations of the key-role role of *Daphnia* in the composition and dynamics of lake plankton contrasted to the minor role of macrozooplankton in the shallow marine waters (Riemann and Bosselmann 1984, Riemann and Christoffersen 1993). M. Søndergaard, professor at the laboratory from 1990, continued the other main research directions evaluating the production of dissolved organic carbon and its utilization by free-living and attached bacteria (Søndergaard 1991, 1993). Outside the Freshwater Biological Laboratory the combined influence of resource and predator control on the lake biota has been demonstrated in large-scale Danish studies combining comparative analyses of many lakes and controlled manipulation studies at different spatial and temporal scales (Jeppesen et al. 1991).

Stream studies – The ecological studies of streams began much later than the lake studies, and they have involved fewer researchers, though the public interest and the ecological importance of the two freshwater environments were equally strong.

The lowland regions in Denmark are drained by a dense network of small streams. Most of the streams have been regulated and are subject to weed cutting to ensure optimal drainage of agricultural fields. The River Susaa project 1978-1981 represented an interdisciplinary ecological study on the regulation of oxygen and nitrogen dynamics and the distribution, composition and biomass of microalgae, macrophytes, invertebrates and fish in this common stream type (Iversen et al. 1984, Jeppesen et al. 1984). The project demonstrated the key-role of submerged macrophytes for the biological structure and stream processes (Sand-Jensen et al. 1988, 1989). Macrophyte surfaces were shown to form an important substratum for microbial processes (e.g. nitrification and degradation of dissolved organic matter) and for attachment of dense and highly productive inverte-

brate communities of filter feeders, grazers, and detritivores (Iversen et al. 1985, Iversen and Thorup 1987, 1988). The direct use of macrophytes as food for macroinvertebrates was shown in later studies (Sand-Jensen and Madsen 1989, Jacobsen and Sand-Jensen 1992). Many common macroinvertebrates, traditionally classified as shredders, actually feed on many different food items and grow well on fresh macrophyte tissue (Jacobsen and Sand-Jensen 1994, 1995). Utilization of macrophyte tissue was widespread and particularly important when other food sources (e.g. benthic microalgae, high-quality terrestrial leaves) were in short supply. Stream macrophytes strongly modified the flow and retention of particles on local spatial scales (plant patches) and larger scales (stream reaches), and influenced the nutrient dynamics of entire watersheds (Sand-Jensen 1997). The spatial coupling of patch growth and fine-scale patterns of flow, sedimentation and nutrient retention has recently been demonstrated with improved analytical techniques (Sand-Jensen and Madsen 1992, Sand-Jensen and Mebus 1996).

Relatively undisturbed springs in forested areas were intensively studied as well. One initial goal was to determine the life cycles of macroinvertebrates in relation to temperature regime and food sources (Thorup 1970, 1973, Iversen 1973, 1976). Another early goal was to quantify conversion and efficiency of the utilization of litter by the large shredder communities (Iversen 1979, 1980, Iversen et al. 1982). Later studies, which included unshaded brooks, were directed at examining the population dynamics and the food sources of the trout populations, and their predatory influence on the macroinvertebrate communities (Mortensen 1977, Andersen et al. 1992, 1993). Clearly, the stream research has moved from the early focus on macroinvertebrates to the study of various biotic interactions such as those between fish and invertebrates, invertebrates and benthic microalgae, and macrophytes and benthic macroalgae. The integrated concepts developed during the River Susaa investigation have been a platform for later ecological studies in Danish streams, and this project greatly improved the equipment for eco-physiological experiments at the laboratory.

Studies of brackish waters – E. Steemann Nielsen's short period (1969-1977) as professor at the laboratory initiated the research in brackish waters because of his main interest in marine environments. This research direction was possible due to the combined chemical-botanical expertise which developed at the laboratory, but was not available at any other Danish university institution at that time. The combination of brackish and freshwater research allowed suitable comparisons and transfer of concepts between the two habitats.

The predominantly shallow, soft-bottom coastal waters in Denmark were formerly covered by eelgrass (*Zostera marina*) meadows. The first brackish studies demonstrated the highly dynamic and productive nature of the eelgrass populations (Sand-Jensen 1975, Wium-Andersen and Borum 1980) and the influence of epiphytic algae on photosynthesis and growth of eelgrass (Sand-Jensen 1977, Borum 1985). The decline of the eelgrass cover followed the gradual spread of cultural eutrophication from freshwaters to brackish waters. Eutrophication stimulated the growth of planktonic and epiphytic microalgae, inducing strong shading effects on macrophytes (Borum 1987). Thus an important goal was to determine how plant communities changed from a dominance of eelgrass and slow-growing perennial macroalgae at low nutrient load to a dominance of phytoplankton and fast-growing ephemeral macroalgae at high nutrient loading. These changes were interpreted as differences in nutrient uptake and storage capacity of the

plant types (Borum 1983, 1996, Sand-Jensen and Borum 1991). The combined primary production of all plant communities in the shallow coastal waters did not change during increasing eutrophication (Borum and Sand-Jensen 1996). However, a progressive spatial and temporal decoupling was observed between oxygen evolution during photosynthesis and oxygen consumption during decomposition, leading to higher risks of oxygen depletion (Borum 1997).

The interest in eelgrass communities was developed along several lines. One aspect was the study of regulation of eelgrass populations at different spatial scales encompassing the demography of shoots within established meadows, the dynamics of small patches and the large-scale dynamics of all meadows within entire estuaries (Olsen and Sand-Jensen 1994). Another aspect was the study of nutrient dynamics emphasizing the ability of eelgrass to conserve nutrients by reallocation within the shoot system. Also, eelgrass influenced sediment processes through nutrient uptake and enhanced sedimentation of particles (Borum et al. 1989, Pedersen and Borum 1992). A final line of studies related light utilization, photosynthesis, and population dynamics of different temperate, subtropical and tropical seagrass species to their growth and survival strategies (Duarte et al. 1994).

A comparative approach was used to evaluate the size dependence of light utilization, photosynthesis, nutrient uptake and growth rate among unicellular and multicellular algae (Nielsen and Sand-Jensen 1990, Enriquez et al. 1993, 1996). This work demonstrated the existence of robuste wide-scale relationships between functional plant traits and the size, or relative surface area, of the plant tissue (Duarte et al. 1995). These scaling studies have helped define important selective processes within the plant kingdom and interpret the changes in composition of plant communities along different environmental gradients of light availability, nutrient availability and physical disturbance (Markager and Sand-Jensen 1992, 1994).

The academic challenge in science and teaching

Narrow and wide-scale scientific approaches both contribute to the progress of scientific understanding, and both approaches have been pursued throughout the 100 years' existence of the Freshwater Biological Laboratory. Questions within specialized ecological disciplines, however, need to be formulated with an understanding of the functioning of the natural communities. Ecophysiological studies, here mentioned only briefly, have always been performed with reference to the natural environmental conditions. This approach has helped formulate questions and set a realistic experimental scene, thereby making the results relevant for evaluating natural ecological phenomena.

Understanding of ecosystem processes is part of the training and heritage of most freshwater ecologists. We believe that this background has been important for the ability of freshwater ecologists to take up new research directions and to assimilate the ideas of colleagues working within different specialized topics. Moreover, the broad background has made it easier to advise research students in many different ecological projects. About 260 students have graduated from the Freshwater Biological Laboratory during the last 20 years. It is probably no coincidence that a large proportion of them have obtained positions at various institutions for applied research and environmental planning on state and county levels.

The two main obligations of university ecologists are teaching and research. However, there is an additional academic responsibilty to inform the public about the relevant scientific findings and to question unwise decisions that would harm aquatic environments. This responsibility evolved naturally as laboratory scientists observed our freshwater and brackish environments undergoing an alarming deterioration during the 100 years' existence of the laboratory. We can easily observe fundamental changes of environmental quality on land and tend to forget that changes have often been more profound and critical below the water surfaces. Since the origin of the Freshwater Biological Laboratory, scientists have been engaged in the environmental problems (e.g. Wesenberg-Lund et al. 1917). However, it was not until the mid-1960's that the information on water pollution problems really penetrated to the political and administrative fora, and strong actions were taken to protect the quality of inland and marine coastal waters. Hopefully, this will also lead to a future situation where the environmental quality of the aquatic environments improve in concert with an ever-rising scientific and public understanding of the ecology and the value of aquatic ecosystems.

Literature cited

Borum, J. 1997. Ecology of Danish coastal waters. Chapter 7 *in* K. Sand-Jensen and O. Pedersen, editors. Freshwater Biology. Priorities and Development in Danish Research. Gad, Copenhagen, Denmark.

Hamburger, K., C. Lindegaard, and P. C. Dall. 1997. Metabolism and survival short of oxygen. Chapter 11 *in* K. Sand-Jensen and O. Pedersen, editors. Freshwater Biology. Priorities and Development in Danish Research. Gad, Copenhagen, Denmark.

Iversen, J. 1929. Studien über die pH-Verhältnisse dänischer Gewässer und ihren Einfluss auf die Hydrophytenvegetation. Botanisk Tidsskrift 40: 277-333.

Jeppesen, E., P. Kristensen, J. P. Jensen, M. Søndergaard, E. Mortensen, and T. Lauridsen. 1991. Recovery resilience following a reduction in external loading of shallow, eutrophic Danish lakes: Duration, regulating factors and methods for overcoming resilience. Memorie dell'Instituto Italiano di Idrobiologia 48: 127-148.

Lindegaard, C., P. C. Dall, and P. M. Jónasson. 1997. Long-term patterns of the profundal fauna in Lake Esrom. Chapter 4 *in* K. Sand-Jensen and O. Pedersen, editors. Freshwater Biology. Priorities and Development in Danish Research. Gad, Copenhagen, Denmark.

Olsen, S. 1950. Aquatic plants and hydrospheric factors. I. Aquatic plants in SW-Jutland, Denmark. II. The hydrospheric types. Svensk Botanisk Tidsskrift 44: 1-34 and 332-373.

Sand-Jensen, K. 1997. Macrophytes as biological engineers in the ecology of Danish lowland streams. Chapter 6 *in* K. Sand-Jensen and O. Pedersen, editors. Freshwater Biology. Priorities and Development in Danish Research. Gad, Copenhagen, Denmark.

Sand-Jensen, K., and M. Søndergaard. 1997. Plants and environmental conditions in Danish *Lobelia*-lakes. Chapter 5 *in* K. Sand-Jensen and O. Pedersen, editors. Freshwater Biology. Priorities and Development in Danish Research. Gad, Copenhagen, Denmark.

Søndergaard, M. 1997. Bacterial plankton and dissolved organic carbon in lakes. Chapter 9 *in* K. Sand-Jensen and O. Pedersen, editors. Freshwater Biology. Priorities and Development in Danish Research. Gad, Copenhagen, Denmark.

Steemann Nielsen, E. 1944. Dependence of freshwater plants on quantity of carbon dioxide and hydrogen ion concentration. Illustrated through experimental investigations. Dansk Botanisk Arkiv 11: 1-25.

Steemann Nielsen, E. 1947. Photosynthesis of aquatic plants with special reference to the carbon-sources. Dansk Botanisk Arkiv 12: 5-71.

A complete list of publications from the Freshwater Biological Laboratory is found in Chapter 14 of this book.

3 Eutrophication and plant communities in Lake Fure during 100 years

By Kaj Sand-Jensen

Lake Fure had transparent water, oxygen-rich bottom waters and diverse plant communities at the first intensive investigations of the lake from 1897 to 1914, performed from the first freshwater laboratory located on the lake bank. During the next 70 years, annual phosphorus loading increased about 30-fold following urbanisation of the watershed. Dense phytoplankton communities reduced lake transparency, and anoxic bottoms waters developed in the lake. Despite the careful early investigations, a golden scientific opportunity to document quantitatively the limnological changes was wasted. Strong actions to protect the once precious lake were not taken until 1969. The 100 years of history, nonetheless, reveal new information on remarkable changes in the community structure of phytoplankton and submerged macrophytes, documented here for the first time.

Lake Fure 100 years ago

Lake Fure has an area of 9.4 km², is 38 m deep, and is located just 20 km North of central Copenhagen. One hundred years ago the lake was still surrounded by forests, fields and small villages, and it received no domestic sewage (Fig. 3.1). The lake had a high water quality and included species-rich plant and animal communities. The water was so transparent that you could observe fish swimming at a depth of 8-10 m (Wesenberg-Lund 1904). Oxygen was always present in the water, even in the hypolimnion at depths exceeding 30 m during summer stratification (Brønsted and Wesenberg 1911-12). The cold and oxygen-rich bottom waters were a prerequisite for the presence at that time of the rare crustaceans *Mysis relicta, Pallasea quadrispinosa* and *Pontoporeia affinis* (Berg et al. 1958).

Eutrophication of Lake Fure

One hundred years ago the small villages surrounding Lake Fure had only 4,400 inhabitants (Hovedstadsrådet 1985). Between 1900 and 1975 the population in the watershed increased 8-fold to 35,000 people, and the proportion of built on areas increased from 4 to 35% (Figs. 3.1 and 3.2). Nutrient loading of the lake with phosphorus increased even further because water closets and sewer pipelines were installed, leading to a more direct phosphorus input to the lake. In addition, phosphorus-rich detergents were introduced in the households in the 1950's, and their application approximately doubled the amount of phosphorus in domestic sewage. No wonder the phosphorus input to Lake Fure increased dramatically (Fig. 3.2).

According to a retrospective evaluation (Hovedstadsrådet 1985), the annual external phosphorus load to Lake Fure was only 1.3 ton P in the year of 1900 (0.14 g P m⁻² lake

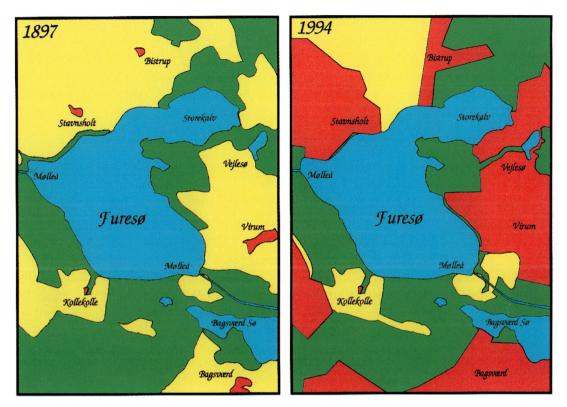

Figure 3.1. Map of Lake Fure and the immediate surroundings in 1897 and 1994. Forests are green, open fields are yellow and built on areas are red. Graphics by O. Pedersen.

surface). However, the loading rate had already increased to 2.4 ton P in 1911, and the first signs of eutrophication had been recorded since the submerged angiosperm, *Potamogeton compressus*, apparently a poor competitor under more eutrophic conditions (Sand-Jensen 1995), had disappeared from the deep slopes between 3 and 7 m depth in the main basin (Petersen 1917). Clear warnings about the detrimental influence of domestic sewage on lake quality were already expressed in the first detailed investigation of the lake environment and the flora and fauna published in 1917 (Wesenberg-Lund et al. 1917). However, environmental problems were to become much worse, and 52 years would elapse before action was be taken to protect the lake against the excessive nutrient input.

The phosphorus load had increased to about 10 ton P by 1950 (Fig. 3.2). Clear signs of eutrophication in the pelagic waters in the form of occasional phytoplankton blooms and reduced water clarity had been reported in the late 1940's (Olsen and Larsen 1948), but overall, Lake Fure had managed to tolerate the increased loading relatively well, thanks to an efficient binding of phosphorus in the sediments to calcium carbonates, aluminium and iron oxy-hydroxides (Olsen 1958 a,b). Until the early 1950's the bottom waters remained oxic and iron-bound phosphorus stayed in the sediments. With the accelerated phosphorus loading during the 1950's and 1960's, however, extensive phytoplankton blooms developed (Fig. 3.3), leading to anoxic bottom waters and release of large amounts of iron-bound sediment phosphorus (Steemann-Nielsen 1973, Hovedstadsrådet 1985).

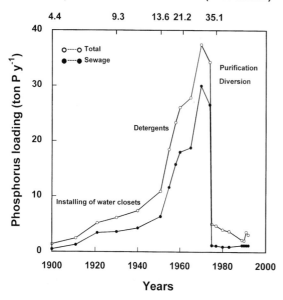

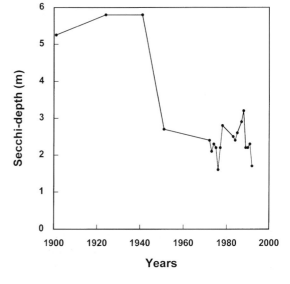

Figure 3.2. From 1900 to 1970 the annual phosphorus loading of Lake Fure has increased 30-fold by sewage discharge, due to increasing population in the watershed, installation of water closets, and use of phosphorus-rich detergents. Since 1970, phosphorus loading has been reduced by diversion and tertiary treatment of the sewage. Sewage and total phosphorus inputs are presented. Total phosphorus input includes point and non-point sources. For comparative purposes the annual loading rates can be recalculated per unit surface area (9.4 km^2 in Lake Fure) or water volume (127x10^6 m^3). Compiled by Sand-Jensen (1995) after Hovedstadsrådet (1985) and Københavns Amt (1991, 1994).

Figure 3.3. Mean Secchi-transparency in Lake Fure during summer (May-October) has declined from 5-6 m before 1940 to 1.5-3.2 m over the last 30 years. No recent systematic changes are evident. Compiled by Sand-Jensen (1995) from several sources.

By 1969, annual phosphorus loading had reached a maximum of 37.4 ton P, or 30-fold higher than the initial level in 1900 (Fig. 3.2). A combination of diversion and tertiary treatment of the sewage subsequently reduced the phosphorus load to 3-4 ton P annually from 1975 and onwards. This profound reduction has not, so far, had any marked influence on the phytoplankton biomass, the transparency or the depletion of oxygen in the hypolimnion (Hovedstadsrådet 1985, Københavns Amt 1994). During the long period from 1900 to 1970 an amount of about 700 ton P, in excess of the loading level in 1900, has been carried into the lake, and large phosphorus pools have accumulated in the sediments. The internal gross release from the sediment to the water column every summer is now 15 ton P which is markedly higher than the present annual external loading (Københavns Amt 1994). Reduction of the phosphorus concentrations in the lake water, therefore, depends on a gradual reduction of the sediment pools by immobilisation and burial in the sediment and a higher loss of phosphorus via the outlet than received via the inlets. This reduction will be slow because the hydraulic retention time is about 16 years, and it is hard to obtain low phosphorus concentrations in the inlet water today because most of the watershed is cultivated or built on.

Transparency and phytoplankton in Lake Fure

Secchi-disc transparency has probably been the most useful measure for documenting changes in lake eutrophication during the 100-150-year history of limnological science. Secchi-disc measurements are easy to perform and reproduce, and the method has remained the same. Time-series of phytoplankton chlorophyll are usually of much shorter duration. One of the classical and most comprehensive time-series of chlorophyll measurements from Lake Washington, USA did not become continuous until 1961 (Edmonson 1993). In Lake Fure frequent chlorophyll measurements did not start until the mid-1970's, while measurements of Secchi-transparency have been performed for the last 100 years.

The mean Secchi-transparency during summer in Lake Fure was between 5 and 6 m from 1901 until 1940 (Fig. 3.3). By 1951, the mean summer value had declined to 2.7 m, demonstrating that the trophic status of the lake had already changed. Since the early 1970's, the mean summer transparency has fluctuated between 1.5 and 3.2 m. The lowest values have been reported in particularly warm summers leading to more extensive nutrient release to the surface waters from the shallow sediments and to denser phytoplankton blooms (Steemann-Nielsen 1973). No overall trend in Secchi-transparency is apparent from the last 20 years (Fig. 3.3), and the varying reports on improvements or deterioration of water quality every summer in the local newspapers are clearly unwarranted.

The long-term developement of phytoplankton biomass has not been recorded in Lake Fure, but we can provide a crude estimate of the development by using the long-term record of Secchi-transparency (Fig. 3.3) and assuming that the contemporary relationships beween Secchi-transparency and chloropyll concentration can be applied for the entire time period (Sand-Jensen 1995). This

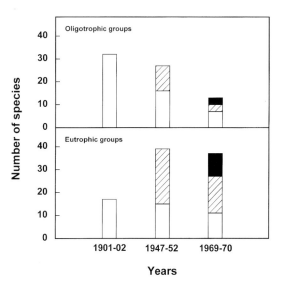

Figure 3.4. Changes in number of phytoplankton species in Lake Fure during the last 100 years within taxonomic groups considered typical of Danish oligotrophic (upper panel) or eutrophic lakes (lower panel). Oligotrophic groups included chrysophytes, desmids and dinoflagellates. Eutrophic groups included blue-greens and chlorococcalean green algae. Not all algal species in the lake were included. The white columns comprise species already observed in 1901-1902, the hatched columns comprise species first observed in 1947-1952, and the black columns comprise species first observed in 1969-1970. Compiled by Sand-Jensen (1995) from Wesenberg-Lund (1904), Nygaard (1958) and Hovedstadsrådet (1985).

estimate suggests that the previous mean summer (May-October) concentrations of chlorophyll were about 2 mg m^{-3}, or 10-fold lower than the mean levels of 15-20 mg m^{-3} measured from the 1970's and onwards (Hovedstadsrådet 1985, Københavns Amt 1991, 1994).

Unlike the biomass, the species number and composition of phytoplankton have been studied in Lake Fure on several occasions during the 100 years. The number of phytoplankton species reported from Lake Fure has remained approximately constant between 70 and 85 over the past 100 years, when the same taxonomical classification was applied, but the species composition has changed dramatically (Fig. 3.4). The number of species within particularly nutrient-demanding groups typical of eutrophic lakes (e.g. blue-greens and chlorococcalean green algae, Nygaard 1949) has increased in Lake Fure from 17 to 39, while the number of species typical of oligotrophic lakes (e.g. chrysophytes, desmids and some dinoflagellates) has declined from 32 to 13 (Fig. 3.4). Nygaard's Compound Quotient (1949) defined as the numerical ratio of eutrophic nutrient-demanding phytoplankton species relative to oligotrophic species has, therefore, also increased with time from summer values of 1.5-1.8 in 1901 and 2.9-3.8 in 1947-1951 to a maximum of 5.3-12.0 in 1969-1978 (Wesenberg-Lund 1904, Nygaard 1958, Hovedstadsrådet 1985).

The species pool of phytoplankton has experienced profound changes during the past 100 years, as original species have disappeared and new species appeared, also within individual algal groups. Among diatoms and blue-greens, for example, species characteristic of oligotrophic and mesotrophic lakes have vanished, and species typical of eutrophic and hypereutrophic lakes have appeared (Nygaard 1958, Hovedstadsrådet 1985). The diatoms *Attheya zachariasii*, *Aulacoseira italica* and *Tabellaria fenestrata*, for example, have disappeared and *Stephanodiscus hantzschii* appeared. The blue-green algae *Oscillatoria prolifica* and *O. rubescens* have disappeared and species of *Microcystis* and *Anabaena* have become very abundant. Also the chrysophyte *Dinobryon sertularia* and species of dinoflagellates within the genera *Peridinium* and *Gymnodinium*, which are able to live in the metalimnion during summer and to migrate vertically through the metalimnion to optimise utilisation of light and nutrients, have disappeared. These changes in species composition are logical, considering that the euphotic zone 100 years ago penetrated into the thermocline and allowed growth of cold-adapted and verti-

cally migrating algae. Today the euphotic zone extends down to between 1.5 and 8 m depth during summer, and phytoplankton growth is not possible in the deeper waters within or below the thermocline. The ability to compete for light has presumably become more important for the success of the present phytoplankton species than it was at the turn of the century.

Macrophyte vegetation in Lake Fure
The diversity of submerged macrophytes in Lake Fure before the recent cultural eutrophication commenced was remarkable and higher than reported for any other Scandinavian lake (Rørslett 1991). Petersen and Raunkiær (1917) found a total of 35 species of submerged macrophytes in 1911. The vegetation was dominated by 17 species of angio-

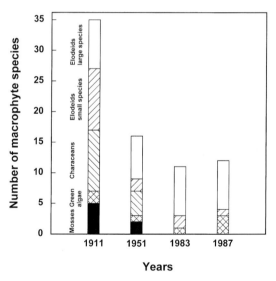

Figure 3.5. Number of species of submerged macrophytes in Lake Fure studied on four occasions during the last 100 years. Characeans, water mosses and small elodeid species of angiosperms have disappeared, while 7-8 species of tall, robust elodeid species of angiosperms have survived. The original species of green algae were *Nostoc pruniforme* and *Vaucheria* sp., whereas the present species are the pollution indicators *Cladophora glomerata*, *Enteromorpha flexuosa* and *Rhizoclonium profundum*. Data from Petersen (1917), Petersen and Raunkiær (1917), Christensen and Andersen (1958) and Københavns Amt (1991, 1994).

sperms and 10 species of characeans, while 5 species of aquatic mosses occurred more sporadically (Fig. 3.5). The high species diversity in Lake Fure was probably due to the transparent water of high alkalinity, the size of the lake, and the variability of habitats. The lake included both exposed and protected sites and sediments with gravel, sand or mud which allowed growth of many different species. Recent comprehensive compilations of the diversity of submerged macrophyte species in 600 Scandinavian lakes have confirmed that species diversity increases with lake size, and that the diversity is highest in clear alkaline lakes and lowest in acid or turbid lakes (Rørslett 1991).

The submerged vegetation was widely distributed in the lake in 1911. Many species grew from shallow water to the depth limit of macrophyte growth at 7-8 m (Petersen 1917, Petersen and Raunkiær 1917). The large elodeid angiosperm species within the genera *Myriophyllum*, *Potamogeton* and *Ranunculus* dominated the vegetation outside the reed belt to depths of about 5 m, but several small angiosperms, mosses and characeans formed a mixed carpet of vegetation below the canopy of large elodeid species. At depths between 5 and 8 m, characeans, and in particular *Nitellopsis obtusata,* and elodeid angiosperms were equally abundant (Fig. 3.6). At greater depths than 8 m, macrophytes were extremely scarce, and a few specimens of green filamentous macroalgae only have been observed to grow attached to mussel shells collected from 12 m depth.

Forty years later, at the second large investigation of Lake Fure in 1951, many of the original submerged macrophyte species had disappeared (Christensen and Andersen 1958). Among the original 35 species only 16 were found (Fig. 3.5). This decimation of submerged plants continued in the coming years as eutrophication progressed (Fig. 3.2 and 3.3). In 1983-1993 only 10 of the original species were found, whereas 3 new

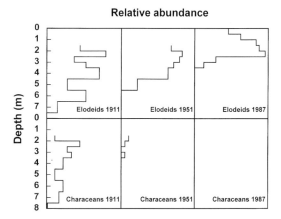

Figure 3.6. Depth distribution of the relative abundance of elodeid angiosperms and characeans in Lake Fure in 1911, 1951 and 1987. Relative abundance was calculated for each depth interval (0.5-1.0 m) as the number of samples containing the plant group in relation to the total number of samples. Between 155 and 270 samples were taken with a rake along many depth transects evenly distributed across the lake. Compiled by Sand-Jensen (1995) from Petersen (1917), Christensen and Andersen (1958) and Københavns Amt (1991).

species of green macroalgae (*Cladophora glomerata*, *Enteromorpha flexuosa* and *Rhizoclonium profundum*), typical of polluted waters, had appeared (Fig. 3.5). The macrophytes which had vanished were mainly small angiosperms, mosses and characeans, which originally covered the deepest parts in the lake or formed the vegetation carpet below the canopy of large elodeid species (Petersen 1917). Hence, all 10 species of characeans and 5 species of mosses originally growing in the lake are no longer present. Eight large species of canopy-forming elodeids have survived in the turbid lake, probably because they reach a height of 1.5-2.0 m and, therefore, in part can compensate for reduced light penetration with depth by growing towards the water surface (Adams and McCracken 1974). The areal distribution of all the original submerged macrophyte species has been severely restricted, however, because of the reduced light penetration in the lake.

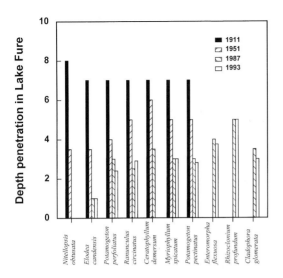

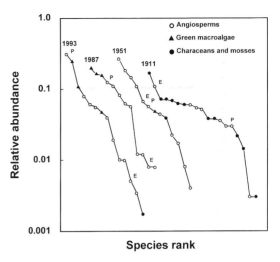

Figure 3.7. Depth limit (m) of selected submerged macrophytes in Lake Fure in 1911, 1951, 1987 and 1993. Shown are the charophyte: *Nitellopsis obtusata*, the six species of large elodeid angiosperms: *Myriophyllum spicatum, Potamogeton pectinatus, Ranunculus circinatus, Potamogeton perfoliatus, Elodea canadensis, Ceratophyllum demersum* and the green macroalgae: *Cladophora glomerata, Enteromorpha flexuosa* and *Rhizoclonium profundum*. Compiled from Petersen (1917), Christensen and Andersen (1958) and Københavns Amt (1991, 1994).

Figure 3.8. Ranked abundance of submerged species of macrophytes in Lake Fure in 1911, 1951, 1987 and 1993. The species are ranked in order of decreasing abundance, and the graphs from the four years have been displaced from each other for clarity. Charophytes and mosses have disappeared over the years, while certain green macroalgae (*Cladophora glomerata, Enteromorpha flexuosa* and *Rhizoclonium profundum*), which indicate pollution, have become abundant. More species were present in 1911, and they had a more even abundance than at the later investigations. *Potamogeton pectinatus* (marked P) has increased, while *Elodea canadensis* (marked E) has declined in rank during the period. Compiled by Sand-Jensen (1995) from Petersen (1917, and unpubl. notes), Petersen and Raunkiær (1917), Christensen and Andersen (1958) and Københavns Amt (1991, 1994).

In 1911 both the elodeid and characean vegetation penetrated to 7-8 m (Fig. 3.6). By 1951, the depth limit had been reduced to 5.5 m for elodeids and 3.5 m for characeans which had also declined more in abundance. At the two latest inventories in 1987 and 1993 the elodeid vegetation had been restricted to the upper 3.5 m while the characeans had died out completely (Fig. 3.6). The overall pattern is the same for all species of elodeid angiosperms and characeans (Fig. 3.7). The previous bottom cover of small angiosperms, characeans and mosses in shallow water has also disappeared most likely by shading from the denser growth of tall elodeid macrophytes and phytoplankton in the water column.

Relative abundance of submerged species

Several results suggest that although the area occupied by macrophytes in Lake Fure is less than it was in the past, the biomass is greater where vegetation does occur. Such a change could have increased the competition for light and space among the surviving macrophytes and, in addition to the restriction of vegetated areas caused by the turbid water, could have contributed to the decline in species diversity.

The submerged species were widely distributed in 1911, and they formed mixed communities of many species, but apparently lower areal density and biomass than in 1951 and later. In 1911 a total of 15 species could be recorded from within a small area (Petersen and Raunkiær 1917), but open space was also available between the plants as a large proportion (77%) of the 270 samples were devoid of plants (Petersen 1917). In 1951 a significantly smaller proportion (36%) of the samples was free of plants (Christensen and Andersen 1958), and in 1987 and 1993, virtually all samples contained plants within the bottom areas receiving sufficient amounts of light, though only 1-2 species per sample were found (Københavns Amt 1991, 1994, Sand-Jensen 1995).

Also the dominance has become more prominent among the surviving species in Lake Fure in 1951, 1987 and 1993 than in the earlier studies in 1911, when a larger number of species had a more equal abundance (Fig. 3.8). My interpretation of this pattern is that certain species have disappeared because of the restriction of suitable growth habitats, and other species have been outcompeted by the robust elodeid species and the recent invasion of mat-forming, fast-growing macroalgae (*Cladophora, Enteromorpha* and *Rhizoclonium*) which have proliferated

Figure 3.9a. The open reed belt of short individuals in Lake Fure in 1900 (a) forms a sharp contrast to the dense and tall reed belt observed in several places in 1994 (b). Photographs by C. Wesenberg-Lund and F. Pedersen.

Figure 3.9b.

because of greater nutrient availability (Sand-Jensen 1995).

The influence of lake fertilization is also apparent for the reed plants which are dominated by *Phragmites australis* and *Scirpus lacustris*. Photographs taken nearly 100 years ago reveal that the reed plants at that time were short and sparse and grew on a coarse sediment, while on many sites today, the reed belt is dense and tall and covers nutrient-rich muddy sediments (Fig. 3.9).

The submerged species which have increased in relative abundance over the years are also those considered as representatives of highly eutrophic waters (e.g. Holmes and Newbold 1984). Among the elodeid angiosperms, *Potamogeton pectinatus* was number 15 in the rank of abundances in Lake Fure in 1911, and it was first in rank in 1993 (Fig. 3.8). *P. pectinatus* is able to grow rapidly in late spring from starch-filled rhizome bulbils in the sediment and to form a leaf canopy immediately below the water surface (Bijl et al. 1989). Moreover, an investigation of herbarium specimens from 1951 suggests that

they are smaller than the individuals collected today, in accordance with the development of more nutrient-rich sediments (Sand-Jensen 1995). *Potamogeton crispus* has also become relatively more abundant in Lake Fure over time (Fig. 3.10). Both *P. pectinatus* and *P. crispus* have increased in relative abundance among submerged species in Danish lakes and streams as eutrophication has progressed since the turn of the century (Pedersen 1976)

A special history is associated with the development of *Elodea canadensis* in Lake Fure. After the accidental introduction of the species to Europe, it was first discovered in the Lake Fure water system in 1886 (Christensen and Andersen 1958). After the characean *Nitellopsis obtusata*, it was the most abundant submerged plant in Lake Fure in 1911, but according to observations, it had by then already declined from its peak abundance in 1900 (Petersen and Raunkiær 1917). During the following 100 years, the distribution of *E. canadensis* has gradually diminished and it is today only ranked as number 12 (Figs. 3.8 and 3.10). This pattern of initial prolific growth followed by a subsequent decline over many years after the invasion of species such as *E. canadensis,* has been observed many times, but the causes of this behaviour are still unclear (Sculthorpe 1967).

Recent invasion of fast-growing macroalgae
A new dimension was recently added to the submerged macrophyte vegetation in Lake Fure by the abundant growth of three species of green macroalgae, *Cladophora glomerata, Enteromorpha flexuosa* subspecies *pilifera* and *Rhizoclonium profundum* (Fig. 3.8; Københavns Amt 1991, 1994). The combined abundance of the three macroalgae among all samples of submerged macrophytes was high both in 1987 (52%) and 1993 (40%). These values represent the percentages of samples with the presence of the macroalgae, but in terms of biomass, their

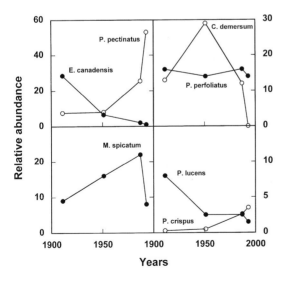

Figure 3.10. Changes in the relative abundance (%) of selected species within the collective group of angiosperms from 1911 to 1993. Data from the same sources as in legend to Fig. 3.8.

contribution is likely to be even higher (Københavns Amt 1994). The green macroalgae were distributed in dense mats in 1987 and 1993 which normally extended to the same or to greater depths (3-5 m) than the rooted angiosperms (Fig. 3.7). Thus, the earlier dominance of characeans at the lower depth boundaries of macrophyte growth in Lake Fure has now been replaced by the growth of green macroalgae with *Rhizoclonium profundum* extending to greatest depth. It is likely that the shift reflects the increased nutrient richness of the sediment and the water over the 100 years period.

The recovery of the former type of macrophyte vegetation at depth is not necessarily reversible, if permanently improved water transparency develops in Lake Fure in the future. Because sediment conditions have changed and dense blankets of macroalgae have formed, reinvasion of mosses, characeans and rooted angiosperms at the reilluminated bottom areas at depth may be impeded. The possible recovery phases are unknown, however, and difficult to interprete because no information, to my knowledge, is available from lakes of the same type and loading history as Lake Fure.

Lost opportunities

Two restrictions are common in environmental evaluations of anthropogenic influences on ecosystems in general and lakes in particular. Background information on the undisturbed lake ecosystem is usually not available, or it is insufficient. Nevertheless, the ecosystem studies which are planned to continue for many years to match the time scales of many chemical and biological properties, often turn out to be of short-time duration for various political, financial and personal reasons. In the case of Lake Fure, several careful background studies were indeed available (e.g. Wesenberg-Lund 1904, Brønsted and Wesenberg-Lund 1912, Wesenberg-Lund et al. 1917), though many were

qualitative studies that have left us with the problem of making certain deductions of the changes over time. Among the biological studies, only the macrophyte studies were quantitative, or can later be reanalysed quantitatively, using notebook information (Petersen 1917) kept at the Botanical Museum in Copenhagen (Sand-Jensen 1995; Fig. 3.8 is an example).

Though the time perspective of the Lake Fure studies is appealing the frequency of the studies is, with hindsight, disappointingly low (Fig. 3.3) considering the scenic beauty, the rich fishery and the recreational interest in this lake located in a region with many wealthy and influential people. The first main investigator of Lake Fure (Wesenberg-Lund 1904), however, returned to his main interest in the biology and behaviour of freshwater invertebrates, and the second main investigator (Berg et al. 1958) studied a variety of different lakes or streams. No initiatives were taken to ensure continuous research on Lake Fure. Indeed the second major limnological study of Lake Fure (Berg et al. 1958) was started on the initiative of the Danish Society of Sanitary Engineers (1950) which suspected that extensive input of sewage harmed the lake (Rastrup 1948). Strangely, however, the second study of Lake Fure between 1950 and 1954, revealing alarming deterioration of water quality, did not lead to changes in management policy. Efficient actions to remove the excessive nutrient loading were not taken until 15 years later after excellent international information on lake eutrophication had become available (e.g.Vollenweider 1964, 1968, Thomas 1968, Edmonson 1969). Thus, important early scientific knowledge was not optimally utilised in the management of Lake Fure and in setting the policy of sewage purification and discharge. In that respect, it is disappointing to read the summary of the second Lake Fure investigation (Berg et al. 1958), predicting that the deterioration of Lake Fure is expected to continue and also to

Figure 3.11. *Potamogeton lucens* is one of largest submerged macrophytes in alkaline Danish lakes. It can become 2-3 m tall and the leaves reach 10-20 cm. *Potamogeton lucens* has disappered from many lakes because of eutrophication, but it is still present in Lake Fure.

gradually involve more Danish lakes. This prediction was fulfilled to a greater extent than the authors and the Society of Danish Engineers had imagined, reminding us that responsibility for ecosystem development extends beyond the knowledge of the academic society and requires implementation of political and administrative actions (Vallentyne 1974).

Gained scientific information

The pristine quality of Lake Fure was lost, and it is only vaguely apparent today in a small number (< 10) of Danish lakes. An early scientific opportunity to document the mech-

anisms and statistics of lake eutrophication was wasted, though the proper mixture of limnologists and engineers quantitatively analysing the loading rate of nitrogen and phosphorus was already in place in 1950-1954 (Berg et al. 1958).

Changes in nutrient dynamics, phytoplankton biomass, water transparency and oxygen conditions in lakes with progressive eutrophication are common knowledge. Hopefully, the changes in distribution, species diversity and community structure of the submerged macrophytes in Lake Fure documented here include important new information. The gradual and slow changes of the macrophyte communities involve the complete loss of mixed, relatively open, multilayered communities and the die-back of small slow-growing elodeids, mosses and characeans. The slow and gradual changes are worth noticing because they suggest time lags and accumulating effects over time due to changes of the sediments and the relatively slow numerical responses of macrophyte communities. The surviving macrophyte species are robust, canopy-forming elodeids forming denser and less diverse communities on gradually richer muddy sediments restricted to shallower waters. Thus, increased shading, both in the water column and within the macrophyte communities, was probably involved in the observed changes. Remarkably, the recent dominance of mat-forming green algae was apparently established after the external phosphorus load had been markedly reduced. This development emphasizes that the instantaneous relationships between nutrient loading rate and macrophyte community structure are weak, at least in this deep lake.

During the 100 years long history of macrophyte development in Lake Fure, some taxonomic groups and species either declined or increased uniformly, while others showed intervening peaks (e.g. *Ceratophyllym demersum,* Fig. 3.10) in a mechanistically explainable manner related to changes in light avalability, sediment richness and competition in the macrophyte community. These findings will require future wide-scale comparisons of the history of macrophyte development in several lakes in order to become fully established as general patterns among the more well-known features of lake eutrophication.

Literature cited

Adams, M. S., and M. D. McCracken. 1974. Seasonal production of the *Myriophyllum* component of the littoral of Lake Wingra, Wisconsin. Journal of Ecology 62: 457-467.

Berg, K., K. Andersen, T. Christensen, F. Ebert, E. Fjerdingstad, C. Holmquist, K. Korsgaard, G. Lange, J. M. Lyshede, H. Mathiesen, G. Nygaard, S. Olsen, C. V. Otterstrøm, U. Røen, A. Skadhauge, and E. Steemann Nielsen. 1958. Furesøundersøgelser 1950-1954. English summary: Investigations of Fure Lake 1950-1954. Folia Limnologica Scandinavica 10: 1-189.

Bijl, L. van der, K. Sand-Jensen, and A.-L. Hjermind. 1989. Photosynthesis and canopy structure of a submerged plant, *Potamogeton pectinatus,* in a Danish lowland stream. Journal of Ecology 77: 947-962.

Brønsted, J. N., and C. Wesenberg-Lund. 1911-12. Chemisch-Physikalische Untersuchungen der dänischen Gewässer nebst Bemerkungen über ihre Bedeutung für unsere Auffassung der Temporalvariation. Internationale Revue der gesamten Hydrobiologie 4: 251-290, 437-492.

Christensen, T., and K. Andersen. 1958. De større vandplanter i Furesø. Folia Limnologica Scandinavica 10: 114-128.

Edmondson, W. T. 1969. Cultural eutrophication with special reference to Lake Washington. Mitteilungen Internationale Vereinigung für Theoretische und Angewandte Limnologie 17: 19-32.

Edmondson, W. T. 1993. Eutrophication effects on the food chains of lakes. Memorie dell'Istituto Italiano di Idrobiologia 52: 113-132.

Holmes, N. and C. Newbold. 1984. River plant communities – Reflectors of water and substrate chemistry. Focus on nature conservation. No 9. Nature Conservancy Council, Shrewsbury, England.

Hovedstadsrådet. 1985. Furesø 1900-2020. Vandkvalitetsinstituttet, ATV for Hovedstadsrådet. Copenhagen, Denmark.

Københavns Amt. 1991. Furesøens plantesamfund 1987. Miljøserie 22. Copenhagen, Denmark.

Københavns Amt. 1994. Overvågning af søer 1993. Miljøserie 57. Copenhagen, Denmark.

Nygaard, G. 1949. Hydrobiological studies on some Danish ponds and lakes. Part II. The quotient hypothesis and some new or little known phytoplankton organisms. Kongelige Danske Videnskabernes Selskabs Skrifter 7: 1-293.

Nygaard, G. 1958. Furesøens planteplankton. Folia Limnologica Scandinavica 10: 109-113.

Olsen, S. 1958a. Phosphorus adsorption and isotopic exchange in lake muds. Verhandlungen Internationale Vereinigung für Theoretische und Angewandte Limnologie 13: 915-922.

Olsen, S. 1958b. Fosfatbalancen mellem bund og vand. Forsøg med radioaktivt fosfor. Folia Limnologia Scandinavica 10: 39-96.

Olsen, S., and K. Larsen 1948. Er Furesøen ved at ændre karakter? Et bidrag til diskussionen om søens eutrofiering. Lystfiskeri-Tidende 60: 467.

Pedersen, A. 1976. Najadaceernes, Potamogetonaceernes, Ruppiaceernes, Zannichelliaceernes og Zosteraceernes udbredelse i Danmark. English summary. Botanisk Tidsskrift 70: 203-262.

Petersen, J. B. 1917. Bemærkninger til plantekortene over Bastrup Sø, Farum Sø, Bagsværd Sø og Lyngby Sø. Pages 39-57 (Kapitel III) *in* C. Wesenberg-Lund et al., editors. Furesøstudier. Kongelige Danske Videnskabernes Selskabs Skrifter, Naturvidenskab og Mathematisk afdeling, 8 Række III. Bianco Lunos Bogtrykkeri, Copenhagen, Denmark.

Petersen, J. B., and S. Raunkiær. 1917. Furesøens vegetationsforhold. Pages 58-77 (Kapitel IV) *in* C. Wesenberg-Lund et al., editors. Furesøstudier. Kongelige Danske Videnskabernes Selskabs Skrifter, Naturvidenskab og Mathematisk afdeling, 8 Række III. Bianco Lunos Bogtrykkeri, Copenhagen, Denmark.

Rastrup, J. A. C. 1948. Furesø-problemer. Ingeniøren 57: 373-380.

Rørslett, B. 1991. Principal determinants of aquatic macrophyte richness in northern European lakes. Aquatic Botany 39: 171-193.

Sand-Jensen, K. 1995. Furesøen gennem 100 år. Naturens Verden 5: 176-187.

Sculthorpe, C.D. 1967. The biology of aquatic vascular plants. Edward Arnold, London, England.

Steemann Nielsen, E. 1973. Hydrobiologi. Polyteknisk Forlag, Lyngby, Denmark.

Thomas, E. A. 1968. Die Phosphattrophierung des

Zürichsees und anderer Schweizer Seen. Mitteilungen Internationale Vereinigung für Theoretische und Angewandte Limnologie 14: 231-242.

Vallentyne, J. R. 1974. The algal bowl – lakes and man. Ottawa, miscellaneous special publication 22. Department of the Environment.

Vollenweider, R. A. 1964. Über oligomiktrische Verhältnisse des Lago Maggiore und einiger anderer insubrischer Seen. Memorie dell'Istituto Italiano di Idrobiologia 17: 191-206.

Vollenweider, R. A. 1968. Scientific fundamentals of the eutrophication of lakes and flowing waters, with particular reference to nitrogen and phosphorus as factors in eutrophication. Paris, Report from the Organisation for Economic Cooperation and Development. DAS/CSI/68.27, 192 pp, Annex 21 pp., Bibliography 61 pp.

Wesenberg-Lund, C. 1904. Studier over de danske søers plankton. Specielle del. Tekst. Copenhagen, Denmark.

Wesenberg-Lund, C., M. J. Sand, J. Boye-Petersen, A. S. Raunkiær, and C. M. Stenberg. 1917. Furesøstudier. En bathymetrisk, botanisk, zoologisk undersøgelse af Mølleaaens søer. French summary. Det Kongelige Danske Videnskabernes Selskabs Skrifter, Naturvidenskabelige og Mathematiske afdeling, 8 Række III. Bianco Lunos Bogtrykkeri, Copenhagen, Denmark.

4 Long-term patterns of the profundal fauna in Lake Esrom

By Claus Lindegaard, Peter C. Dall and Pétur M. Jónasson

Eutrophication of Lake Esrom has presumably increased phytoplankton production up through this century. The subsequent nutrient reduction since the late 1960's has not improved water transparency nor lowered primary production. A surplus of organic sedimentation has continued to increase oxygen depletion in the profundal zone. Our long-term surveillance of the zoobenthos community documents the increased environmental stress and the resulting shifts in composition of the fauna from a chironomid to an oligochaete dominated macroinvertebrate community.

Long-term studies are essential in describing slow and complex ecological processes, rare events, and phenomena of high annual variability (Likens 1989). However, long-term ecological studies are expensive and personally costly and are therefore rare. A few lakes have been studied over a long time-span and provide long-term data sets. Some of the well known examples are Lake Zürich (Thomas 1979), Lake Windermere (Lund 1978), Lake Washington (Edmondson 1991) and Lake Tahoe (Goldman 1993), where data on nutrient concentration and water transparency or phytoplankton production have documented the process of eutrophication. Long-term data sets on benthic macroinvertebrates are known from a few Swedish lakes (Wiederholm 1988) and from the profundal zone of Lake Esrom, where the rich macroinvertebrate community has fascinated limnologists at the Freshwater Biological Laboratory since 1932. We are therefore able to outline changes in the profundal fauna of Lake Esrom during the past 65 years; a period when human activities have influenced the lake and its surroundings in different ways.

Lake Esrom has a volume of 0.233 km³, which is the largest in Denmark. This volume is modest compared with many European lakes, but the lake is, nevertheless, large enough to have a relatively slow response to changes in environmental loading and stresses, due to the small catchment area and the long water residence time of 16 years. Thus, long-term surveillance of the biocoenoses in the lake can reveal intrinsic, natural year to year variability in community structure, unaffected by external changes, as well as trends in the response to major long-term changes in nutrient loading.

This paper describes the eutrophication process in Lake Esrom and its effects on nutrient level, pelagic primary production and the profundal environment. We discuss the impact on the profundal fauna in view of its natural variability as opposed to the documented deterioration in oxygen availability, suspected fish predation and the results of long-term monitoring of population dynamics within the invertebrate community.

Lake Esrom and its eutrophication

Lake Esrom is dimictic, 22 m deep and has a surface area of 17.3 km². The surrounding area is morainic, with marlpits rich in calcium carbonate in the farmland on the eastern

Figure 4.1. Photo of Lake Esrom in Northern Zealand viewed from the south. The village Nødebo is located on the SW shore of the lake and the town Fredensborg to the east. Note the extensive forest Gribskov to the west and the intensively farmed areas to the east. To the north is the Kattegat and behind that the Swedish coastline, (Photo 1996 by J. K. Winther/BIOFOTO).

Figure 4.2. The profundal mud flats in Lake Esrom are dominated by the larval tubes of *Chironomus anthracinus*. In the foreground, a juvenile fish is foraging (Photo by Bio/consult-Naturfocus, Denmark).

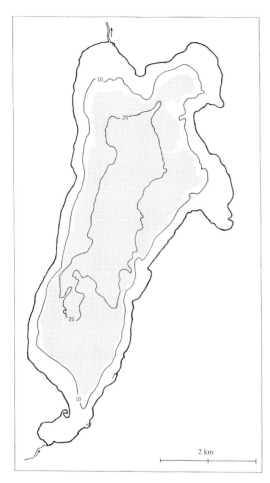

Figure 4.3. Bathymetric map of Lake Esrom showing the extension of the profundal zone.

side of the lake, and sandy areas covered by deciduous and coniferous forests on the western side (Fig. 4.1). The lake basin is a glacier-carved kettle hole providing the lake with an extensive and uniform profundal zone (Fig. 4.2). The area below 15 m depth covers 9 km², or 52% of the lake area (Fig. 4.3).

The catchment area of Lake Esrom is only 59 km². Water is supplied from small brooks and ditches, but inflow of ground water and precipitation are the main sources. Half of the catchment area is arable land from which fertilizers can reach the lake, and the immediate vicinity of the lake houses some 10,000 inhabitants, whose sewage until 1971 was led to the lake. The sewage loading was then reduced to 3,800 person equivalents (p. e.), further reduced to 1,200 p. e. in 1987 and to 200 p. e. in 1992.

The first sewage plant in the catchment area was established in 1914, and the use of artificial fertilizers increased rapidly after 1920, indicating the start of modern eutrophication. The external loading with sewage peaked in 1975 with 20 t of phosphorus and 80 t of nitrogen. By 1993 the total external loading had been reduced to 1.5 t of phosphorus and 40 t of nitrogen. This estimate

Figure 4.4. Concentration of phosphorus in surface water of Lake Esrom during summer (black bars) and winter (shaded bars) from 1929 to 1990.

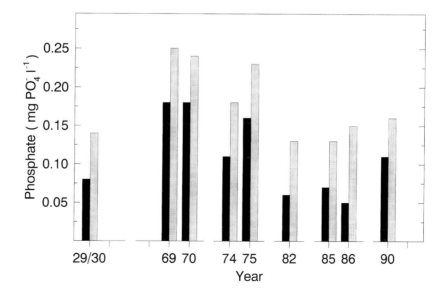

includes 30 t of nitrogen from precipitation on the lake surface.

The winter concentration of orthophosphate increased from 140 µg l⁻¹ in 1930 (Nygaard 1938) to a maximum of 250 µg l⁻¹ in 1969 (Jónasson et al. 1974) after which it decreased to about 150 µg PO_4^--Pl⁻¹ in the 1980's (Fig. 4.4). During all the years the concentration of orthophosphate never declined below 50 µg l⁻¹ in the surface water during summer. Total inorganic nitrogen concentration (NO_2^-+ NO_3^-+ NH_4^+) in the same period ranged between 300 and 400 µg l⁻¹ during winter and dropped to very low summer levels, suggesting that nitrogen was the limiting nutrient.

The orthophosphate concentration in the inlets at present averages about 70 µg l⁻¹, which indicates that the 1930-level of 140 µg l⁻¹ in the lake water was already enhanced above the background level. It is unknown whether phosphorus originally was the limiting factor for phytoplankton production, but it is possible that the original chemical composition in Lake Esrom was similar to that of the other large lakes in northern Zealand, where phosphorus was the limiting nutrient (Sand-Jensen 1997).

Phytoplankton responses

Water transparency as an indication of phytoplankton biomass suggests that eutrophication increased during the first half of this century, followed by a more steady situation since the 1950's (Fig. 4.5). The annual median Secchi-depth was higher in 1933 (5.0 m) than later in 1956-1975 (3.6 m) and 1985-1995 (3.7 m). The year to year variation since 1956 has, however, exceeded the long-term changes, and the 1933-transparency is not significantly different from the transparency in 1972 or 1990. A further evaluation of the long-term changes of transparency was made using the number of hours with sunshine in the years concerned. The year 1933 had 15% more hours with sunshine that the average from 1955 to 1990. This result shows that it was not low irradiation which caused the higher transparency in 1933 as compared with later. Although no correlation between transparency and hours of sunshine was found, we believe that transparency was actually higher and the primary production lower in 1933 as compared to later measurements.

In August 1901 Wesenberg-Lund (1904) made a single Secchi-depth measurement of 5 m in Lake Esrom. The transparency in August 1933 was 3.0 m (Berg 1938). The transparency in August normally varied

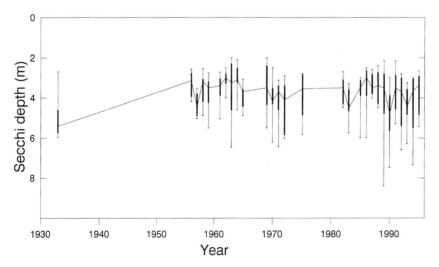

Figure 4.5. Compilation of Secchi-disc measurements in Lake Esrom. Results are presented as annual median (graph), 0.25-0.75 quantile intervals (heavy bars) and 0.1-0.9 quantiles (thin bars). The latter intervals includes 80 % of the range in Secchi-depths measured in each particular year.

between 2 m and 3 m during 1955-1995, but up to 4.5 m was recorded in single years. Despite the limited data these August values also suggest a decrease in transparency between 1901 and 1933. We thus assume that transparency has indeed decreased during the first half of this century.

The species composition of the phytoplankton in Lake Esrom was almost the same in 1961-1963, 1986 and 1990, and apparently not very different from the composition in 1901 (Wesenberg-Lund 1904, Jónasson and Kristiansen 1967, Christoffersen and Olrik 1996). It appears, however, that some of the rare species considered 'clean water' species have diminished in abundance or vanished. Thus, apparently 12 taxa among the Chrysophyceae, Dinophyceae and Desmidiaceae have disappeared from 1961-1963 to 1990 (Frederiksborg Amt 1991).

The annual succession in composition of the major taxonomic groups of algae is rather similar in all four investigations during this century. The spring maximum of diatoms is followed by rhodomonads and green algae. The summer and early autumn is dominated by blue-greens and a second maximum of diatoms. Diatoms also dominate during winter, but rhodomonads and chrysophyceans are abundant too. The bloom of blue-greens varies form year to year, depending on late summer temperatures and vertical stability of the water column (Frederiksborg Amt 1991).

Annual phytoplankton production averaged 240 g C m^{-2} y^{-1} in the years 1955-1982, and year to year variation was found to correlate to annual solar radiation (Jónasson 1992). This level, however, is substantially higher than the 70-100 g C m^{-2} y^{-1} proposed by Riemann and Mathiesen (1977) to be the common pre-cultural level of production in Danish lakes. Lakes with this level of production exhibit average transparencies of 6-7 m when not influenced by humic material. This transparency level exceeds the measurements in Lake Esrom during this century.

The higher transparency in 1933 than later indicates an increase in phytoplankton biomass and production during 1930-1955. We propose, in addition, that modern eutrophication had already started earlier, probably around the year 1900. Transparency varied little only from 1955 to 1995 (Fig. 4.5), and we conclude, despite the substantial increase and subsequent reduction in nutrient loading, that the phytoplankton biomass and production were constant during this period, and that year to year variation in production was due to meteorological differences in insolation. A rather constant concentration of the limiting inorganic nitrogen during the past 40 years supports this suggestion.

Oxygen conditions in the hypolimnion document eutrophication

Oxygen condition in the hypolimnion has responded to the past eutrophication of Lake Esrom. Oxygen measurements in 1908 showed that the hypolimnion contained oxygen throughout the stratification period. In 1933 an oxygen concentration below 1 mg O$_2$ l^{-1} was found during just one week in early October before the autumn overturn. The records from 1955 to 1975 showed oxygen depletion in the profundal zone for 2-3 months each summer. During the 1990's the oxygen depletion lasted 3-4 months (Fig. 4.6). The period from establishment of the thermocline until the oxygen concentration in the bottom water becomes close to anoxia (< 0.2 mg O$_2$ l^{-1}, defined as microxia) has decreased from about 75 days in the late 1950's to the present level of about 50 days (Fig. 4.7). Concurrently, the thickness of the microxic water layer has increased and now includes 6-8 m above the lake bottom in late summer.

The shortening of the well-oxygenated period can be explained by increased oxygen demand in the profundal sediment caused by a steady accumulation of biodegradable

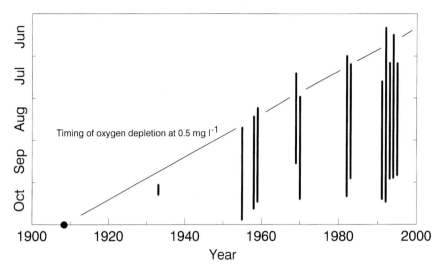

Figure 4.6. The increasing period of oxygen depletion below 0.2 mg O_2 l^{-1} in the bottom layers of Lake Esrom during the 20th century. Vertical bars show the total duration of the microxic period in the respective years.

organic matter. Because we cannot demonstrate enhanced phytoplankton production since 1955 this explanation may imply that accumulation of organic matter started already early in this century. With a constant phytoplankton production since the 1950's, the accumulation of sediment organic matter may still progress, resulting in a slow ongoing reduction of the period from establishment of thermal stratification until microxic conditions develop in the hypolimnion (Fig. 4.7).

Benthic faunal sampling

A response of the fauna to eutrophication is to be expected in all communities, but since the first quantitative studies in Lake Esrom took place in 1932-1934, when the increase in primary production had presumably already occurred, at least in part, it has not been possible to demonstrate any significant long-term changes in the littoral fauna from 1930's to the 1990's. The progressive deterioration of the oxygen conditions in the hypolimnion are, however, likely to have influenced the ambient fauna. Therefore, we

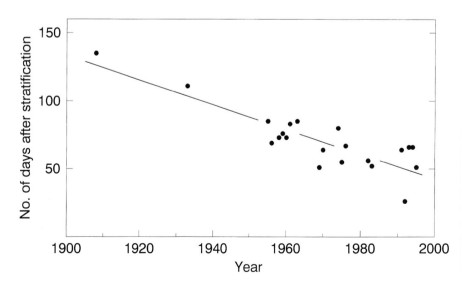

Figure 4.7. The advancing timing of oxygen depletion (< 0.2 mg O_2 l^{-1}) in days after the establishment of the summer stratification in Lake Esrom during the 20th century.

focus our discussion on changes in bathymetric distribution of the sublittoral fauna and changes in abundances of the profundal species.

The choice of bottom sampler, sieving and sorting procedures determine the accuracy of quantitative measurements. Therefore, it may be questionable to compare long-term results obtained with different techniques and by different persons. However, we have compared the methods involved and made the necessary adjustments. A 225 cm²-Ekman sampler and a 200 μm sieve were employed in all studies except in 1932-1936 when a 600 μm sieve was used (Berg 1938). Small animals are lost through the coarse sieve. Approximately 50% of the Tubificidae are lost, and only 4th instar larvae of *Chironomus anthracinus* are fully retained (Jónasson 1955). To include the 1930-data in the comparison we have adjusted the abundance estimates or compare only the large specimens which have been retained by all sieving techniques applied.

Bathymetric distribution of the sublittoral fauna

The species richness of invertebrates in Lake Esrom decreases markedly from more than hundred taxa in the littoral to about 40 species in the sublittoral zone (Jónasson 1978, Dall et al. 1984). Many of the sublittoral species are filtrators exploiting suspended organic matter circulating across the bottom by wind-induced currents in the epilimnion and internal seiches in the metalimnion. The thermocline is located at shallower depth or within the sublittoral zone during summer stratification. Hence, the sublittoral organisms will periodically be exposed to oxygen deficits. Today, the microxic water layer reaches a thickness of 6-8 m, which means that animals located at depths greater than 14 m can experience microxic or anoxic conditions.

In 1989 all sublittoral species disappeared at 13-14 m depth, and only the true profundal species penetrated further down (unpublished data). One species, *Dreissena polymorpha* (Pallas) displayed a slightly wider bathymetric distribution in 1933 than in 1989. Berg (1938) found a few specimens at 14-15 m depth, while presently it does not occur below 13 m (Hamburger et al. 1990). Otherwise, the records from 1932-34 showed the same depth limits except for a few sublittoral species which were observed sporadically at greater depth but at low densities (< 20 individuals m⁻² for *Valvata piscinalis* (O. F. Müller) and *Tanytarsus* sp.). In the 1930's the microxic bottom layer was very thin and existed for a few weeks only. Thus, the high accordance in bathymetric distribution between the previous and present benthic fauna was surprising. Perhaps substratum and/or food limitations may be more important for the bathymetric distribution than the oxygen conditions. An alternative explanation could be that only the true profundal species are able to survive microxic conditions for hours or days. The latter suggestion is supported by experiments with a number of oligochaetes and chironomids from the littoral and sublittoral zone, which – although they are able to regulate the oxygen uptake at decreasing oxygen concentration – survived only a few days' exposure to experimental anoxia (Hamburger et al., in press).

The profundal fauna

The profundal fauna in Lake Esrom includes only six species of macroinvertebrates (Table 4.1), four of which are able to live in soft mud, to survive severe oxygen deficits and to grow at severe limitations of food quantity and quality. The main species – the oligochaete *Potamothrix hammoniensis* (Michaelsen) and the midge *Chironomus anthracinus* Zetterstedt – have haemoglobin and can maintain a high oxygen uptake down to 1 and 3 mg O_2 l⁻¹ respectively. When oxygen concentration drops further, they stop feeding, reduce metabolism and survive by anaerobic

Table 4.1. Average abundance of profundal zoobenthos and the ratio between oligochaetes (O) and oligochaetes plus sedentary chironomids (O+C) in Lake Esrom during 1933 to 1995. Data compiled from Berg (1938) and Jónasson (1972, 1984). The abundance of *Potamothrix hammoniensis* in 1933 may have been underestimated (see text).

Species	1933 no. m^{-2}	1954-60 no. m^{-2}	1962-71 no. m^{-2}	1972-73 no. m^{-2}	1982 no. m^{-2}	1990-95 no. m^{-2}
Potamothrix hammoniensis	3000	10480	9410	19150	3210	19150
Chironomus anthracinus	4780	13000	5070	10030	1130	5460
Procladius pectinatus	20	190	220	80	50	90
Chaoborus flavicans	1390	2020	1380	2420	3260	4660
Pisidium spp.	230	2700	890	650	170	1330
Total	7710	28390	16950	32330	7820	30690
O/O+C	0.39	0.45	0.65	0.66	0.74	0.78

degradation of glycogen (Hamburger et al. 1997). The two *Pisidium* species (*P. casertanum* Poli and *P. subtruncatum* Malm) lack haemoglobin and are unable to regulate respiration as efficiently as *C. anthracinus* and *P. hammoniensis*. At 2-3 mg O_2 l^{-1} their oxygen uptake is reduced by 50% relative to air saturation. They also form glycogen stores and survive the microxic period partly by anaerobic degradation of glycogen (cf. Holopainen 1987). *Procladius pectinatus* Kieffer leaves the profundal zone during summer stratification, and *Chaoborus flavicans* Meigen hides from predators in the profundal mud during daytime, but migrates to the oxygen-rich pelagic zone for feeding during night. When evaluating the possible impact of eutrophication and oxygen conditions on the profundal environment we have, therefore, focused on the permanent inhabitants: *P. hammoniensis*, *C. anthracinus* and the two *Pisidium* species.

Oxygen depletion during the stratification period and food availability are main controlling factors for the abundance and population dynamics of the profundal community. The population dynamics of the profundal species and their adaptations to survive the severe environments in Lake Esrom are described in detail by Jónasson (1972, 1978), and partly summarised by Hamburger et al. (1997).

The life cycle of *C. anthracinus* is uni- or bivoltine. A new generation starts in April-May, and the larvae grow quickly and reach 2nd and/or 3rd instar before oxygen depletion during summer stratification stops the growth. After the autumn overturn, the larvae resume growth and immediately moult to 3rd and/or 4th instar larvae. Growth continues during autumn until temperature decreases to about 4°C. During spring, the mature 4th instar larvae pupate and the adults emerge and swarm. In the sublittoral and the upper profundal zone, the entire population emerges after one year, while part of the 4th instar larvae remain in the deep profundal, continue to grow and emerge in the following spring. If this remaining population surpasses about 2,000 larvae m^{-2}, it prevents recruitment of a new generation by eating the newly-settled eggmasses produced by that part of the population which emerged as adults (Jónasson 1972, Lindegaard et al. 1987). Our longterm data represent the deep profundal area and, consequently, we normally encounter recruitment every second year. The partial hemivoltine life cycle in the deep profundal reflects the longer microxic period which prevents feeding during summer when temperature and food are otherwise favourable.

The life span of *P. hammoniensis* in the profundal zone is 5-6 years, and reproduction takes place when the worms are 3-4 years old (Jónasson and Thorhauge 1972). Cocoons are found in the bottom sediments from late spring to early autumn. Summer stratification probably arrests the development of embryos as suggested by peaks in worm density just after autumn overturn (Jónasson and Thorhauge 1976). Growth of *P. hammoniensis* is fastest during spring and autumn, and the individuals loose weight during summer stratification at low oxygen concentrations (Thorhauge 1976). Cocoons are eaten by 4th instar larvae of *C. anthracinus*, which is reflected by low density of *P. hammoniensis* in years with many *C. anthracinus* (Jónasson and Thorhauge 1972, 1976)

Growth, age and population dynamics of the *Pisidium* species are quite similar to those of *P. hammoniensis* (Holopainen and Jónasson 1983, 1989b). They become 4-5 years old. Growth stops during summer stratification and at low winter temperatures. Eggs are laid in late April, and embryos are released in December or April, six or twelve month later, depending on the hypolimnion temperature.

Long- and short-term fluctuations in abundances

The population density of *C. anthracinus* can vary within a year from almost zero after the emergence of the adults in April-May to more than 40,000 larvae m^{-2} when the larvae hatch in June. High mortality of the young instars reduces the number of wintering 3rd and/or 4th instar larvae to typically 5-10.000 individuals m^{-2}. The original counts, without any corrections for differences in sampling and sorting processes, show a very regular pattern from 1954-1962 with partly uni- and hemivoltine generations and recruitment of very high initial densities every second year (Fig. 4.8). From 1962 this pattern disappears: the average density declines, some generations become almost univoltine, and the high peaks of newly-hatched larvae appear only occasionally. This change in abundance and

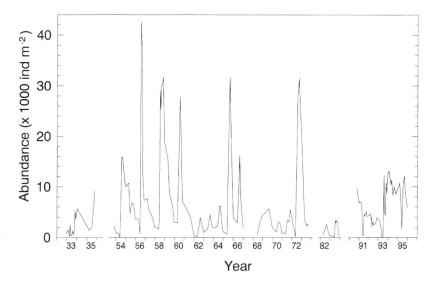

Figure 4.8. Changes in abundance of *Chironomus anthracinus* in the deep profundal zone of Lake Esrom during 1932-1995.

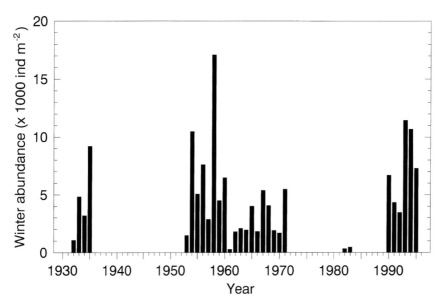

Figure 4.9. Winter abundances (October-March) of *Chironomus anthracinus* in the deep profundal zone of Lake Esrom during 1932-1995. No samples are available for the years 1936-52, 1972-1981, 1984-1989.

population dynamics in the 1960's has been related to the large input of untreated sewage from 1961 to 1971 (e. g. Jónasson 1984, 1993). The very small generations in 1982 and 1983, cannot be explained by sewage input, which had strongly declined by then. The density of the other profundal species was also low during 1982 and 1983 and some sort of sampling error cannot be excluded.

To compensate for differences in sieving procedure and large fluctuations in numbers of small instars, we made a year to year comparison of the abundances during autumn and winter when all larvae (large 3rd or 4th instar) are easy to count and cohort density has stabilized (Fig. 4.9). We still observe slightly higher abundances during the 1950's, but the differences between the four periods (i. e. the 1930's, 1950's, 1960's and 1990's) are small and not significant.

P. hammoniensis does not show large year-to-year variations due to its 5-6-year lifespan,

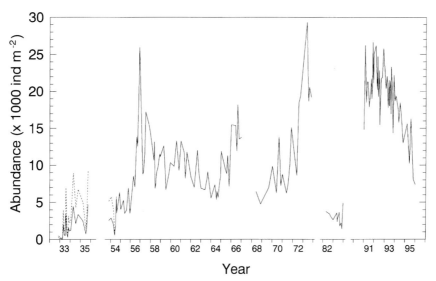

Figure 4.10. Changes in abundance of *Potamothrix hammoniensis* in the deep profundal of Lake Esrom during 1932-1995. Stippled curve shows bundances after correction for animals lost by applying a 600 μm sieve during 1932-1935.

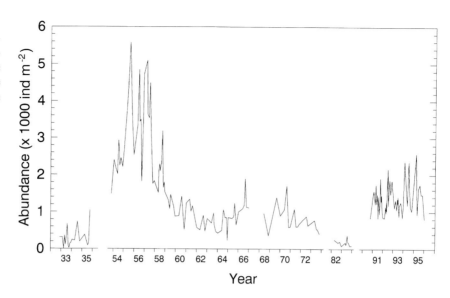

Figure 4.11.
Changes in abundance of the *Pisidium* species in the deep profundal zone of Lake Esrom during 1932-1995.

and because the individuals do not necessarily die after breeding (Fig. 4.10). However, the interspecific competition with *C. anthracinus* is suggested to cause low numbers of worms in years with many chironomid larvae (Jónasson and Thorhauge 1972, 1976). These coupled fluctuations are not always realised, but may be seen in the 1990's where an increase in *C. anthracinus* to about 10,000 larvae m-2 is followed by a decrease in *P. hammoniensis* from about 25,000 to 10,000 individuals m-2. Despite fluctuations in densities of *P. hammoniensis* it seems that the abundance has increased since the 1930's and early 1950's (Table 4.1).

Wiederholm (1980) showed that a ratio between numbers of oligochaetes (O) and oligochaetes plus sedentary chironomids (C), O/O+C correlates with the content of chlorophyll *a* in the great Swedish lakes. Therefore, permanent changes in this ratio over decades may describe fluctuations in trophic level within a lake. The ratio O/O+C in Lake Esrom increased from 0.40-0.45 in the 1930's and 1950's to 0.65 during 1962-1973 and finally to 0.78 for 1990-1995 (Table 4.1). The increase in the O/O+C ratio may reflect the increased duration of the microxic period.

Several conditions may account for the

increase in the abundance of *P. hammoniensis* compared to the constancy, or minor decrease, in the abundance of *C. anthracinus*. Firstly, *P. hammoniensis* may benefit from the possibly increasing amount of accumulated organic matter in the sediment, because it feeds selectively on the bacteria which degrade the organic matter (Jónasson 1993). Secondly, *P. hammoniensis* is better at maintaining oxygen uptake at decreasing ambient oxygen concentration (Hamburger et al. 1997). Thirdly, *P. hammoniensis* survives an anoxic condition longer than *C. anthracinus* (Hamburger et al. 1997).

Three species of *Pisidium* occurred in the profundal of Lake Esrom during the 1950's. One species, *P. henslowanum* (Sheppard), was rare (< 100 individuals m-2) and disappeared from the profundal zone in 1962 (Holopainen and Jónasson 1983). The abundance of the remaining two *Pisidium* species decreased markedly from the 1950's to 1982. During the 1990's the *Pisidium* species have partly recovered, and their abundance is about half of the level in the 1950's (Fig. 4.11). Whether *P. henslowanum* has recolonized the profundal zone in the 1990's is unknown. The low abundances in the 1930's are ascribed to shortcomings in the sorting process.

Natural variability

We suppose that significant changes in abundance and life history from one decade to the next can be ascribed to overall changes in the lake environments (e. g. the slow but progressive deterioration of the hypolimnetic oxygen condition), whereas year-to-year changes within a period may be caused by meteorological factors such as variability in annual temperature, insolation and wind climate. The latter factors influence the physical and chemical regime in the lake in single years and, thereby, the biological structure; weather also affects swarming and egg laying of *C. anthracinus*. Finally, biological interactions such as predation by fish and interspecific competition among the profundal species can cause considerable annual variation.

We have recorded that the duration of thermal summer stratification varies from three to six months, depending on insolation and wind climate, particularly in spring and autumn. Thermal stratification affects the profundal temperature and the length of the microxic period, which can theoretically reach five months. The ability to survive microxia increases with the larval size of the *C. anthracinus* (Hamburger et al. 1997). Thus, an early establishment of the thermocline will markedly shorten the growth period in early summer, resulting in a population of small larvae prior to the initiation of the microxic period. This situation is critical for the survival of the larvae.

The pupation of *C. anthracinus* takes place when the water temperature in spring reaches 5-7° C. After mild winters without ice-cover this temperature is reached in early April, often exposing swarming adults to cold, rainy and windy weather which can hamper mating and egg laying. Cold winters, on the other hand, cause late emergence in mid-May with improved odds for breeding success. The small recruitment in 1991 was apparently caused by a windy and cold mid-April. Calm and sunny weather resulted in a successful breeding in May 1993, when eggmasses with more than 100,000 individuals m^{-2} were found in the profundal zone. In the laboratory, eggmasses showed an almost 100% success in hatching small 1st instar larvae. Nevertheless, the resulting population in the lake a few weeks later was only about 10,000 individuals m^{-2}. This result might suggests that sediment condition have deteriorated compared to the 1950's, when the large initial densities were recorded (Fig. 4.8).

A large eel population in Lake Esrom is considered the main predator on profundal *C. anthracinus* larvae. Jónasson (1972) observed a high mortality before and after the microxic period and related this to predation, mainly by eel. A similar pattern of mortality has not be recognized in the 1990's, when population densities of *C. anthracinus* during autumn were almost constant. This finding may be correlated to the drastic decrease of the eel population measured as the yearly catches, which have declined from 10-20 t y^{-1} during 1950-1990 to about 5 t y^{-1} in 1991-1994 and recently to less than 1 t y^{-1}. The high chironomid densities during autumn and winter in the 1990's may thus be a consequence of reduced predation from a diminishing eel population. A predation similar to the previous 1950's level could have diminished the present population density of *C. anthracinus*, and thus revealed the increased stress on the population, i.e. reduced oxygen availability (cf. Fig. 4.7) and less successful recruitments (Fig. 4.8).

Concluding remarks

Natural year to year variation heavily influences abundance and life history of the profundal fauna. Superimposed on this variation was a long-term decline in density of the *Pisidium* species from 1932 to 1995, and increased numbers of *P. hammoniensis* during that same period. We consider the *C. anthracinus* population as almost constant in numbers, although the large initial population densities from the early 1950's have not been observed during the investigations in the 1990's. We ascribe the progressive deterioration of the hypolimnetic oxygen regime during summer stratification to be responsible for the long-term trend. The observed changes in the profundal fauna of Lake Esrom, however, have been comparatively small, and the present community still includes numerous individuals compared to other eutrophicated lakes.

Several factors are probably responsible for the preservation of a rich profundal fauna in Lake Esrom. The catchment area is small, and loading with nitrate from agricultured areas has therefore been limited. Sewage input of phosphorus had limited effect, because nitrogen always was or soon became the main limiting nutrient for phytoplankton growth. The present primary production is about 240 g C m^{-2} y^{-1}, which is believed to exceed the pre-1900 level about 2-fold. The increased sedimentation of biodegradable organic matter in the profundal zone has prolonged the duration of microxic conditions. Such a prolonged microxic and anaerobic period as found in Lake Esrom normally restricts macroinvertebrate life in the profundal zone and leads to the disappearance of *C. anthracinus* (e. g. Thienemann 1954, Hofmann 1971).

Hamburger et al. (1997) have described how three adaptations to survive profundal oxygen depletion are successfully combined by *C. anthracinus* in Lake Esrom. They conclude, nevertheless, that a microxic period of two to four months in Lake Esrom exceeds the period during which *C. anthracinus* is able to survive anoxia, and they suggest that the profundal animals in Lake Esrom are rarely exposed to complete anoxia and that wind induced circulations, facilitated by the morphometry of the large profundal zone, supply sufficient oxygen to maintain a definitely low, but mainly aerobic, metabolism during summer stratification.

We have records of the profundal fauna back to the 1930's. At that time the primary production had presumably already increased and oxygen conditions deteriorated. The species composition in the profundal zone was almost the same as today: rich in specimens but poor in species. With a primary production at the supposed pre-cultural level (70-100 g C m^{-2} y^{-1}) and without microxic conditions during summer stratification, the profundal fauna composition might have been richer in species, and a number of the present sublittoral species (e. g. the crustaceans *Pallacea quadrispinosa* Sars, *Asellus aquaticus* L., the chironomid *Tanytarsus bathophilus* Kieffer and the oligochaete *Psammoryctides barbatus* (Grube)) were likely permanent members of the profundal zone. A palaeoecological investigation of especially the crustacean and chironomid subfossil remnants in the profundal sediment of Lake Esrom is required to reveal the ancient species composition of the profundal community and date the time of likely changes.

Literature cited

Berg, K. 1938. Studies on the bottom animals of Esrom Lake. Det Kongelige Danske Videnskabernes Selskabs Skrifter, Naturvidenskabelig og Mathematisk Afdeling 9, 8: 1-255.

Christoffersen, K., and K. Olrik. 1996. Giftige alger i Esrum Sø og Å. Vand og Jord 3: 21-23.

Dall, P. C., C. Lindegaard, E. Jónasson, G. Jónasson, and P. M. Jónasson. 1984. Invertebrate communities and their environment in the exposed littoral zone of Lake Esrom, Denmark. Archiv für Hydrobiologie Supplementum 69: 477-524.

Edmonson, W. T. 1991. The uses of ecology: Lake Washington and beyond. University of Washington Press, Seattle, U. S. A.

Frederiksborg Amt. 1991. Esrum Sø. Plante- og dyreplankton 1990. Rapport udført for Frederiksborg Amt af Miljøbiologisk Laboratorium APS, Humlebæk, Denmark.

Goldman, C. R. 1993. The conservation of two large lakes: Tahoe and Baikal. Verhandlungen der Internationale Vereinigung für Theoretische und Angewandte Limnologie 25: 388-391.

Hamburger, K., P. C. Dall, and P. M. Jónasson. 1990. The role of *Dreissena polymorpha* Pallas (Mollusca) in the energy budget of Lake Esrom, Denmark. Verhandlungen der Internationale Vereinigung für Theoretische und Angewandte Limnologie 24: 621-625.

Hamburger, K., P. C. Dall, and C. Lindegaard. 1994. Energy metabolism of *Chironomus anthracinus* (Diptera: Chironomidae) from the profundal zone of Lake Esrom, Denmark, as a function of body size, temperature and oxygen concentration. Hydrobiologia 294: 43-50.

Hamburger, K., P. C. Dall, and C. Lindegaard. 1995. Effects of oxygen deficiency on survival and glycogen content of *Chironomus anthracinus* (Diptera, Chironomidae) under laboratory and field conditions. Hydrobiologia 297: 187-200.

Hamburger, K., C. Lindegaard, and P. C. Dall. 1996. The role of glycogen during the ontogenesis of *Chironomus anthracinus* (Chironomidae, Diptera). Hydrobiologia 318: 51-59.

Hamburger, K., C. Lindegaard, and P. C. Dall. 1997. Metabolism and survival of benthic animals short of oxygen. Chapter 11 *in* K. Sand-Jensen and O. Pedersen, editors. Freshwater Biology – Priorities and Development in Danish Research. Gad, Copenhagen, Denmark.

Hamburger, K., C. Lindegaard, and P. C. Dall (in press). Strategies of respiration and glycogen metabolism in oligochaetes and chironomids from habitats exposed to different oxygen deficits. Verhand-

lungen der Internationale Vereinigung für Theoretische und Angewandte Limnologie 26.

Hofmann, W. 1971. Die postglaziale Entwicklung der Chironomiden- und *Chaoborus*-Fauna (Dipt.) des Schöhsees. Archiv für Hydrobiologie Supplementum 40: 1-74.

Holopainen, I. J. 1987. Seasonal variation of survival time in anoxic water and the glycogen content of *Sphaerium corneum* and *Pisidium amnicum* (Bivalvia, Pisidiidae). American Malacological Bulletin 5: 41-48.

Holopainen, I. J., and P. M. Jónasson 1983. Long-term population dynamics and production of *Pisidium* (Bivalvia) in the profundal of Lake Esrom, Denmark. Oikos 41: 99-117.

Holopainen, I. J., and P. M. Jónasson 1989a: Bathymetric distribution and abundance of *Pisidium* (Bivalvia: Sphaeriidae) in Lake Esrom, Denmark, from 1954 to 1988. Oikos 55: 324-334.

Holopainen, I. J., and P. M. Jónasson 1989b. Reproduction of Pisidium (Bivalvia, Sphaeriidae) at different depths in Lake Esrom, Denmark. Archiv für Hydrobiologie 116: 85-95.

Hovedstadsrådet. 1989. Esrom Sø 1900-1988. – Recipientovervågning nr. 40. Hovedstadsrådet, Copenhagen, Denmark, (in Danish).

Jónasson, P. M. 1955. The efficiency of sieving techniques for sampling freshwater bottom fauna. Oikos 6: 183-207.

Jónasson, P. M. 1972. Ecology and production of the profundal benthos in relation to phytoplankton in Lake Esrom. Oikos Supplementum 14: 1-148.

Jónasson, P. M. 1978. Edgardo Baldi Memorial lecture. Zoobenthos of Lakes. Verhandlungen der Internationale Vereinigung für Theoretische und Angewandte Limnologie 20:13-37.

Jónasson, P. M. 1984. Decline of zoobenthos through five decades of eutrophication in Lake Esrom. Verhandlungen der Internationale Vereinigung für Theoretische und Angewandte Limnologie 22:800-804.

Jónasson, P. M. 1993. Lakes as a basic resource for development: the role of limnology. Memorie dell'Istituto Italiano di Idrobiologia 52:9-26.

Jónasson, P. M., and F. Thorhauge. 1972. Life cycle of *Potamothrix hammoniensis* (Tubificidae) in the profundal of a eutrophic lake. Oikos 23:151-158.

Jónasson, P. M., and F. Thorhauge. 1976. Population dynamics of *Potamothrix hammoniensis* in the profundal of Lake Esrom with special reference to environmental and competitive factors. Oikos 27:193-203.

Jónasson, P. M., E. Lastein, and A. Rebsdorf 1974. Production, insolation, and nutrient budget of eutrophic

Lake Esrom. Oikos 25: 255-277.

Likens, G. E., (ed.) 1989. Long-term studies in ecology. Springer-Verlag New York Inc., New York, U. S. A.

Lindegaard, C., P. C. Dall, and S. B. Hansen 1993. Natural and imposed variability in the profundal fauna of Lake Esrom, Denmark. Verhandlungen der Internationale Vereinigung für Theoretische und Angewandte Limnologie 25: 576-581.

Lund, J. W. G. 1978. Changes in the phytoplankton of an English lake, 1945-1977. Hydrobiological Journal 14: 10-27.

Nygaard, G. 1938. Hydrobiologischen Studien über dänische Teiche und Seen. Archiv für Hydrobiologie 32: 523-692.

Riemann, B., and H. Mathiesen 1977. Danish research into phytoplankton primary production. Folia Limnologica Scandinavica 17: 49-54.

Sand-Jensen, K. 1997. Eutrophication and plant communities in Lake Fure during 100 years. Chapter 3 *in* K. Sand-Jensen and O. Pedersen, editors. Freshwater Biology – Priorities and Development in Danish Research. Gad, Copenhagen, Denmark.

Thienemann, A. 1954. *Chironomus*. Leben, Verbreitung und wirtschaftlische Bedeutung der Chironomiden. Die Binnengewässer 20: 1-834.

Thomas, E. A. 1979. Planktonleben und Stoffkreisläufe; physikalishe und chemische Einflüsse. Pages 61-87 *in* Der Zürichsee und seine Nachbarseen. Office du Livre. Buchverlag Neue Züricher Zeitung. Zürich, Switzerland.

Thorhauge, F. 1976. Growth and life cycle of *Potamothrix hammoniensis* (Tubificidae, Oligochaeta) in the profundal of eutrophic Lake Esrom. A field and laboratory study. Archiv für Hydrobiologie 78: 71-85.

Wesenberg-Lund, C. 1904. Plankton investigations of the Danish lakes. Part I-II. Gyldendalske Boghandel, Copenhagen, Denmark.

Wiederholm, T. 1980. Use of benthos in lake monitoring. Journal of Water Pollution Control Federation 52: 537-547.

Wiederholm, T. 1988. Changes in the profundal Chironomidae of Lake Mälaren during 17 years. Spixiana Supplement 14: 7-15.

5 Plants and environmental conditions in Danish *Lobelia*-lakes

By Kaj Sand-Jensen and Morten Søndergaard

Danish *Lobelia*-lakes are small oligotrophic softwater lakes located in uncultivated sandy regions. The lakes have been intensively studied because of their high water quality and unique vegetation of small rooted vascular plants of the isoetid growth form. The isoetids have a remarkable ability to exploit CO_2 and nutrients by root uptake in the sandy nutrient-poor sediments and to influence sediment processes through extensive root release of O_2. The isoetid species photosynthesize and grow slowly, but remain active throughout the year resulting in an important role in organic carbon production and nutrient cycling in the lakes. The few *Lobelia*-lakes remaining in Denmark are threatened by acidification and nutrient enrichment of the main water inputs through precipitation and groundwater, and we foresee a continued deterioration of their quality in the future.

Lobelia dortmanna is a common representative of the isoetid growth form and has given the name *Lobelia*-lakes to the nutrient-poor, softwater lakes in North America and Europe (Wium-Andersen and Andersen 1972a,b). The isoetids are small plants with a rosette of short stiff leaves, a dense root system, and large air lacunae in leaves and roots that allow efficient intra-plant transport of CO_2 and O_2 (Sand-Jensen and Prahl 1982, Raven et al. 1988). The isoetids are adapted to growth in exposed, nutrient-poor sediments, and they dominate the vegetation in the shallow water of oligotrophic lakes in temperate and subarctic regions (Sculthorpe 1967, Sand-Jensen and Søndergaard 1979). The isoetid physiology generates a strong dependence and influence on the sediment environment and the overall nutrient dynamics in the lakes (Christensen and Sørensen 1986, Pedersen et al. 1995).

The Danish *Lobelia*-lakes are restricted to sandy uncultivated regions of mid- and western Jutland (Iversen 1929). The lakes are mainly surrounded by forest or heather vegetation and are often without surficial water input (Sand-Jensen and Lindegaard 1996). This location and hydrology minimises the external nutrient input and maintains the transparent and oligotrophic status which is a prerequisite for the dominance of the isoetid vegetation. Many of the Danish *Lobelia*-lakes have lost their isoetid vegetation during the past hundred years because of eutrophication by domestic sewage and drainage from fertilised fields. The few remaining *Lobelia*-lakes deserve a high priority in nature conservation because of their high water quality, remarkable vegetation and status as reference lakes for a relatively low human impact. The Danish *Lobelia*-lakes are, nonetheless, threatened by increased input of acids and nutrients and increased recreational activity (Sand-Jensen 1980, Sand-Jensen and Søndergaard 1981, Riis et al. 1996).

The research interest in Danish *Lobelia*-lakes probably originated from their high water quality and scenic beauty. Also, studies on the ecology and physiology of submerged macrophytes have a long Danish tradition (e.g. Petersen 1917, Iversen 1929, Steemann Nielsen 1947, Olsen 1950), which stimulated the interest in the remarkable freshwater isoetid plants, whose photosynthesis and inorganic carbon sources became a subject for research (Nygaard 1958, Steemann Nielsen 1960). More than any other publication, however, Wium-Andersen's finding of CO_2 uptake by the roots of *Lobelia dortmanna* – the first demonstration in the plant kingdom of the sediment as the main inorganic carbon source – stimulated the national and international interest in the ecophysiology of isoetid species.

In this review we discuss four aspects of the environmental conditions and the ecophysiology of the isoetid vegetation in *Lobelia*-lakes: 1) Distribution and abundance of macrophytes in the lakes, 2) Ecophysiology of the isoetids, focusing on the regulation of photosynthesis and the exchange of CO_2 and O_2 between the plants and the sediments, 3) Primary production of phytoplankton, epiphytes and macrophytes in the lakes, and 4) Changes of plant communities in the Danish *Lobelia*-lakes in the context of the environmental threats associated with acidification and eutrophication. We focus on the isoetid communities because they have been our main area of research and have generated significant international interest. Most of the examples presented are from investigations in Lake Grane Langsø and Lake Kalgaard.

Distribution and abundance of the isoetid vegetation

Most Danish *Lobelia*-lakes are small oligotrophic softwater seepage lakes. The three

most common of the isoetid species are *Isoetes lacustris*, *Littorella uniflora* and *Lobelia dortmanna*, but other widespread species are *Carex oederi*, *Elatine hexandra* and *Juncus bulbosus*. Very rare freshwater isoetid species in Denmark include *Isoetes echinospora* and *Subularia aquatica* (Moeslund et al. 1990). *Isoetes lacustris* and *Lobelia dortmanna* are confined to the most oligotrophic softwater lakes, while *Littorella uniflora* is more widely distributed in mesotrophic alkaline lakes. Occasionally, *Littorella uniflora* maintains small populations on exposed sediments in shallow water of alkaline lakes which have been eutrophied during the past hundred years (Sand-Jensen and Søndergaard 1981).

The vegetation of *Lobelia*-lakes is quantitatively dominated by isoetid species. The emergent reed plants are often absent, and if present at the land-water interface, they usually form short, open stands of low shoot density because of the nutrient-poor sediments. Most of the isoetid species are amphibious organisms which flower in air and grow well above and below the water (Pedersen and Sand-Jensen 1992). Isoetids can, therefore, form a homogeneous vegetation in the transition zone from land to water, and they can grow within the open reed stands. Floating-leaved macrophytes also represent a sparse cover and density in the Danish *Lobelia*-lakes, though they can apparently reach high biomasses in some foreign oligotrophic lakes (e.g. Hutchinson 1975, Moeller 1975). In Lake Kalgaard, the emergent and floating-leaved macrophytes occupy only 20% of the potentially available lake bottom between 0 and 2 m, and their density is very low (Sand-Jensen and Søndergaard 1979). The emergent and floating-leaved macrophytes may form denser stands in some *Lobelia*-lakes at sites exposed to local nutrient enrichment from small inlets and seepage from fertilised fields close to the lakes.

Submerged macrophytes in lakes often form a distinct vertical zonation controlled

Table 5.1. Depth limits (m) of submerged macrophytes in Grane Langsø in 1955 (Nygaard 1958) and 1994 (Riis et al. 1996).

Species	1955	1994
Lobelia dortmanna	3	1.5
Littorella uniflora	4	1.5
Isoetes lacustris	5	3.0
Drepanocladus exannulatus	11	10.5
Sphagnum subsecundum	10	10.0
Nitella flexilis	11	extinct

by differences in the resistance to physical exposure, the light requirements and the competitive ability among species (Spence 1967). Isoetids have their photosynthetic tissue restricted to a 3-15 cm narrow zone above the sediment, and they have a poor ability to compete for light, compared with elodeid species which develop tall ramifying canopies and can regulate their vertical position and density of leaves dependent on the water transparency (Adams and McCracken 1974). Isoetids, on the other hand, are well protected against the physical exposure in shallow water, because the leaves are placed in a basal rosette (Farmer and Spence 1986). *Lobelia dortmanna* in particular can form leaves closely compressed against the sediment to resist strong wave action, whereas *Isoetes lacustris* have more erect and fragile leaves and are absent from the most shallow

Table 5.2. Depth distribution and average biomass of isoetids within the colonized area in Lake Kalgaard. Data from Sand-Jensen and Søndergaard (1979).

Species	Depth (m)	Biomass (g org. DW m⁻²)
Lobelia dortmanna	0 – 1.5	5
Littorella uniflora	0 – 2.5	112
Isoetes lacustris	1.35 – 4.5	66

and exposed sites (Sand-Jensen and Søndergaard 1979).

The common isoetids *Littorella uniflora* and *Lobelia dortmanna* penetrate to about 3 and 4 m respectively in the clearest Danish *Lobelia*-lakes (Table 5.1, Nygaard 1958). The two species normally coexist in shallow water, but *Littorella uniflora* has a higher density and biomass and penetrates deeper than *Lobelia dortmanna* (Table 5.1 and 5.2). *Isoetes lacustris* is usually absent from shallow water but grow deepest among the isoetids. The depth limit of *Isoetes lacustris* is about 5 m in the clearest Danish lakes (Table 5.1), but the species has been observed at 8-m depth in some oligotrophic English lakes (Hutchinson 1975). In transparent *Lobelia*-lakes, mosses or characeans occupy the largest depths below the isoetid vegetation (Nygaard 1958, Moeller 1975, Sand-Jensen and Søndergaard 1981). During the 1950's, for example, species of *Drepanocladus, Sphagnum* and *Nitella* formed an almost complete cover on the bottom of Grane

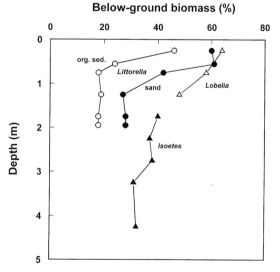

Figure 5.1. Percentage below-ground biomass of *Isoetes lacustris, Littorella uniflora* and *Lobelia dortmanna* at depth in Lake Kalgaard during summer. Redrawn from data in Sand-Jensen and Søndergaard (1979) with permission from Blackwell Science Ltd.

Langsø at depths from 8 to 11 m (Table 5.1, Nygaard 1958).

A unique morphological feature of the isoetids is the high percentage of plant biomass allocated to the roots (Wetzel 1983, Raven et al. 1988). The below-ground biomass of *Lobelia dortmanna* in Lake Kalgaard is 50-60%, with the highest percentages occurring in plants from shallow water (Fig. 5.1, Sand-Jensen and Søndergaard 1979). The highest percentage of below-ground biomass of *Littorella uniflora* is also found in shallow water and on the most nutrient-poor sandy sediments with low organic content (Fig. 5.1). The percentage of below-ground biomass of *Littorella uniflora* decreases steeply from about 60% to 20% as the organic content of the sediment increases from close to 0 to more than 2% of the dry weight (Sand-Jensen and Søndergaard 1979). *Isoetes lacustris* is absent from the most sandy and nutrient-poor sediments in shallow water, and the species has a percentage of below-ground biomass of about 40%, decreasing only slightly with depth (Fig. 5.1). Sand-Jensen and Søndergaard (1979) proposed that the reduced allocation of biomass to the roots of *Littorella uniflora*, as the organic content increased in the sandy sediments, was due to the higher availability of inorganic nutrients and CO_2. This explanation was supported by the higher plant size and biomass of *Littorella uniflora* as the organic content increased in the sediments from less than 1% to 2-5% of the dry weight among different shallow sites in Lake Kalgaard (Sand-Jensen and Søndergaard 1979). Further support for this explanation was gained by experiments with *Littorella uniflora* from sandy sites in Lake Hampen. The plants were nutrient-limited and responded to phosphorus addition by enhanced growth (Christiansen et al. 1985).

Another important feature of Danish *Lobelia*-lakes is the dense, carpet-like cover of the isoetids. The shoot density often exceeds 5000 m^{-2}, and the mean biomass for the entire *Littorella uniflora* zone can surpass 100 g DW m^{-2} (Table 5.2). The closed isoetid cover of high biomass is probably due to the fact that the Danish *Lobelia*-lakes are small, physically protected and have more stable and nutrient-rich sediments compared with many foreign *Lobelia*-lakes (Nygaard 1958, Moeller 1975, Sand-Jensen and Søndergaard 1979). The high biomass is a prerequisite for the quantitative importance of the isoetids in the carbon and nutrient cycling of the Danish *Lobelia*-lakes (Sand-Jensen 1980).

Ecophysiology of the isoetid vegetation

The isoetids inhabit the most nutrient-poor lake sediments and show all the characteristics of stress-selected organisms (*sensu* Grime 1977) to acquire and maintain the limited nutrient resources (Boston 1986, Farmer and Spence 1986). The isoetids are perennial species with an evergreen biomass. They photosynthesise and grow throughout the year, and seasonal variations in photosynthetic capacity are much smaller than for other submerged macrophytes (Sand-Jensen 1978, Søndergaard and Bonde 1988). The specific growth rates are very low in isoetids, in accordance with the low photosynthetic rates and the higher respiratory costs associated with the high proportions of roots and stem of the plant biomass (Moeller 1978, Sand-Jensen and Søndergaard 1978, Nielsen and Sand-Jensen 1991, Madsen et al. 1993). There is, however, considerable variability in resource requirements and metabolic rates among the different isoetids species. Rates of photosynthesis, respiration and growth are higher for *Littorella uniflora* than for *Lobelia dortmanna* (Sand-Jensen and Søndergaard 1978, Madsen et al. 1993). Isoetid species also differ markedly in their structure and exchange of CO_2 and O_2 with the sediments (Sand-Jensen et al. 1982, Boston et al. 1987, Raven et al. 1988). Finally, isoetids vary in their capacity to fix carbon in the dark through C-4 metabolism (Richardson et al. 1984, Madsen 1985).

Isoetid photosynthesis at depth

It is a general conclusion that deep-growing plants need to optimise light capture for the minimum investment in biomass and to reduce respiratory costs (Sand-Jensen and Madsen 1991, Markager and Sand-Jensen 1994). In that respect, isoetid species are poorly adapted to shaded environments because of the thick leaves placed in the basal rosette and the large biomass in the stem and the roots. A comparison of the depth distribution of different submerged macrophytes in lakes of variable transparency world-wide showed that isoetids require more light at the depth limit than elodeid angiosperms, characeans and mosses (Middelboe and Markager 1997). The elodeid angiosperms have erect stems, thinner leaves and smaller root systems in order to facilitate light capture, and characeans and mosses have no roots and often very low respiratory rates (Nygaard and Sand-Jensen 1981, Riis and Sand-Jensen 1997).

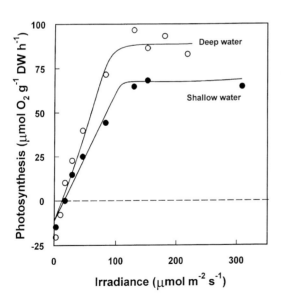

Figure 5.2. Photosynthesis-irradiance curves of entire plants of *Littorella uniflora* collected from shallow (0.2 m) and deep water (2.3 m) in Lake Kalgaard during summer. Redrawn from Søndergaard and Bonde (1988) with permission from Elsevier Science B. V.

Isoetid species experience substantial self-shading among overlapping leaves. They can acclimatise to sun and shade conditions at depth by varying the number and inclination of leaves, the relative allocation of biomass to leaves and roots, and the pigmentation and photosynthetic capacity of leaves. However, the metabolic plasticity is generally much lower among the stress-selected isoetid species than among other freshwater angiosperms (Sand-Jensen 1978, Søndergaard and Bonde 1988). Pigmentation and photosynthesis of *Littorella uniflora* have been studied within its depth limits (0-2.3 m) in Lake Kalgaard (Søndergaard and Bonde 1988). The deep water plants from 2.3 m showed considerably higher leaf chlorophyll concentration, presumably due to larger antennae for light capture judging from measurements of higher chlorophyll b/a ratios (0.39) than in the shallow water plants from 0.2 m (0.30). This difference was accompanied by higher initial slopes and maximum rates of the photosynthesis curves for deep water than for shallow water plants (Fig. 5.2). The deep water plants also had much higher leaf/root ratios (3.3) than the shallow water plants (1.5), resulting in lower light compensation points at depth (Søndergaard and Bonde 1988).

It is likely that the increasing sediment content of organic matter and nutrients at depth stimulated chlorophyll formation, electron transport capacity and Rubisco activity and, thus, the photosynthetic performance of the deep water plants relative to the most shallowly growing plants from the coarse sediments in the surf zone (Søndergaard and Bonde 1988). Differences in pigmentation and photosynthesis among plants were smaller in laboratory experiments across light gradients with otherwise similar environmental conditions (Table 5.3). The most prominent influence of decreasing light was that the proportions of leaves increased and the respiratory rates declined to maintain a photosynthetic surplus on a 24-hour basis.

Table 5.3. Morphology and physiology of *Littorella uniflora* and *Isoetes lacustris* after 10-14 weeks' growth in the laboratory at different irradiances in a 14-10 hour light-dark cycle. Photosynthesis was measured on detached leaves at saturating irradiance and CO_2 concentrations. Respiration was measured in the dark. All values are means of 3 5 replicates at 15 °C. Different markers are indicators of significant differences (p < 0.05).

Species	Irradiance (μmol m^{-2} s^{-1})	Leaf biomass (% of total)	Chlorophyll (mg g^{-1} DW)	Photosynthesis (mg O_2 mg^{-1} chl h^{-1})	Respiration (mg O_2 mg^{-1} chl h^{-1})
Littorella	172	45[a]	5.4[b]	1.22[b]	0.24[a]
uniflora	52	59[b]	7.2[c]	0.76[c]	0.10[b]
	39	56[b]	8.5[c]	0.79[c]	0.10[b]
	22	58[b]	7.6[c]	0.76[c]	0.11[b]
	6[*]	63[b]	7.9[c]	0.67[d]	0.09[b]
Isoetes	172		4.7[a]	1.45[a]	0.090[b]
lacustris	52		4.9[a]	1.18[b]	0.075[c]
	39		5.4[b]	0.95[b]	0.075[c]
	22		5.4[b]	1.11[b]	0.055[c]
	6		5.9[b]	0.92[c]	

[*] The plants lost weight and became senescent with time

Populations of *Isoetes lacustris* grow deeper in Lake Kalgaard, between 1.5 and 4.5 m, than *Littorella uniflora* (Sand-Jensen and Søndergaard 1979). The same distinct depth zonation is also found in other mid-Jutland lakes (Nygaard 1958, Sand-Jensen and Søndergaard 1981). *Littorella uniflora* forms a denser, more homogeneous cover by vegetative propagation by runners than *Isoetes lacustris* and *Lobelia dortmanna*, which disperse by macrospores and seeds respectively (Table 5.2, Sand-Jensen and Søndergaard 1979). The depth limit of *Littorella uniflora* is often very distinct and varies among lakes according to differences in light penetration (Sand-Jensen and Søndergaard 1981). This pattern suggests that the depth limit reflects the minimum amount of light required to ensure growth and survival of the population (Sand-Jensen and Madsen 1991). The minimum light requirement is apparently much lower for *Isoetes lacustris* than for *Littorella uniflora*. Photosynthesis-light relationships are not significantly different between the two isoetid species, but markedly lower dark respiration rates of *Isoetes lacustris* may account for the ability of this species to penetrate much deeper into the lakes than *Littorella uniflora* (Sand-Jensen 1978, Table 5.3).

Use of sediment CO_2 among isoetids
Measurements of ^{14}C-fixation of *Lobelia dortmanna* in split-chambers have shown that the majority of CO_2 assimilated is taken up by the roots (Wium-Andersen 1971). Both structural, physiological and environmental conditions are responsible for this large root uptake of CO_2. *Lobelia dortmanna* has a particularly high surface area and gas permeability of the roots relative to the leaves, and large, continuous air lacunae ensure efficient intra-plant gas transport over the short distance between roots and leaves (Sand-Jensen and Prahl 1982). In the natural lake environment, the roots are surrounded by sediments high in CO_2, while the leaves are surrounded by lake water close to air equilibration (about 0.015 mM CO_2). Carbon dioxide concentrations in the sediment pore-water typically range from 0.3 mM to 3 mM depending on

the variable rates of bacterial mineralization and seepage of CO_2-rich groundwater (Wium-Andersen and Andersen 1972a, Sand-Jensen and Søndergaard 1979, Pedersen et al. 1995).

Experiments by Nygaard (1958) already showed that respiration exceeded photosynthesis when *Lobelia dortmanna* was kept in air-equilibrated water of low CO_2 concentration. The CO_2-rich sediment is, therefore, required for the plants to achieve a net photosynthetic gain in the light. The CO_2 compensation point of *Lobelia dortmanna* is high (about 0.080 mM, Sand-Jensen 1987) and, like other isoetid species, it is unable actively to utilize HCO_3^- (Madsen and Sand-Jensen 1991). Moreover, *Lobelia dortmanna* does not temporarily fix CO_2 in C-4 acids in the dark, like *Littorella uniflora* and species of *Isoetes* do, to supplement the supply of CO_2 in the light (Keeley 1981, Madsen 1985). Thus, *Lobelia dortmanna* is strongly dependent on the CO_2 supply from the sediment, and concentrations above 3 mM CO_2 in the pore-water are required to saturate photosynthesis (Pedersen et al. 1995). This CO_2 requirement is so high that rates of photosynthesis of *Lobelia dortmanna* should change with the spatial and temporal variations of CO_2 concentrations in the sediments.

Other common isoetid species such as *Littorella uniflora* and *Isoetes lacustris* also use the sediment as the main CO_2 source for photosynthesis, though the resistance to CO_2 uptake is higher from the roots than from the leaves for the same external concentration of CO_2 (Søndergaard and Sand-Jensen 1979a, Sand-Jensen 1983, Boston et al. 1987, Raven et al. 1988). Leaves of *Littorella uniflora* and *Isoetes lacustris* have a thinner cuticle and smaller transport resistance than leaves of *Lobelia dortmanna* (Sand-Jensen unpublished data). The gas transport also has to pass thin tissue barriers between leaves and roots in *Littorella uniflora* and *Isoetes lacustris*, whereas the aerenchyma is continuous in

Lobelia dortmanna. Finally, the surface area of roots relative to leaves is usually lower for *Littorella uniflora* and *Isoetes lacustris* than for *Lobelia dortmanna*. The surface area of roots relative to leaves does, however, depend on the light availability and the sediment content of nutrients and CO_2 (Fig. 5.1, Madsen unpublished data). Studies on *Littorella uniflora* showed that the plants were smaller – but the surface area of roots relative to leaves was higher – on sandy sediments poor in nutrients and CO_2 than on richer sediments (Søndergaard and Sand-Jensen 1979a). The small plants with high relative root surface area had a higher affinity for sediment CO_2 during photosynthesis than the large plants with low relative root surface area. Despite considerable variability in leaf and root development, *Littorella uniflora* and *Isoetes lacustris* receive the majority of CO_2 from the sediment because the CO_2 concentrations here are several-fold higher than in the lake water (Søndergaard and Sand-Jensen 1979a, Boston 1986).

Many isoetids are amphibious species which form continuous populations from below to well above the mean water level (Sculthorpe 1967). The water level can fluctuate by more than 1 m between dry summers and wet winters because many of the *Lobelia*-lakes are seepage lakes. Hence, *Littorella uniflora* and *Lobelia dortmanna* often grow exposed to air during summer. *Littorella uniflora* can form extensive monospecific populations on sandy flats which dry up every summer. Utilization of sediment CO_2 continues in both species on land where they maintain many of the submergent characteristics (e.g. short leaves, large lacunae, sparse stomata, Nielsen et al. 1991, Pedersen and Sand-Jensen 1992). Land forms of *Littorella uniflora* produce new and thinner leaves with apical stomata and can obtain a net carbon gain during photosynthesis in air (Nielsen et al. 1991). Provided high CO_2 concentrations (about 0.5 mM) are maintained in a water-

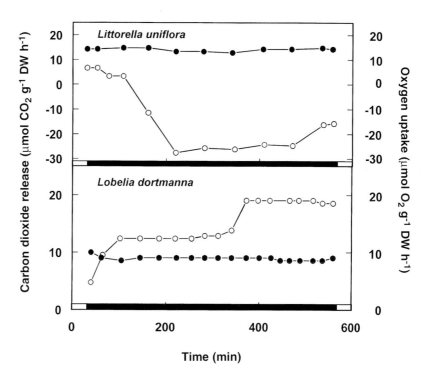

Figure 5.3. Uptake and release of CO$_2$ (open circles) and O$_2$ (closed circles) in *Littorella uniflora* and *Lobelia dortmanna* during a dark period. *Littorella uniflora* releases CO$_2$ in the early part and takes up CO$_2$ during the remaining part of the dark period. *Lobelia dortmanna* releases CO$_2$ throughout the dark period. Both species consume O$_2$ in the dark. Redrawn from Madsen (1985) with permission from Elsevier Science B. V.

saturated soil, estimates suggest that the plant would still obtain about 80% of the CO$_2$ supply from the hydrosoil (Nielsen et al. 1991). The dependence of the hydrosoil for CO$_2$ supply is even higher for land forms of *Lobelia dortmanna*, which develops leaves with a thicker cuticle and without stomata and derive more than 95% of the CO$_2$ supply from the wet soils (Pedersen and Sand-Jensen 1992).

Dark fixation of CO$_2$ among isoetids
Isoetids such as *Littorella uniflora* and species of *Isoetes* show a substantial dark fixation of CO$_2$ into C-4 acids (Keeley 1981, Keeley and Bowes 1982, Richardson et al. 1984, Madsen 1985). This temporal separation between C-3 fixation reactions in the light and C-4 fixation reactions in the dark resemble the crassulacean metabolism (CAM) of some terrestrial plants. Unlike the terrestrial CAM-plants, the isoetids maintain a net CO$_2$ influx during most of the light and dark periods, and CO$_2$ fixation is much higher

in the light than in the dark (Fig. 5.3, Madsen 1985). *Lobelia dortmanna* does not share these features and releases CO$_2$ at night in approximately equal molar amounts as the consumption of O$_2$ (Fig. 5.3, Madsen 1985).

The dark assimilation of CO$_2$ by isoetids is associated with accumulation of malate in the cell sap, which becomes progressively more acidic at night (Keeley 1982, Keeley and Morton 1982). During the day, the malate concentration and the acidity decline again. The observed daytime decline of malate is associated with a higher net fixation of CO$_2$ in organic substances than can be accounted for by external CO$_2$ uptake alone (Madsen 1987). The dark fixation of CO$_2$ in some isoetid species can improve their carbon balance on a 24-hour basis, particularly because their CO$_2$ uptake in the light does not appear to be hampered by the internal decarboxylation of CO$_2$. The ability of isoetids to utilise sediment CO$_2$, and of some of them to fix CO$_2$ in the dark, can be viewed as general adaptations to ensure the supply of CO$_2$ in an environment of limiting CO_2. The modes of CO_2 utilization are integrated in the overall morphology, anatomy and physiology of the plants to influence multiple biological processes. One associated consequence of a high CO_2 supply often mentioned is the possibility of reducing the investments in the primary carboxylating enzyme Rubisco, while still maintaining a high CO_2 fixation capacity (Raven et al. 1988). A possible reduction of the concentration of Rubisco is by no means trivial in an oligotrophic environment, since this enzyme is the most abundant in the plants and requires large proportions of the combined nitrogen pool (Raven 1984).

Oxygenation and microbial processes of isoetid sediments

Isoetid species are known to release large proportions of their photosynthetic oxygen from the roots (Sand-Jensen and Prahl 1982, Sand-Jensen et al. 1982, Christensen et al. 1994). Root release of oxygen influences the rhizosphere close to the root surfaces under conditions of low oxygen efflux and/or high

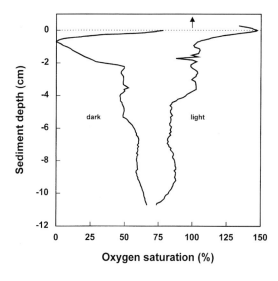

Figure 5.4. Depth profiles of O_2 in the pore-water of a sandy sediment with *Lobelia dortmanna* after 12 hours of light and 12 hours of darkness. Unpublished data by Pedersen and Sand-Jensen (1997).

sediment oxygen consumption rates. The influence of oxygen release reaches further away from the roots, and may include the entire root zone, when dense populations of isoetids inhabit coarse sediments of low organic content and low oxygen consumption rates (Wium-Andersen and Andersen 1972b, Pedersen et al. 1995).

Photosynthesis and respiration of *Lobelia dortmanna* on sandy sediment result in very pronounced changes of dissolved oxygen and CO_2 in the pore-water because of the dominant gas exchange across the root surfaces and the low microbial activity in the sediments (Pedersen et al. 1995). Photosynthesis during the day results in O_2 concentrations close to saturation in the root zone and O_2 penetration to great sediment depth, while O_2 consumption by plant respiration and microbial processes leads to much lower O_2 concentrations at night (Fig. 5.4). At the depth of maximum root density (about 1-2 cm into the sediment), the pool of dissolved O_2 is completely consumed by plant and bacterial respiration at night (Fig. 5.4). Deeper into the sediment, rates of bacterial O_2 consumption are much lower due to the lack of degradable organic substrates, and O_2 amplitudes between day and night are much dampened because of lower root density and exchange rates of O_2 (Fig. 5.4, Pedersen et al. 1995). The depth-integrated pools of O_2 and CO_2 in the sediments follow complementary diel pulses, because *Lobelia dortmanna* roots release O_2 and absorb CO_2 in the light but absorb O_2 and release CO_2 in the dark (Fig. 5.5). Because CO_2 is depleted in the root zone, and photosynthesis of *Lobelia dortmanna* is CO_2 limited, most of the CO_2 produced by microbial degradation and plant respiration will subsequently be assimilated by the plants and result in an almost equivalent amount of O_2 being released from the roots (Fig. 5.5). Thus, within certain limits, plant and sediment processes work in concert in these nutrient-poor sediments.

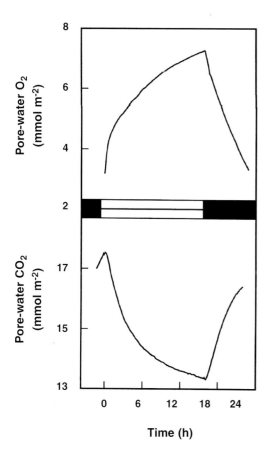

Figure 5.5. Diel changes in the depth-integrated pools of O_2 and CO_2 in the root zone of a sandy sediment inhabited by *Lobelia dortmanna*. From Pedersen et al. (1995) with permission from the Ecological Society of America.

The oxygenation of sediments has important implications for biogeochemical processes. Root release of oxygen to the sediments leads to high redox potentials and the presence of nitrate and oxidized iron and manganese in the sandy sediments (Wium-Andersen and Andersen 1972b, Tessenow and Baynes 1978, Christensen and Sørensen 1986). The formation of nitrate by nitrification in the oxic parts of the root zone will subsequently allow denitrification and the permanent loss of nitrate in the anoxic parts of the root zone in the dark, or in the permanently anoxic strata deeper into the sediment

(Christensen and Sørensen 1986, Risgaard-Petersen and Jensen 1997). Denitrification in sediments inhabited by isoetids may lead to a gradual decline of the total nitrogen pool in the sediment, if not balanced by new nitrogen supplied by sedimentation of particles, nitrogen fixation or production of roots based on leaf uptake of nitrogen from the water (Flindt 1994).

The presence of oxygen in the sediment should allow efficient aerobic degradation of the organic matter by bacteria and fungi. VA-mycorrhiza fungi are associated with the roots of isoetid species (Søndergaard and Lægaard 1977), and their abundance apparently increases with the oxygen status of the sediment, because the extent of infection increases with higher plant density, higher redox potential and lower organic content (Wigand et al. unpublished data). VA-mycorrhiza fungi are believed to assist in the acquisition of those mineral nutrients (e.g. iron, manganese and phosphorus) whose solubilities decline under oxidized conditions (Harley and Smith 1983, Bolan 1991). Sediment oxygenation leads to low solubility of phosphorus because of co-precipitation with and adsorption to oxyhydroxy-forms of iron and manganese (Christensen and Andersen 1996, Christensen et al. 1997). This mechanism is important because the majority of phosphorus in sandy sediments inhabited by isoetids is found in the root biomass or bound to oxidized iron (Christensen and Andersen 1996). The vegetated isoetid sediments, therefore, contain higher phosphorus pools that the unvegetated sediments, and vegetated sediments also show much higher retention of dissolved phosphate from solutions slowly percolated through the pore-water than non-vegetated sediments do (Fig. 5.6, Christensen and Andersen 1996). The nutrient-poor status of the isoetid sediments therefore requires efficient mechanisms of the plants to take up the nutrients and maintain them by reallocation within the evergreen biomass (Moeller

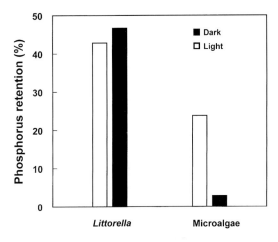

Figure 5.6. Relative retention of phosphate from solutions percolated through sediments from Lake Kalgaard inhabited by *Littorella uniflora* and sediments without isoetids, but with surface growth of benthic microalgae. Data from Christensen and Andersen (1996).

1978). Reallocation of phosphorus from old to new leaves has been shown to occur in *Littorella uniflora* (Christiansen et al. 1985). Particular risks of phosphorus deficiency for the isoetids may, nonetheless, arise on organic sediments with high supply rates of reduced iron, which is oxidized and precipitated as thick coatings directly on the root surfaces. Thick iron coatings can probably inhibit root uptake of phosphorus and other nutrient elements (Christensen and Sand-Jensen unpublished data). Interspecific interactions between roots are perhaps strong, because species with particularly high O_2 release and efficient nutrient uptake may be able to reduce the growth and survival of other more nutrient demanding species. It remains to be tested how nitrogen and phosphorus are supplied to different plant species under varying O_2 and redox conditions, and how the species can influence each other through symbiosis with the sediment fungi. Tests of the importance of VA-mycorrhiza for phosphorus acquisition of isoetids by the use of fungicides have not been successful so far,

because the treatment disturbs the normal root performance (Christensen unpublished data).

Growth of isoetids and primary production in Lobelia-lakes
Growth of isoetids
Leaf-marking methods are easy to apply to isoetids for measuring the long-term growth rates in the natural populations with a minimum of experimental disturbance (Moeller 1978, Sand-Jensen and Søndergaard 1978). These methods are preferable to traditional measurements of photosynthesis and respiration by short-time exchange rates of O_2 and CO_2 in tissue samples. The latter methods raise particular problems for isoetids, because O_2 and CO_2 are recycled within the aerenchyma and exchanged with the roots, and integration of short-time metabolic measurements to yield reliable daily or weekly net growth rates is virtually impossible (Nygaard 1958, Sand-Jensen and Prahl 1982). The leaf-marking technique applied to *Lobelia dortmanna* has the additional advantage that root production can be determined simultaneously because two roots are produced for every new leaf formed (Sand-Jensen and Søndergaard 1978).

The leaf growth rates have been measured in natural populations of *Littorella uniflora* and *Lobelia dortmanna* at 0.5-m depth in Lake Kalgaard during a full year (Fig. 5.7, Sand-Jensen and Søndergaard 1978). The leaf turnover rate of both species followed the seasonal variation of irradiance and temperature with maximum summer rates at 0.82% d[-1] for *Littorella uniflora* and 0.65 d[-1] for *Lobelia dortmanna*. The average turnover rate of the leaf pool was 0.41% d[-1] for *Littorella uniflora* and 0.21% d[-1] for *Lobelia dortmanna*. These rates correspond to a mean leaf life-time of about 245 days for *Littorella uniflora* and 475 days for *Lobelia dortmanna* (Sand-Jensen and Søndergaard 1978). The coupled production and mortality of the

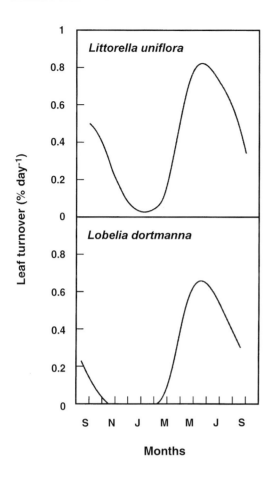

leaves resulted in a relatively constant leaf number and biomass over the year, though with slightly higher values in summer than during the winter (Sand-Jensen and Søndergaard 1978). The leaf growth rate of *Isoetes lacustris* has not been measured in Danish *Lobelia*-lakes, but the annual mean rate was 0.27% d^{-1} in a Swedish lake (Erikson 1973). These findings support the general impression of very low growth rates and long tissue life-time among the isoetid species in accordance with similar patterns for stress-selected species in nutrient-poor terrestrial habitats (Grime 1977, Moeller 1978). In contrast, the leaf turnover rates are often 10-fold higher among elodeid species inhabiting eutrophic freshwater habitats, and their leaf longevity is only about one month (Adams and McCracken 1974, Nielsen and Sand-Jensen 1991).

The main role of phytoplankton and isoetids in primary production in Lobelia-lakes

We have determined the organic production of phytoplankton, isoetids and their epiphytic algae in Lake Kalgaard in order to evaluate their overall contribution to primary production in the lake (Søndergaard and Sand-Jensen 1978). As expected, the annual production rates were low per m^2 of lake surface for all communities, due to the oligotrophic

Figure 5.7. Leaf turnover rate of *Littorella uniflora* and *Lobelia dortmanna* in shallow water in Lake Kalgaard during 1977. Redrawn from Sand-Jensen and Søndergaard (1978) with permission from E. Schweitzerbart'sche Verlagsbuchhandlung.

Table 5.4. Daily and annual averages of net primary production among phytoplankton, epiphytes, and isoetids in Lake Kalgaard. Data compiled from Søndergaard and Sand-Jensen (1978, 1979b)

Plant community/ Species	Primary production		Relative contribution
	Daily (mg C m^{-2} d^{-1})	Annual (g C m^{-2} year)	(% of total)
Phytoplankton	66.1	24	52
Epiphytes	2.3	1	2
Isoetids			
Lobelia dortmanna	0.5	0.2	<0.5
Littorella uniflora	29.4	10.7	23
Isoetes lacustris	27.7	10.1	22
Total	126.0	46.0	

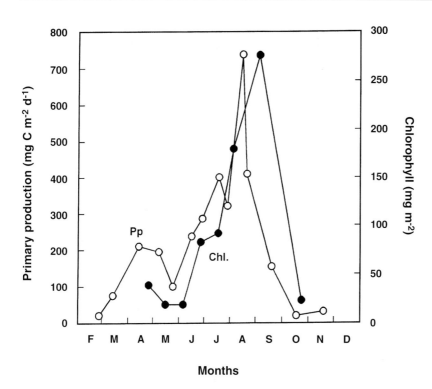

Figure 5.8. Seasonal changes in depth-integrated production (Pp) and biomass of phytoplankton (as chlorophyll *a*) in Lake Kalgaard during 1977. Data from Søndergaard and Sand-Jensen (1979b and unpublished).

nature of the lake system (Table 5.4). An important finding was the equal contribution of phytoplankton and isoetids to the total primary production within the lake, while the contribution of epiphytic algae was very small (Table 5.4). A similar pattern was found in Grane Langsø (Nygaard and Sand-Jensen 1981), but here the dense populations of deep-growing mosses and characeans contributed grossly to submerged macrophyte production and to the oxygen maximum in the hypolimnion, while this vegetation component was unimportant in Lake Kalgaard.

The phytoplankton development in Lake Kalgaard and Grane Langsø is characterized by prominent sub-surface biomasses and production maxima in the metalimnion or the upper-hypolimnion during summer (Søndergaard and Sand-Jensen 1979b, Nygaard and Sand-Jensen 1981). The transparent epilimnion and the accumulation of nutrients in the hypolimnion of oligotrophic lakes promote this formation of deep-water phytoplankton

maxima (Schindler and Holmgren 1971). The deep-water maximum in Lake Kalgaard between 6 and 9 m reached a chlorophyll concentration of 90 mg m^{-3} in August, or 20-fold higher concentrations than in the epilimnion, and represented a major contribution to the integrated production in the water column (Søndergaard and Sand-Jensen 1979b). The depth-integrated rates of phytoplankton primary production measured *in situ* very closely followed the depth-integrated chlorophyll biomass during the year (Fig. 5.8). Annual phytoplankton production in the water column was 50 g C m^{-2} in the deepest part of the lake (10 m), which included the deep phytoplankton maximum. However, the mean lake depth is 4.6 m, and large areas of the lake do not include the metalimnetic maximum. The phytoplankton production averaged for the entire lake surface was only 24 g C m^{-2} y^{-1}, which places Lake Kalgaard in an oligotrophic category of production rates (Wetzel 1983).

Only a few lakes show such a large contribution by submerged macrophytes to the total primary production as Lake Kalgaard, Grane Langsø and, presumably other *Lobelia*-lakes do (Søndergaard and Sand-Jensen 1978, Nygaard and Sand-Jensen 1981, Wetzel 1983). Although submerged macrophytes in oligotrophic hardwater Lawrence Lake have been suggested to account for 50% of the primary production (Rich et al. 1971), new, higher estimates of primary production by benthic microalgae reduce the contribution of macrophytes to less than 20% (Wetzel and Søndergaard 1997). The production of benthic microalgae in lakes is usually not known, but it may well be important on non-vegetated shallow sediments in many lakes. In Lake Kalgaard, however, the dense isoetid cover should preclude substantial growth of benthic microalgae. Therefore, we consider the high quantitative importance of isoetids for primary production in Lake Kalgaard as unique for freshwater lakes.

Long-term development of Lobelia-lakes

Lobelia-lakes are oligotrophic softwater lakes and, therefore, extremely vulnerable to human impact by eutrophication and acidification. Evaluation of the long-term development of *Lobelia*-lakes in Denmark is very depressing. One hundred years ago, Danish *Lobelia*-lakes were common in Jutland, and isoetids species were also present, though not dominant, in many alkaline lakes throughout the country (Bågøe and Ravn 1895-1896, Iversen 1929, Rostrup and Jørgensen 1961). Progressive eutrophication of lakes since the turn of the century stimulated the growth of phytoplankton, epiphytes, elodeid angiosperms and reed plants and led to a decline, or a complete die-off, of isoetid species due to their poor ability to compete for light with the other plant communities. Small surviving isoetid populations are still found occasionally on coarse sediments in shallow parts of eutrophic lakes in which growth of other

rooted plants is impaired by physical exposure (Sand-Jensen and Søndergaard 1981).

Evaluations of long-term changes in lake ecosystems are usually based on measurements of lake transparency and water chemistry (e.g. Lindegaard et al. 1997, Sand-Jensen 1997). These measurements do not always provide a satisfactory account of the ecological changes, particularly not in small, shallow lakes in which benthic organisms may be more important for the biological interactions and the dynamics of carbon and nutrient flow than the pelagic organisms (Søndergaard and Sand-Jensen 1978, Wetzel 1983). It is noteworthy that most of the Danish *Lobelia*-lakes belong to this category. The median size of Danish lakes is only 0.22 km^2, and the median of mean water depth in these lakes is only 2.0 m (Miljøstyrelsen 1990). One of the early signs of eutrophication and beginning deterioration of the Danish *Lobelia*-lakes is the stimulation of biomass accumulation by epiphytic algae on the submerged plant surfaces rather than the increase of phytoplankton biomass (Sand-Jensen and Søndergaard 1981). Epiphytic algae are less vulnerable to loss by grazing and sedimentation than phytoplankton, and the epiphyte biomass responded several-fold stronger than the phytoplankton biomass to eutrophication (Sand-Jensen and Søndergaard 1981). The accumulation of epiphytic algae can severely inhibit growth of the submerged plants, ultimately reducing both their horizontal development and depth penetration (Sand-Jensen and Søndergaard 1981, Sand-Jensen and Borum 1984, Riis 1996). The influence of an increased nutrient input to the *Lobelia*-lakes can, therefore, develop relatively unnoticed, if only measurements of transparency and water chemistry are conducted (Riis et al. 1997). Increased plant growth as a response to eutrophication may take place mainly in the littoral zone with the increased nutrient pools accumulating mainly in the sediments rather than in the water column. It

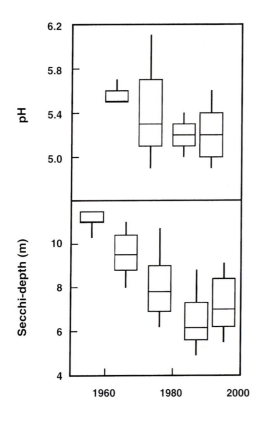

Figure 5.9. Long-term development of surface pH (above) and Secchi-depth transparency (below) in Lake Grane Langsø. The horizontal lines show the median, the boxes show the 25-75 percentiles and the vertical lines show the range of all measurements. The vertical lines are placed in the middle of the measuring periods. Unpublished data by Riis and Sand-Jensen (1997).

Figure 5.10. A: flowering *Lobelia dortmanna* growing on a sandy, nutrient-poor sediment exposed to air during a hot, dry summer. Photograph Ole Geertz-Hansen.

is a serious warning to long-term monitoring programmes that isoetid communities in Grane Langsø, for example, have deteriorated alarmingly, despite the fact that statistically significant long-term changes in nutrient concentrations in the water could not be detected (Riis and Sand-Jensen unpublished data). One reason for the inability to detect the pelagic changes is probably the common analytical insufficiencies (i.e. changes of methods over time, too few measurements annually), but another probable reason is that the pelagic communities are less susceptible than the benthic communities to the early impact.

The low alkalinity (<0.1 meqv. l^{-1}) of most Danish *Lobelia*-lakes makes them susceptible to anthropogenic acidification. Grane Langsø had a very low alkalinity (0.022 meqv. l^{-1}) already in the 1950's and belongs to a group of *Lobelia*-lakes which has experienced a decline of pH during the last forty years (Rebsdorf 1981, Riis 1996). Acidification of Grane Langsø has taken place along with a progressive decline of transparency and major alterations in the composition and depth distribution of macrophytes. In the 1950's, Secchi-transparency exceeded 10 m in Grane Langsø, mean pH was 5.6, isoetids extended to 5.5 m and characeans covered the deepest parts of the lake at 11.5 m (Table 5.1, Fig. 5.9, Riis 1996). During the 1960's and 1970′s, mean pH declined to 5.2 and

Figure 5.10. C: a sandy *Lobelia*-sediment with precipitates of oxyhydroxides of iron and manganese along the roots as a result of oxygen release. Photograph Ole Pedersen.

Figure 5.10. B: *Lobelia dortmanna* is the character plant of oligotrophic softwater *Lobelia*-lakes. *Lobelia dortmanna* is also a type species for the isoetid growth form characterized by short, thick, evergreen leaves in a rosette and an extensive root system, which establishes the intimate relationship between plant and sediment processes. Photograph Ole Pedersen.

Figure 5.10. D: a transverse section of a leaf of *Lobelia dortmanna* showing the two large air lacunae, which ensure efficient transport of CO_2 and O_2 between the leaves and the roots and allow the main gas exchange during photosynthesis and respiration to take place across the root surfaces with the sediment. The leaf surface is covered by a thick cuticle which impedes gas exchange with the lake water. Photograph Ole Pedersen.

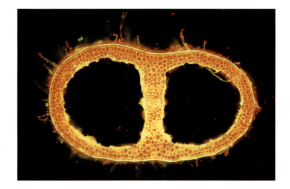

annual pH fluctuation increased from 0.3 to 0.9 pH units, probably as a result of enhanced rates of organic production and decomposition in the lake (Fig. 5.9, Riis and Sand-Jensen unpublished data). Median Secchi-transparency also declined from about 10.0 m in the 1950's to 6.7 m in the 1990's (Fig. 5.9) due to higher concentrations of dissolved humic material which followed a blow-down of trees in the surrounding pine plantation, and to a likely increase of phytoplankton biomass and production, which cannot be verified because of insufficient data (Riis et al. 1996). The long-term changes have been much more pronounced in the littoral plant communities in Grane Langsø. Dense populations of filamentous green algae in shallow water and mosses in deep water have spread in the lake, while the areal distribution and depth penetration of isoetids have diminished dramatically, and the former characean populations in deep water have disappeared. Increased atmospheric deposition of sulphuric acid and nitric acid are important for the observed changes in the lake ecosystem, but causal relationships and models for nutrients and protons cannot be established because the amount and composition of groundwater input and the production and decomposition processes in the littoral zone and the sediments have not been quantified.

A slow but progressive acidification and eutrophication has apparently taken place in the most oligotrophic softwater *Lobelia*-lakes in accordance with the decline of pH and the increase of nutrient concentrations in precipitation and in groundwater from uncultivated regions in Denmark (Rebsdorf 1981, Kristensen et al. 1990). Hence, it will probably be difficult to maintain the status of the Danish *Lobelia*-lakes in the future. The chemical threats are accompanied by physical threats associated with increased recreational interests in the few transparent lakes which are still present in this country. The isoetid species are very susceptible to trampling because of their low growth rate and the ease by which the stiff leaves break. Unregulated fishing and bathing activity during hot summers has resulted in loss of vegetation from large regions of several lakes. The vegetation recovers slowly because of the slow growth of the individuals. We may, therefore, reach an unfortunate situation where frequent periods of disturbance preclude full recovery during intervening favourable periods. Damage to the isoetid vegetation will also have a cascading negative influence on the remaining vegetation. Resuspension is increased, and nutrients, formerly incorporated in relatively closed nutrient cycles between isoetid plants and sediments (Pedersen et al. 1995), become engaged in pelagic nutrient cycles resulting in a larger biomass of phytoplankton and less transparent waters. Pools of phosphorus retained in the isoetid biomass and bound to oxidized iron in the vegetated sediments in the undisturbed *Lobelia*-lakes are several-fold higher than the maximum pools observed in the lake water (Sand-Jensen 1980).

We would prefer to be optimistic, considering the possibility for maintaining the high water quality and the widespread distribution of the remarkable isoetid vegetation in Danish *Lobelia*-lakes in the future. However, the protection of the few *Lobelia*-lakes remaining in Denmark requires a stop to the fertilization of cultivated fields and forest plantations within the catchment areas and a reduction of the recreational activity in the lakes and their immediate surroundings. We hope the willingness to introduce the required protection will increase in the future. Otherwise we foresee that the ongoing decline of the number *Lobelia*-lakes and their cover of isoetid vegetation will continue.

Literature cited

Adams, M. S, and M. D. McCracken. 1974. Seasonal production of the *Myriophyllum* component in the littoral of Lake Wingra, Wisconsin. Journal of Ecology 62: 457-465.

Baagøe, J., and J. K. Ravn. 1895-1896. Exkursion til jydske søer og vandløb i sommeren 1895. Botanisk Tidsskrift 20: 288-326.

Bolan, N. S. 1991. A critical review on the role of mycorrhizal fungi in the uptake of phosphorus by plants. Plant Soil 134: 180-207.

Boston, H. L. 1986. A discussion of the adaptations for carbon acquisition in relation to the growth strategy of aquatic isoetids. Aquatic Botany 26: 258-270.

Boston, H. L., M. S. Adams, and T. P. Pienkowski. 1987. Utilization of sediment-CO_2 by selected North American isoetids. Annals of Botany 60: 485-494.

Christensen, K. K., and F. Ø. Andersen. 1996. Influence of *Littorella uniflora* on phosphorus retention in sediment supplied with artificial porewater. Aquatic Botany 55: 183-197.

Christensen, K. K., F. Ø. Andersen, and H. S. Jensen. 1997. Comparison of iron, manganese, and phosphorus retention in freshwater littoral sediment with growth of *Littorella uniflora* and benthic micro-algae. Biogeochemistry, in press.

Christensen, P. B., N. P. Revsbech, and K. Sand-Jensen. 1994. Microsensor analysis of oxygen in the rhizosphere of the aquatic macrophyte *Littorella uniflora* (L.) Aschers. Plant Physiology 105: 847-852.

Christensen, P. B., and J. Sørensen. 1986. Temporal variation of denitrification in plant-covered sediment from Lake Hampen, Denmark. Applied Environmental Microbiology 51: 1174-1179.

Christiansen, R., N. J. S. Friis, and M. Søndergaard. 1985. Leaf production and nitrogen and phosphorus tissue content of *Littorella uniflora* (L.) Aschers. in relation to nitrogen and phosphorus enrichment of the sediment in oligotrophic Lake Hampen, Denmark. Aquatic Botany 23: 1-11.

Eriksson, F. 1973. Makrofyterna och deras produktion i sjön Vitalampa. Scripta Limnologica Upsaliensis 346. University of Upsala, Sweden.

Farmer, A. M., and D. H. N. Spence. 1986. The growth strategies and distribution of isoetids in Scottish freshwater lochs. Aquatic Botany 26: 247-258.

Flindt, M. 1994. Omsætning af organisk stof, kvælstof og fosfor i Roskilde Fjords sedimenter belyst eksperimentelt og ved matematisk modelering. Ph.D-thesis. Freshwater Biological Laboratory, University of Copenhagen, Copenhagen, Denmark.

Grime, J. P. 1977. Plant strategies and vegetation processes. John Wiley & Sons, New York, USA.

Harley, J. L., and S. E. Smith. 1983. Mycorrhizal symbiosis. Academic Press, London, England.

Hutchinson, G. E. 1975. A treatise on limnology. Volume III. John Wiley & Sons, New York, USA.

Iversen, J. 1929. Studien über die pH-Verhältnisse dänischer Gewässer und ihren Einflüss auf die Hydrophytenvegetation. Botanisk Tidsskrift 40: 277-353.

Keeley, J. E. 1981. *Isoetes howelii*: a submerged aquatic CAM plant? American Journal of Botany 68: 420-424.

Keeley, J. E., and G. Bowes. 1982. Gas exchange characteristics of the submerged crassulacean acid metabolism plant, *Isoetes howelii*. Plant Physiology 70: 1455-1458.

Keeley, J. E., and B. A. Morton. 1982. Distribution of diurnal acid metabolism in submerged aquatic plants outside the genus *Isoetes*. Photosynthetica 16: 546-553.

Kristensen, P., B. Kronvang, E. Jeppesen, P. Græsbøll, M. Erlandsen, Aa. Rebsdorf, A. Bruhn, and M. Søndergaard. 1990. Ferske vandområder – vandløb, kilder og søer. Vandmiljøplanens Overvågningsprogram. Danmarks Miljøundersøgelser. Miljøministeriet, Copenhagen, Denmark.

Lindegaard, C., P. C. Dall, and P. M. Jónasson. 1997. Long-term pattern of the profundal fauna in Lake Esrom. Chapter 4 *in* K. Sand-Jensen and O. Pedersen, editors. Freshwater Biology. Priorities and Development in Danish Research. Gad, Copenhagen, Denmark.

Madsen, T. V. 1985. A community of submerged aquatic CAM plants in Lake Kalgaard, Denmark. Aquatic Botany 23: 97-108.

Madsen, T. V. 1987. Interactions between internal and external pools of CO_2 in the photosynthesis of the aquatic CAM plants *Littorella uniflora* (L.) Aschers. and *Isoetes lacustris* L. New Phytologist 106: 35-50.

Madsen, T. V., and K. Sand-Jensen. 1991. Photosynthetic carbon assimilation in aquatic macrophytes. Aquatic Botany 41: 5-40.

Madsen, T. V., K. Sand-Jensen, and S. Beer. 1993. Comparison of photosynthetic performance and carboxylation capacity in a range of aquatic macrophytes of different growth forms. Aquatic Botany 44: 373-384.

Markager, S., and K. Sand-Jensen. 1994. The physiology and ecology of light-growth relationships in macroalgae. Progress in Phycological Research 10: 210-298.

Middelboe, A. L., and S. Markager. 1997. Depth limits and minimum light requirements of freshwater macrophytes. Freshwater Biology, in press.

Miljøstyrelsen. 1990. Vandmiljø-90. Redegørelse fra Miljøstyrelsen nr. 1/1991. Miljøstyrelsen, Copenhagen, Denmark.

Moeller, R. E. 1975. Hydrophyte biomass and community structure in a small, oligotrophic New Hampshire lake. Verhandlungen Internationale Vereinigung für Theoretische und Angewandte Limnologie 19: 1104-1012.

Moeller, R. E. 1978. Seasonal changes in biomass, tissue chemistry and net production of the evergreen hydrophyte *Lobelia dortmanna*. Canadian Journal of Botany 56: 1425- 1433.

Nielsen, S. L., and K. Sand-Jensen. 1991. Variation in growth rates of submerged rooted macrophytes. Aquatic Botany 39: 109-120.

Nielsen, S. L., Gacia, E., and K. Sand-Jensen. 1991. Land plants of amphibious *Littorella uniflora* (L.) Aschers. maintain utilization of CO_2 from the sediments. Oecologia 88: 258-262.

Nygaard, G. 1958. On the productivity of the bottom vegetation in lake Grane Langsø. Verhandlungen Internationale Vereinigung für Theoretische und Angewandte Limnologie. 13: 144-155.

Nygaard, G., and K. Sand-Jensen. 1981. Light climate and metabolism of *Nitella flexilis* (L.) Ag. in the bottom waters of oligotrophic Lake Grane Langsø, Denmark. Internationale Revue der gesamten Hydrobiologie 66: 685-699.

Olsen, S. 1950. Aquatic plants and hydrospheric factors. I. Aquatic plants in SW-Jutland. Denmark. II. The hydrospheric types. Svensk Botanisk Tidsskrift 44: 1-34 and 332-373.

Pedersen. O., and K. Sand-Jensen. 1992. Adaptations of submerged *Lobelia dortmanna* to aerial life form: morphology, carbon sources and oxygen dynamics. Oikos 65: 89-96.

Pedersen, O., K. Sand-Jensen, and N. P. Revsbech. 1995. Diel pulses of O_2 and CO_2 in sandy sediments inhabited by *Lobelia dortmanna*. Ecology 76: 1536-1545.

Petersen, J. B. 1917. Bemærkninger til plantekortene over Bastrup Sø, Farum Sø, Bagsværd Sø og Lyngby Sø. Pages 39-57 *in* C. Wesenberg-Lund et al., editors. Furesøstudier. Kongelige Danske Videnskabernes Selskabs Skrifter. Naturvidenskab og Mathematisk Afdeling 8, række III. Bianco Lunos Bogtrykkeri, Copenhagen, Denmark.

Raven, J. A. 1984. Energetics and Transport in Aquatic Plants. M. B. L. Lectures in Biology, 4. A. R. Liss, New York, USA.

Raven, J. A., L. L. Handley, J. J. MacFarlane, S. McInroy, L. McKinzie, J. H. Richards, and G. Samuelsson 1988. The role of CO_2 uptake by roots and CAM in acquisition of inorganic C by plants of the isoetid life form: a review with new data on *Eriocaulon decangulare* L. New Phytologist 108: 125-148.

Rebsdorf, Aa. 1981. Forsuringstruede danske søer. Miljøprojekter 38. Miljøstyrelsens Ferskvandslaboratorium. Miljøstyrelsen, Copenhagen, Denmark.

Rich, P. H., R. G. Wetzel, and N V Thuy. 1971. Distribution, production and role of aquatic macrophytes in a southern Michigan marl lake. Freshwater Biology 1: 3-21.

Richardson, K., H. Griffiths, J. A. Raven, and N. M. Griffiths. 1984. Inorganic carbon assimilation in the isoetids, *Isoetes lacustris* L. and *Lobelia dortmanna* L. Oecologia 61: 115-121.

Riis, T. 1996. Forsuring og udvikling af vegetationen i Grane Langsø og Kalgaard Sø. MS-thesis. Freshwater Biological Laboratory, University of Copenhagen, Copenhagen, Denmark.

Riis, T., and K. Sand-Jensen. 1997. Growth reconstruction and photosynthesis of aquatic mosses: influence of light, temperature and carbon dioxide at depth. Journal of Ecology, in press.

Riis, T., K. Sand-Jensen, and T. Jørgensen. 1996. Danmarks reneste sø gennem 40 år. Vand og Jord 5: 199-203.

Risgaard-Petersen, N., and K. Jensen. 1997. Nitrification and denitrification in the rhizosphere of the aquatic macrophyte *Lobelia dortmanna* L. Limnology and Oceanography, in press.

Rostrup, E., and C. A. Jørgensen. 1961. Den Danske Flora. Gyldendal, Copenhagen, Denmark.

Sand-Jensen, K. 1978. Metabolic adaptation and vertical zonation of *Littorella uniflora* (L.) Aschers. and *Isoetes lacustris* L. Aquatic Botany 4: 1-10.

Sand-Jensen, K. 1980. Økologisk balance i danske *Lobelia*-søer. Naturens Verden 7: 222-227.

Sand-Jensen, K. 1987. Environmental control of bicarbonate use among freshwater and marine macrophytes. Pages 99-112 *in* R. M. M. Crawford, editor. Plant Life in Aquatic and Amphibious Habitats. Blackwell Science, Oxford, England.

Sand-Jensen, K. 1997. Eutrophication and plant communities in Lake Fure during 100 years. Chapter 3 *in* K. Sand-Jensen and O. Pedersen, editors. Freshwater Biology. Priorities and Development in Danish Research. Gad, Copenhagen, Denmark.

Sand-Jensen, K., and J. Borum. 1984. Epiphyte shading and its effect on diel metabolism of *Lobelia dortmanna* L. during the spring bloom in a Danish lake. Aquatic Botany 20: 109-119.

Sand-Jensen, K., and C. Lindegaard. 1996. Økologi i søer og vandløb. Gad, Copenhagen, Denmark.

Sand-Jensen, K., and T. V. Madsen. 1991. Minimum

light requirements of submerged freshwater macrophytes in laboratory growth experiments. Journal of Ecology 79: 749-764.

Sand-Jensen, K., and C. Prahl. 1982. Oxygen exchange with the lacunae and across the leaves and roots of the submerged vascular macrophyte, *Lobelia dortmanna* L. New Phytologist 91: 103-120.

Sand-Jensen, K., and M. Søndergaard. 1978. Growth and production of isoetids in oligotrophic Lake Kalgaard, Denmark. Verhandlungen Internationale Vereinigung für Theoretische und Angewandte Limnologie 20: 659-666.

Sand-Jensen, K., and M. Søndergaard. 1979. Distribution and quantitative development of aquatic macrophytes in relation to sediment characteristics in oligotrophic Lake Kalgaard, Denmark. Freshwater Biology 9: 1-11.

Sand-Jensen, K., and M. Søndergaard. 1981. Phytoplankton and epiphyte development and their shading effect on submerged macrophytes in lakes of different nutrient status. Internationale Revue der gesamten Hydrobiologie 66: 529-552.

Sand-Jensen, K., C. Prahl, and H. Stokholm. 1982. Oxygen release from roots of submerged aquatic macrophytes. Oikos 38: 249-354.

Schindler, D. W., and S. K. Holmgren. 1971. Primary production and phytoplankton in the Experimental Lake Area, north-western Ontario, and other low-carbonate waters, and a liquid scintillation method for determining ^{14}C activity in photosynthesis. Journal of the Fisheries Research Board Canada 28: 189-201.

Sculthorpe, C. D. 1967. The biology of aquatic vascular plants. Edward Arnold, London, England.

Steemann Nielsen, E. 1947. Photosynthesis of aquatic plants. Dansk Botanisk Arkiv 12: 5-71.

Steemann Nielsen, E. 1960. Uptake of CO_2 by plants. Pages 70-84 *in* W. Ruhland, editor. Encyclopedia of Plant Physiology. Springer Verlag, Berlin, Germany.

Søndergaard, M., and G. Bonde. 1988. Photosynthetic characteristics and pigment content and composition in *Littorella uniflora* (L.) Aschers. in a depth gradient. Aquatic Botany 32: 307-319.

Søndergaard, M., and S. Lægaard. 1977. Vesicular-arbuscular mycorrhiza in some aquatic vascular plants. Nature 268: 232 233.

Søndergaard, M., and K. Sand-Jensen. 1978. Total autotrophic production in oligotrophic Lake Kalgaard, Denmark. Verhandlungen Internationale Vereinigung für Theoretische und Angewandte Limnologie 20: 667-673.

Søndergaard, M., and K. Sand-Jensen. 1979a. Carbon uptake by leaves and roots of *Littorella uniflora* (L.) Aschers. Aquatic Botany 6: 1-12.

Søndergaard, M., and K. Sand-Jensen. 1979b. Physico-chemical environment, phytoplankton biomass and production in oligotrophic softwater Lake Kalgaard, Denmark. Hydrobiologia 63: 241-252.

Tessenow, U., and Y. Baynes. 1978. Redoxchemische Einflüsse von *Isoetes lacustris* L. im Littoralsediment des Feldsees (Hochschwarzwald). Archiv für Hydrobiologie 82: 20-48.

Wetzel, R. G. 1983. Limnology. 2nd edition. Saunders College Publishing, Philadelphia, USA.

Wetzel, R. G., and M. Søndergaard. 1997. The role of submersed macrophytes for the microbial community and dynamics of dissolved organic carbon in aquatic ecosystems. In press *in* E. Jeppesen, Ma. Søndergaard, Mo. Søndergaard, and K. Christoffersen, editors. Springer Verlag, Berlin, Germany.

Wium-Andersen, S. 1971. Photosynthetic uptake of free CO_2 by the roots of *Lobelia dortmanna*. Physiologia Plantarum 25: 245-248.

Wium-Andersen, S., and J. M. Andersen. 1972a. Carbon dioxide content of the interstitial water in the sediment of Grane Langsø, a Danish *Lobelia*-lakes. Limnology and Oceanography 17: 943-947.

Wium-Andersen, S., and J. M. Andersen. 1972b. The influence of vegetation on the redox profile of the sediment of Grane Langsø, a Danish *Lobelia*-lake. Limnology and Oceanography 17: 948-952.

6 Macrophytes as biological engineers in the ecology of Danish streams

By Kaj Sand-Jensen

The ecological role of rooted macrophytes in the predominantly small and nutrient-rich Danish streams is remarkable. The macrophytes are directly responsible for a large production of organic matter and oxygen, but because of the large stature and surface area their indirect role as bio-engineers is equally profound. New studies reveal how macrophytes modify flow, stabilise sediments and promote retention of organic matter and nutrients at local spatial scales within macrophyte patches and large spatial scales encompassing reaches and stream systems. Macrophytes also regulate the abundance of benthic and epiphytic microalgae and stimulate the production of invertebrates and fish. Fundamental differences in species diversity and dynamics of carbon and nutrients, therefore, exist between vegetated and unvegetated streams emphasising the key-role of plant cover and plant architecture in shallow aquatic ecosystems.

Denmark is a small flat country surrounded by the sea. The tallest hill is no more than 173 m above sea level, and no inland locality is more than 60 km from the nearest coast. The landscape is intensely cultivated and drained by a dense network of ditches, canals and natural streams of which more than 95% have been regulated to improve agricultural practices (Madsen 1995). The streams are generally small and shallow and have small slopes (< 0.005 m m^{-1}), low water velocities (< 0.8 m s^{-1}) and sediments dominated by clay, silt and sand (Kern-Hansen et al. 1980, Thyssen et al. 1990). More than 95% of the reaches are less than 10 m wide and 1 m deep. The largest Danish stream, the River Gudenå drains an area of only 2600 km^2 and is placed very far down on the list of the largest rivers in Europe.

Danish streams offer optimal conditions for abundant growth of rooted macrophytes (Sand-Jensen et al. 1989b). The open landscape, the shallow waters, the slow current and the soft sediments allow the plants to colonise most of the stream bottom. The high nutrient concentrations in the sediments and in the flowing water ensure nutrient saturation of plant growth under most circumstances (Kern-Hansen and Dawson 1978). The common CO_2-supersaturation of the stream water (Rebsdorf et al. 1991, Sand-Jensen et al. 1995) stimulates colonisation of amphibious plants from the riparian zone, and the excess CO_2 enhances photosynthesis of both the amphibious and the permanently submerged water plants (Madsen and Maberly 1991, Sand-Jensen et al. 1992).

The key-role of macrophytes in the ecology of streams is due to their large size. The macrophytes commonly have a leaf surface area 4-10 times higher than the sediment area (Kern-Hansen et al. 1980, Bijl et al. 1989).

The macrophytes, therefore, function as biological engineers (*sensu* Jones et al. 1994) because they create habitats and alter the environmental conditions for microalgae, invertebrates and fish (Sand-Jensen et al. 1989b, Sand-Jensen 1995). Hence, the macrophytes influence ecosystem ecology to an extent which far exceeds their direct role in primary production of organic material and oxygen. Whereas microalgae, invertebrates and fish depend on the environmental conditions for their survival and development, submerged macrophytes both depend on and strongly modify the physical environment.

My goal here is to review the key-role of macrophytes in the ecology of small lowland streams starting, with the conceptual formulation by Sand-Jensen et al. (1989b, Fig. 6.1) and proceeding to the recent studies of the influence of macrophytes on flow and sedimentation (Sand-Jensen and Mebus 1996). I follow the concepts in Figure 6.1 by first analysing the species composition and life forms of macrophytes in Danish lowland streams, emphasising the importance of amphibious and secondary water plants in the streams. Secondly, I examine the spatial and temporal patterns of macrophyte distribution and biomass which are regulated mainly by light availability and physical processes. Thirdly, I address the interaction between macrophytes and stream flow on the large scale of stream reaches and on the local scale of individual plant patches, and I illustrate how macrophyte species of different morphology and architecture differ in their resistance to the flow. Fourthly, I examine the influence of macrophytes on the retention of mineral particles, organic material and nutrients as a direct consequence of their ability to reduce stream flow and stabilise sediments. Fifthly, I review the ability of macrophytes to regulate the development of benthic microalgae by influencing light availability and sedimentation and to influence the epiphytes by offering plant surfaces available for colonisation. Finally, I evaluate how macrophytes influence diversity and abundance of macroinvertebrates and fish by forming shelter and habitats and altering the type and magnitude of food sources available.

Macrophytes species and life forms

Streams are in intimate contact with the adjacent terrestrial environment. The extensive contact zone between land and water, partic-

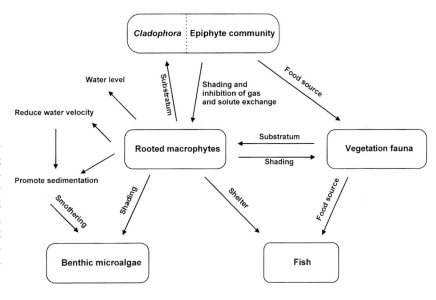

Figure 6.1. Conceptual model describing the key-role of macrophytes in the ecology of small lowland streams. Redrawn from Sand-Jensen et al. (1989b) with permission from Blackwell Science Ltd.

Table 6.1. Classification of rooted aquatic plants in the three main groups of true water plants, amphibious plants and secondary water plants. For each group the five most common species observed submerged at 131 Danish streams sites are listed together with their rank of relative abundance (Riis, Sand-Jensen and Vestergaard. unpublished data.)

Aquatic life form	Species	Rank
True water plants	*Elodea canadensis*	3
Grow under water	*Callitriche platycarpa*	6
and occasionally	*Ranunculus peltatus*	7
with air contact	*Potamogeton crispus*	10
	Callitriche hamulata	11
Amphibious plants	*Sparganium emersum*	1
Grow both in water	*Berula erecta*	2
and in air	*Myosotis palustris*	4
	Veronica anagallis-aquatica	5
	Mentha aquatica	14
Secondary water plants	*Ranunculus repens*	8
Grow on land and	*Solanum dulcamara*	9
occasionally in water	*Glyceria fluitans*	12
	Cardamine amare	13
	Epilobium hirsutum	16

ularly in small narrow streams, allows plants and animals to move from one environment to the other and exploit the mutual benefits of the two environments. Many stream plants are amphibious species which form flowers and seeds on land and grow vegetatively in the stream. Most insects have aquatic larval stages and terrestrial flying imagos which mate and disperse along and among the streams. Also several birds and mammals collect food under water but spend most of the time on land. Consequently, the hydrology, the biological structure and the transport and conversion of carbon and nutrients in streams are strongly dependent on the interaction with the riparian zone. Management of streams has, therefore, to consider both the stream and its surrounding catchment.

Macrophytes in streams are composed of different species and life forms with variable adaptation to life under water or to the transition zone between land and water (Holmes

and Newbold 1984, Sand-Jensen et al. 1992, 1995). The composition and recruitment of stream plants are, therefore, impossible to understand and predict without regarding them as a mixed assemblage of true water plants, amphibious plants and secondary water plants. The term secondary water plants is applied here for terrestrial plants which can tolerate submergence, and not in the evolutionary sense of angiosperms secondarily adapted to life in water from terrestrial ancestors. The three main groups of rooted angiosperms in lowland streams and the most common species are listed in Table 6.1. The true or primary water plants rarely grow above the water. They are poorly protected against evaporation and have a limited ability to grow in height on land due to insufficient structural tissue. The amphibious plants are able to grow both under water and on land. In some cases, the water and land populations grow separately from each other. In other

Table 6.2. Total number of species and observations among true water plants, amphibious plants and secondary water plants at 131 Danish streams sites. Percentage distribution in parenthes (Riis, Sand-Jensen and Vestergaard unpublished data).

Plant category	Number of species		Number of observations	
True water plants	36	(24%)	619	(33%)
Amphibious plants	33	(22%)	887	(47%)
Secondary water plants	83	(54%)	389	(20%)

cases the submerged plants grow out of the water as the season progresses, or the land plants expand from the shore into the water. The leaves are often entire and broad in air but finely dissected with capillary filaments or ribbon-like in water. Flowering and seed formation of amphibious plants can take place on land, or on aerial parts raised above the water. Plants in water can spread by seeds and vegetative fragments along the stream (Butcher 1933, Jones 1955, Andersen and Andersen 1991, Johansson and Nilsson 1993), and they may survive in active growth during winter because the water rarely freezes in many streams, unlike the situation on land (Moeslund et al. 1990). The secondary water plants are predominantly terrestrial species. Some species are found under water only very rarely, whereas others appear more frequently (Sand-Jensen et al. 1992, Sand-Jensen 1995). The transition from amphibious to secondary water plants is, therefore, gradual, and the distinction between the two groups is subjective.

In the Danish flora there are about 1265 herbaceous species of flowering plants (Rostrup and Jørgensen 1961). From the description of their growth habitat, Sand-Jensen et al. (1992) initially identified a total of 171 species of water plants, or 13.5% of all Danish flowering plants, belonging to the true water plants (4.0%), the amphibious plants (3.5%) and the secondary water plants (6.0%). The notion of freshwater plants as a species-rich heterogeneous group is confirmed by comprehensive studies of the stream flora of Great Britain (Holmes and Newbold 1984) and Denmark (Sand-Jensen et al. unpublished data). A recent Danish study included measurements of areal cover, species by species, along 131 stretches uniformly distributed in 20 streams. The macrophyte species found submerged in this selection of Danish streams included 36 species of true water plants, 33 species of amphibious plants and 83 secondary water plants (Table 6.2). Species of true water plants and amphibious plants were found more frequently than secondary water plants, so that 33% of all observed species in the 131 stretches were true water plants, 47% were amphibious species and 20% were secondary water plants. The true water plants and amphibious plants dominated the areal cover in the streams about equally, whereas secondary water plants covered just a few per cent of the stream bottom (Sand-Jensen et al. unpublished data).

The numerous species of secondary water plants, and the quantitative importance of amphibious plants in the submerged flora in streams, reflects the recruitment of plants from the less disturbed permanent flora along the streams. The large number of secondary water plants probably reflects the large diversity of bank-side vegetation and the fact that these plants have evolved to tolerate submergence. Many seeds and propagules of true water plants are also trapped and develop in shallow water among the semi-aquatic plants and subsequently spread to deeper waters (Henry and Amoros 1996). Most submerged macrophytes of Danish lowland streams are highly disturbed by variable flow, cutting and dredging. The environmental conditions and the distribution of plants may also allow a fast and significant invasion of the aquatic habitat from land in the small streams and ensure that adaptation to submerged life is an active on-going process involving many species (Sand-Jensen et al. 1992). The evolutionary changes and directions have not, however, been analysed. Recent changes of management regimes in many Danish streams have further increased the representation of amphibious and secondary water plants in the streams (Moeslund 1995, Sode 1996). These changes involve reduced cutting of the emergent flora in streams that artificially have been kept extremely wide to eliminate any risk of flooding and frequent cutting of a mid-stream flow passage to ensure drainage, but leaving the shallow shore region intact.

The extensive invasion of amphibious and secondary water plants in small streams suggests that it is easier to pass the land-water barrier than was once believed. The stream habitat offers open space, and the flowing water, being rich in nutrients and carbon dioxide, alleviates the restrictions on gas and solute exchange associated with submerged life. The common supersaturation of the stream water with carbon dioxide, due to inflow of CO_2-rich groundwater or drainage water (Rebsdorf et al. 1991, Sand-Jensen et al. 1995), is important for the ability of amphibious and secondary water plants to photosynthesise and obtain a sufficient carbon surplus to permit survival and growth under water (Sand-Jensen et al. 1992). True water plants have higher affinities for carbon

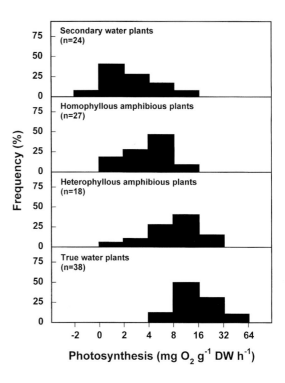

Figure 6.2. Frequency distribution of photosynthetic rates among species of secondary water plants, amphibious plants and primary water plants. All species were collected submerged in Danish streams, and light-saturated rates of photosynthesis were measured at 15 °C in natural stream water supersaturated with respect to carbon dioxide. n = number of measurements. Redrawn from Sand-Jensen et al. (1992) with permission from Blackwell Science Ltd.

dioxide and can, in contrast to secondary water plants and most amphibious plants, often utilise dissolved bicarbonate, which is present at high concentrations in most Danish freshwaters (Sand-Jensen 1983, Rebsdorf et al. 1991). Thus, in the transition from secondary to true water plants there is a gradual increase in the ability of the plants to photosynthesise and grow under water (Fig. 6.2) and to increase the fraction of inorganic carbon that can be extracted from the total inorganic carbon pool (Sand-Jensen et al. 1992). Consequently, the intimate contact between land and water and the high availability of nutrients and carbon dioxide in streams should make this ecotone particularly suitable for the exchange of species and invasion of the aquatic habitat by land plants in an evolutionary perspective.

Spatial and temporal distribution of macrophytes

Macrophytes display different spatial and temporal patterns of distribution in unshaded streams, depending on stream type (i.e. morphometry, hydrology and substrate) and plant species. Submerged rooted macrophytes are often limited by insufficient light and unsuitable substrates and are, therefore, sparse in upstream narrow reaches with coarse substrates and overhanging riparian vegetation as well as in downstream deep reaches (Westlake 1975, Dawson and Kern-Hansen 1979). Maximum plant cover is obtained in medium-sized reaches with stable sediments and stream widths of 1-15 m and depths of 0.2-1.2 m (Westlake 1975).

The macrophytes often develop dense monospecific patches that undergo temporal changes in size and location depending on the flow regime and the incoming irradiance (Butcher 1933, Sand-Jensen and Madsen 1992). The presence of monospecific patches reflects rapid vegetative growth from spatially separated seedlings, trapped shoots and wintering below-ground parts (Andersen and Andersen 1991). The overall physical instability of streams, and the substantial plant loss at high discharges during winter, often remove most of the plants and keep the community in an early successional stage where mixed species assemblages have insufficient time to develop (Dawson et al. 1978, Sand-Jensen et al. 1989b). The small newly colonised patches often experience substantial mortality, and the survival of patches increases as they grow in size and become more firmly rooted (Andersen and Andersen 1991). Shoots in large patches offer mutual protection against physical disturbance and shoots in the centre of the patch are particu-

larly protected. The patches typically grow in size during periods of high light availability and stable flow in summer (Andersen and Andersen 1991, Sand-Jensen and Madsen 1992). The patches expand predominantly laterally and downstream, whereas the upstream front of the patches, which experiences the direct pressure and shear forces of the flow, remains approximately at the same location, though it may move slowly upstream during periods of low flow and downstream during periods of high flow. As the patches increase in size, the spatial variability of flow and substrate conditions also increases along the reaches because the flow velocity is dampened within the patches, whereas the deflected flow is accelerated around the patches and may restrict further lateral expansion (Andersen and Andersen 1991, Sand-Jensen and Madsen 1992). A scattered mosaic of monospecific patches with low flow velocities and sedimentation of fine nutrient-rich particles alternating with deeper flow-channels with coarse substrates hampering further plant expansion is, therefore, a common feature in streams. The resulting physical diversity of flow and substrate types should increase diversity of invertebrates and fish by forming numerous habitats and territories (Mortensen 1977, Nielsen 1986).

The tendency to form a mosaic of plant patches is, however, not constant among species and streams. Plant species which form a dense entangled network of leaves and stems and display limited horisontal spread, such as species of *Callitriche*, have a stronger tendency to form distinct patches and enhance spatial diversity than plant species which form an open canopy and possess efficient horisontal spread by rhizomes and runners, such as species of *Sparganium* (Sand-Jensen and Madsen 1992, Sand-Jensen unpublished data). Stream reaches with unstable sediments (e.g. sand) are also more likely to have a highly scattered distribution of patches, because frequent redistribution of sediments reduces the probability of forming a homogenous cover with time. This distinction between homogenous or heterogeneous plant distributions is important for the physical and biological diversity and the influence of plants on the hydraulic resistance as well as the retention of sediments.

Biomass of macrophytes

The biomass of submerged macrophytes follows a unimodal seasonal course in Danish lowland streams with the maximum in summer (Kern-Hansen et al. 1980). Seasonal changes in biomass to a large extent vary with the stability of discharge. In streams fed by groundwater, the discharge is relatively constant and there is a substantial wintering green biomass (Thyssen et al. 1990). In contrast, in streams with a very variable discharge, most green biomass is lost during high winter discharge and the macrophytes

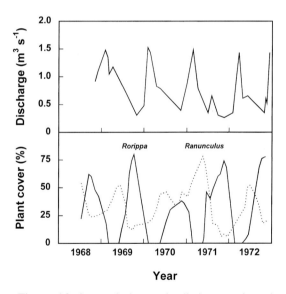

Figure 6.3. Seasonal changes in discharge and areal cover of *Ranunculus pennicillatus* var. *calcareus* (a true water plant) and *Rorippa nasturtium-aquatica* (an amphibious plant) in an English chalk stream. *Ranunculus* colonises the open stream bottom, establishes patches and is overgrown by amphibious *Rorippa* during summer. Redrawn from Dawson et al. (1978) with permission.

exist as starch-filled propagules and rhizomes (Sand-Jensen et al. 1989b, Sand-Jensen and Lindegaard 1996). The warmer winter temperatures in streams fed by groundwater may also allow growth of species which have an evergreen habit.

Streams with stable flow include the chalk streams fed by large aquifers in calcareous catchment areas and the less alkaline streams in catchments with permeable soils of sand or gravel. The first stream type has been intensively examined in southern England and is often dominated by the submerged macrophyte *Ranunculus penicillatus* var. *calcareus,* which is present around the year but has a minimum biomass in winter and a maximum biomass in late spring (Dawson 1978a). The amphibious macrophyte, *Rorippa nasturtium-aquaticum,* invades the *Ranunculus* patches during summer from the permanent riparian populations and the temporary stream populations are later lost during high winter discharge. The two plant species, thereby, form a cyclic successional pattern where *Ranunculus* colonises the open stream bottom, generates patches with reduced flow and subsequently traps floating shoots or allows ingrowth from the banks of *Rorippa* which overgrows the *Ranunculus* plants in late summer (Fig. 6.3). The sequence is reset during winter when most plants decay and are washed out, but *Ranunculus* maintains small populations. Similar cyclic sequences can be observed in many Danish streams. However, the first colonizer is often *Ranunculus peltatus*, or species of *Callitriche*, and the amphibious invaders are often *Berula erecta, Veronica anagallis-aquatica* or *Mentha aquatica* (Sand-Jensen unpublished data). Submerged species may also invade patches of other submerged species, given sufficient time of stable flow. These successional sequences are not necessarily annual. The sequence may last one, two, or more years depending on the frequency of peak discharges which remove

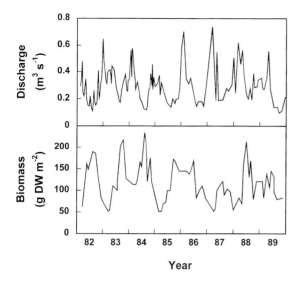

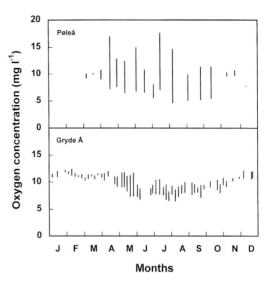

Figure 6.4. Seasonal changes in discharge and macrophyte biomass at a stretch in the River Gryde over several years. River Gryde has a relatively stable flow and has green macrophytes around the year. The main macrophytes are *Ranunculus peltatus*, *Potamogeton natans* var. *submersus* and species of *Callitriche*. Redrawn from Thyssen et al. (1990).

Figure 6.5. Diurnal amplitudes of oxygen concentration during the year in an East-Danish stream where flow is variable (River Pøleå, above) and in a West-Danish stream where flow is stable (River Gryde, below). Data extracted from Kern-Hansen et al. (1980) and Sand-Jensen (orig.).

most of the plant cover and the fine-grained surface sediments (Andersen and Andersen 1991). Several successional phases may also be represented at any given time among different stretches or sites located at different depths and distances from the stream bank.

Danish streams with sandy sediments in permeable catchment areas are very common in the mid and western parts of Jutland. River Gryde has been intensively studied and is a representative of these streams. The seasonal variation in discharge is relatively small and the seasonal changes of the macrophyte biomass are highly regular with minimum winter biomasses of 20-40 g DW m^{-2} and maximum summer biomasses of 150-220 g DW m^{-2} (Fig. 6.4). The individual macrophyte species also show a regular annual growth cycle with species of *Callitriche* maintaining the highest biomass during winter, *Ranunculus peltatus* forming the peak biomass in early summer

and *Potamogeton natans* var. *submersus* having the highest biomass in late summer (Kelly et al. 1983, Thyssen et al. 1990). This regularity of plant development, deriving from the regular discharge pattern, is also reflected in plant photosynthesis and oxygen regimes in the stream. Thus, the gross photosynthesis of the macrophytes can be accurately predicted from the plant biomass and the hyperbolic photosynthesis-light relationships characterising the main seasons (Kelly et al. 1983). The oxygen concentration in streams is determined by the relative rates of photosynthesis, respiration and exchange across the air-water interface and is, therefore, strongly influenced by the presence and metabolic activity of the macrophytes. Fluctuations in oxygen concentration over a day, or a year, are largely predictable and occur within relatively narrow limits in the River Gryde (Fig. 6.5, Thyssen 1981, Kelly et al. 1983).

These streams in sandy regions are, however, inherently unstable if the plant cover is disturbed by extensive weed cutting or by peak discharge leading to erosion of the banks, the stream bed or the plant patches (Thyssen et al. 1990, Andersen and Andersen 1991). Sandy sediments are easily moved by the current, resulting in shifting bed forms and burial or erosion of plant patches. Hence, if the stabilising plant cover is lost, it may take the plants a long time to become re-established. In disturbed streams the macrophyte biomass may, therefore, vary markedly from summer to summer, and it may easily take two years to regain a dense plant cover following extensive losses (Andersen and Andersen 1991).

Streams with large seasonal variations in discharge are found in catchment areas with impermeable clayish soils mainly located in

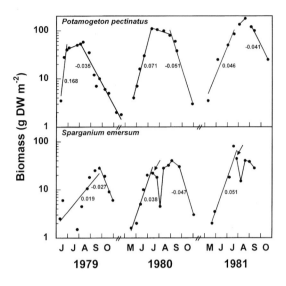

Figure 6.7. Seasonal changes in above-ground biomass of *Potamogeton pectinatus* and *Sparganium emersum* at a stretch in the River Suså over three years. The biomass is plotted on a logarithmic scale, and the numbers show the apparent specific daily rates of biomass increase (ln-units day^{-1}) during early summer and of biomass decline during autumn. Redrawn from Sand-Jensen et al. (1989b) with permission from Blackwell Science Ltd.

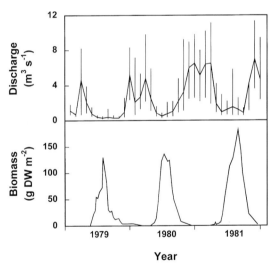

Figure 6.6. Seasonal changes in discharge and macrophyte biomass at a stretch in the River Suså over three years. The discharge gives the mean and the range for each month. The green macrophyte biomass, mainly *Potamogeton pectinatus* and *Sparganium emersum*, is confined to summer when discharge is low. The peak biomass increased gradually with increasing summer discharge during the three years, as biomass and shading by filamentous *Cladophora* declined. Redrawn from Sand-Jensen et al. (1989b) with permission from Blackwell Science Ltd.

East-Denmark (Sand-Jensen et al. 1989b). Winter discharges are often 10-100 times higher than summer discharges (Fig. 6.6). Despite the variable seasonal flow, the development of submerged macrophytes is highly predictable (Fig. 6.6). The high biomass of green plants is confined to a short summer period of low discharge and the plants survive the high discharge and water levels during winter mainly as rhizomes protected in the sediments (Sand-Jensen et al. 1989b, Nørgaard 1992). Starch reserves in the rhizomes allow the macrophytes to form green shoots and grow rapidly in early summer when water levels are low and light conditions are suitable. In late summer and autumn the plant populations decay while transferring storage compounds to the rhizomes or forming special wintering turions and bulbils as observed for *Elodea canadensis, Pota-*

mogeton crispus, P. pectinatus P. perfoliatus and *Sparganium emersum* (Sand-Jensen et al. 1989b, Nørgaard 1992). The regular growth patterns with explosive development of the green biomass during summer, and lack of above-ground biomass for a long period between late autumn and early summer, are well illustrated for *Potamogeton pectinatus* and *Sparganium emersum* in the River Suså (Fig. 6.7). The rapid and exponential rate of increase of the biomass in early summer (0.046-0.168 day^{-1} for *P. pectinatus* and 0.019-0.051 day^{-1} for *S. emersum*) takes place over about 30-40 days. The rate of development is regulated by light availability and is subsequently constrained by self-shading in the accumulating biomass, resulting in excellent fits to a logistic growth model (Fig. 6.7, Sand-Jensen et al. 1989b). Increased shading by intermingled filamentous green algae (e.g. *Cladophora glomerata*) in years of particularly low summer flow (Fig. 6.6) constrains the biomass of rooted macrophytes (Sand-Jensen et al. 1989b).

The marked contrast between the permanent plant cover in streams with stable flow and the concentrated summer cover in streams with variable flow has cascading influences on most other environmental conditions and processes in the streams. Seasonal variability is greater for most parameters in the streams where flow is variable and processes are extremely rapid during the hectic summer, leading to shifting periods of super- and subsaturation with oxygen (Fig. 6.5, Iversen et al. 1984, Jeppesen et al. 1984). These rapidly changing environmental conditions are further enhanced by the higher summer temperatures in the streams with variable flow (e.g. 20-25 $^\circ$C) than in the streams with stable flow (e.g.10-15 $^\circ$C), because the latter receive greater amounts of homothermic groundwater (Iversen et al. 1984, Thyssen et al. 1990). Large amplitudes of discharge, temperature and oxygen concentration are particularly critical to sensitive fish and invertebrates with long life cycles. Salmonids are, for example, not present in streams with variable flow, and these experience high summer temperatures and periodic low oxygen concentrations (Iversen et al. 1984).

Macrophytes and stream flow

A main hydraulic effect of macrophytes in streams is to resist the downstream flow and, thereby, increase the water level. The increased water level is the fundamental reason to cut the plants in order to improve drainage of the surrounding fields and reduce the risk of flooding. Hydrologists, engineers and biologists have, therefore, been particularly interested in being able to predict the relationships between water level, mean flow velocity and hydraulic resistance on the one hand, and macrophyte cover or biomass on the other hand. These attempts have been relatively successful for selected stream reaches, but they have failed to produce a general predictive model of sufficient accuracy (Larsen et al. 1990, 1991). Details of flow and sedimentation within individual plant patches have been studied only recently because of the lack of appropriate methods for measuring flow with sufficient spatial and temporal resolution within the tissue network of plant patches (Sand-Jensen and Mebus 1996, Sand-Jensen et al. 1996).

Flow patterns along stream reaches

The lack of general and accurate models to predict the hydraulic conditions along stream reaches is not surprising, considering the complexity of flow regulation (Chow 1959). For small plant-free channels, an empirical relationship exists between mean water velocity (v, m s^{-1}) over stream stretches and hydraulic radius (R, m; i.e. transversal area divided by wetted perimeter), energy gradient (S, m m^{-1}; i.e. downstream slope of water surface) and hydraulic resistance to the flow (Manning's n):

$$v = n^{-1} R^{2/3} S^{1/2} \quad (1)$$

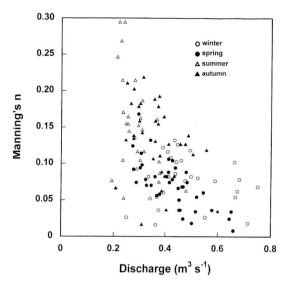

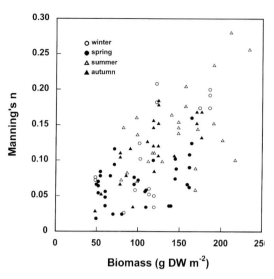

Figure 6.8a. Hydraulic resistance (Manning's n) in the River Gryde as a function of a) discharge and b) macrophyte biomass. Data were derived from measurements at different seasons over several years by Thyssen et al. (1990).

Figure 6.8.b

Manning's n varies depending on form and particle composition of the stream bed, the meandering of the stream and the discharge. Moreover, Manning's exponents are not constants, but represent mean values subject to considerable error. For natural streams with submerged macrophytes, the hydraulic conditions are more complex because the macrophytes form an additional resistance which changes with species, distribution and biomass of the plants.

Plant species with rigid bushy shoots are more resistant to the flow than species with flexible, streamlined shoots (Sand-Jensen and Mebus 1996). Plant species forming dense patches are also more resistant than species forming open patches which are more penetrable to the water flow (Sand-Jensen and Mebus 1996). A continuous plant cover is more resistant to the flow than a mosaic cover of the same mean biomass (Larsen et al. 1990, 1991). A meandering stream with a variable bottom profile, often exaggerated by

scattered plant patches, also exerts greater resistance to the flow than a straight channel of a uniform profile. Attempts to measure the influence of macrophytes on Manning's n may, therefore, reach dubious results even in simple situations where the hydraulic conditions are compared for the same stream reach before and after cutting the plants (Dawson 1978b, Kern-Hansen et al. 1980). In this instance, Manning's n is affected by changes in bed forms following removal of the plants.

The hydraulic resistance and the influence of the plants depends strongly on the discharge. The resistance in the stream channel declines at increasing discharge because the hydraulic radius, the energy gradient and the mean velocity all increase (Fig. 6.8, Larsen et al. 1990, 1991). At high discharge, a larger proportion of the water will pass unaffected above the plant canopy, and this proportion is increased by the enhanced velocity which compresses the plant canopy closer against the stream bed. Consequently, at very high discharges plant resistance will become negligible (Thyssen et al. 1990). Thus, macrophytes reduce the velocity by resisting the flow, but high velocities influence the plants by minimising this resistance (Sand-Jensen and Mebus unpublished data). This compensatory effect is particularly high for flexible submerged plants and is smaller for stiffer amphibious plants growing in shallow water and in the riparian zone (Sode 1996). Amphibious plants with emergent parts may also maintain flow resistance better at high discharge and water level. Bedforms may also show a compensatory effect leading to redistribution of the surface sediment and reduced resistance as discharge and water velocity increase (Kern-Hansen et al. 1980). The desire to establish simple hydraulic models incorporating the influence of macrophytes in an environment of notoriously variable hydraulic resistance has, therefore, not been fulfilled (Larsen et al. 1990, 1991).

Studies in small Danish streams illustrate the inability to obtain accurate relationships between hydraulic resistance and macrophyte biomass. A long-term study in the River Gryde (Thyssen et al. 1990) showed that the hydraulic resistance (Manning's n) increased with plant biomass and declined with increased discharge, but the scatter of data points was substantial (Fig. 6.8). The relationship of hydraulic resistance to plant biomass was even more variable if several small lowland streams were compared. The variable discharge is seasonally coupled to the plant biomass (i.e. the biomass is high and the discharge is low during summer and *vice versa* during winter) which makes it difficult to analytically isolate the influence of the plant biomass on the hydraulic resistance. To overcome this problem Thyssen et al. (1990) abandoned the use of hydraulic resistance and hydraulic radius and established a direct empirical relationship between mean velocity (v, m s^{-1}) along stretches and discharge (Q, m^3 s^{-1}), plant biomass (B, g DW m^{-2}), energy gradient (S, m m^{-1}) and an index for meandering (I), which was set at 1, 2 or 3. For several small Danish streams with discharges ranging from 0.002 to 2.17 m^3 s^{-1} the best predictive model (r^2 = 0.76, p < 0.001) was

$$v = 79.3 \ Q^{0.47} \ B^{-0.58} \ S^{0.69} \ I^{-0.79} \qquad (2)$$

As anticipated, mean water velocity increased with discharge and slope, but decreased with plant biomass and meandering. The mean velocity declined with plant biomass raised to the 0.58 (± 0.06) power and not in direct proportion to the biomass. The retardation of the mean velocity because of high summer biomass was, however, very substantial. The lower mean velocity along reaches should also reduce turbulence and, thereby, reduce reaeration and also promote sedimentation of fine-grained organic material (Thyssen et al. 1987, Sand-Jensen and Mebus 1996). As the mean velocity of the stream decreases, the time taken for a parcel of water to pass along a reach will increase. As a consequence, the likelihood that macrophytes will produce large diel changes in oxygen concentration and critically low oxygen concentrations at night will increase.

It is in many ways simpler and intuitively easier to use relationships of discharge to water level (Q-H) for evaluating the influence of the plants on the flow. The Q-H relationships can also directly evaluate the probability of flooding of cultivated fields along the stream. Several authors (Kern-Hansen et al. 1980, Larsen et al. 1990, 1991, Sode 1996) have described the practical use of Q-H relationships for evaluating the influence of the plant biomass and the probability of flooding.

Flow within plant patches

The local flow pattern within and around individual plant patches is spatially complex but highly reproducible for a given plant species and canopy architecture (Sand-Jensen and Mebus 1996). The flow patterns are important because they determine the physical disturbance of the plants, influence the flux of gases and solutes and regulate sedimentation and erosion. Detailed measurements are, nonetheless, few, due to lack of appropriate methods. The conventional propellers for flow measurements have a horizontal axis rotor which can easily become entangled in the plants, and the advanced acoustic and optical doppler current profilers cannot operate because of interference from the plant tissue. Measurements of dissolution rates of salt tablets are applicable within macrophyte patches (Madsen and Warncke 1983), but the technique is too laborious to reveal details of velocity patterns.

Sand-Jensen and Mebus (1996) used small and physically robust hot-wire anemometers for measuring flow velocity within stream macrophyte patches and close to the sediment with high spatial (0.5 cm) and temporal (0.05 s) resolution. Water velocity dropped markedly from positions upstream to positions within the patches of four common stream macrophytes (Fig. 6.9). The velocity increased somewhat in the most downstream part of the patches and increased much more in the free water behind the patches. The flow reduction was large and uniform within patches of *Callitriche cophocarpa* and *Elodea canadensis,* whereas the reduction was large and variable for *Ranunculus peltatus* and small and variable for *Sparganium emersum* (Fig. 6.9). The different effect of the four species on flow relates to their morphology. *C. cophocarpa* and *E. canadensis* form a relatively dense and homogeneous tissue matrix. *R. peltatus* has long flexible stems and leaves with long capillary filaments which increase the variability of flow within the patches, because *R. peltatus* forms narrow flow-channels which change position within the patches as the long stems move in

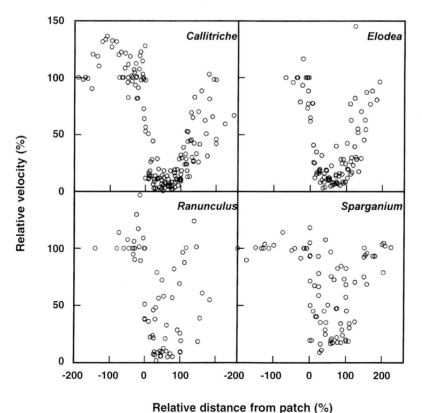

Figure 6.9. Mean velocity measured in longitudinal transects along the mid-axis at the mid-canopy height through patches of *Callitriche cophocarpa, Elodea canadensis, Ranunculus peltatus* and *Sparganium emersum.* Velocity was normalised to the one measured 20-80 cm upstream of each patch, and distance was normalised to the length of each patch. Each panel represents measurements on 6-13 patches measuring 60-200 cm in length. Redrawn from Sand-Jensen and Mebus (1996) with permission from Munksgaard.

Figure 6.10. Vertical profiles of velocity measured in positions upstream (pos. I), along the mid-axis through a patch (pos. II-V) and in the water behind a patch (pos. VI) of *Callitriche cophocarpa*. The location of the patch surface (S) in the positions II-V and the longitudinal extension of the patch (hatched bar on x-axis) are indicated (Sand-Jensen orig.)

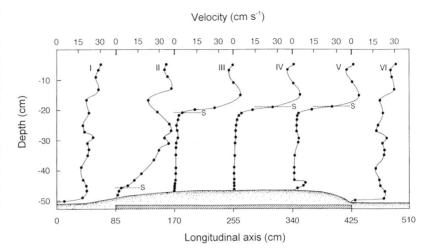

the current. Finally, *S. emersum* forms relatively open patches of long, streamlined ribbon-like leaves which allow the water to pass through with little resistance.

The vertical velocity profiles have been measured for the same four macrophyte species as exemplified for *C. cophocarpa* (Fig. 6.10). The velocity profile measured above the open stream bed upstream of the patch showed small changes with depth until immediately above the stream bed, which constituted the main resistance to the flow. In the most upstream part of the patch the steep vertical gradient had moved to above the canopy surface. In positions further into the patch the velocity gradient was located in the immediate transition from the water to the canopy and was steeper than the gradient encountered above the sediment upstream of the patch. Inside the patch the velocity was strongly dampened (1-8 cm s^{-1}). Hence, the location of major shear forces moved from the unvegetated sediment surface to the patch surface, whereas the interior of the patch and the sediment below the plants was protected against strong shear forces. Downstream of the patch, pronounced vortices form and can generate considerable erosion of the sediment.

The strong turbulence and shear forces at the surface of the patch caused the exposed leaves and stems to flap continuously. Leaf flapping will reduce transport resistance to gas and solute exchange, which can stimulate photosynthesis and growth under nutrient and carbon limitation (Jørgensen and Enevoldsen 1989). Leaf flapping will, however, also increase mechanical wear and has been observed to increase respiration and retard net photosynthesis (Madsen et al. 1993). The formation of the plant patches will, therefore, generate mutual physical protection of the leaves and shoots in the interior of the patches, whereas the physical stress is large on the patch surface (Sand-Jensen and Mebus 1996). The overall consequences of the modified flow and mechanical strain for plant growth and morphogenesis have never been studied. For stream macrophytes exposed as individual shoots to a wide range of velocities, however, the specific plant growth rate has been observed to increase with velocities up to 10-20 cm s^{-1} and, thereafter, decrease up to the highest velocity tested (about 80 cm s^{-1}, Jørgensen and Enevoldsen 1989, Andersen and Andersen 1991, Madsen unpublished data). This pattern presumably reflects the opposing influence of improved gas and solute exchange across gradually thinner diffusion boundary layers (Westlake 1967,

Jenkins and Proctor 1985) *versus* the enhanced mechanical stress with increasing velocity (Madsen et al. 1993).

Several scientists have speculated that the reduced velocity within plant patches may introduce self-limitation of photosynthesis because pH and dissolved oxygen increase, and dissolved inorganic carbon decreases, due to plant metabolism. In all plant patches examined by Sand-Jensen and Mebus (1996), however, the velocity exceeded 1 cm s^{-1} in virtually all positions within patches having a maximum length of 5 m. If, in order to simplify calculations we assume, that the water passes the longest patch as an undisturbed front, the maximum possible passage time will be 500 s. Given the measured plant densities and the photosynthetic capacities, the oxygen released to the water during passage can only amount to a few per cent of the oxygen concentration in the water at air saturation (Sand-Jensen and Mebus 1996). The relative changes in dissolved inorganic carbon are even smaller because the background levels in the Danish streams are typically 2-15-fold higher than oxygen concentrations at air saturation, but concentrations of carbon dioxide could be more affected. For major accumulation of oxygen and depletion of dissolved carbon to take place within plant patches, the patches need to be large, and velocities need to be at least 10-fold lower (i.e. < 0.1 cm s^{-1}) than the lowest velocities normally encountered. Such low flow velocities are only possible in patches located close to the banks in meandering streams protected against the main flow behind the bends. Almost stagnant water may also arise in patches of rooted macrophytes overgrown by dense filamentous algae. Since such patches also experience profound accumulation of organic particles, they may face oxygen supersaturation during the day and oxygen depletion during the night.

Macrophytes and sediments

The relationship between macrophytes and sediments is interactive. Macrophytes influence the sediment composition by affecting sedimentation and erosion and producing organic material. At the same time, the growth and survival of macrophytes depends on sediment nutrition and stability. Our knowledge of the influence of stream macrophytes on sediment composition has grown considerably recently, whereas our knowledge of their dependency on sediment composition is mainly anecdotal.

Stream macrophytes have a specific influence on sedimentation and erosion of inorganic and organic particles within and around plant patches and a general influence over stream reaches. This influence is attributed to three conditions. Firstly, the macrophytes influence velocity and turbulence along stretches and within plant patches. Secondly, the patches generate such a lamellar structure that suspended particles, making contact with the plant tissue, experience local stagnant conditions which facilitate trapping and sedimentation. Thirdly, the plant roots can bind sediment particles closely together so that they become more difficult to erode. The local effects of the plants are examined first, followed by their effects along stream reaches.

Sediment topography and composition within macrophyte patches

The local changes in sediment topography within plant patches depend on the velocity and turbulence close to the sediment and the location of rooted shoots. The near-bed velocity is strongly reduced among the rooted shoots in the upstream and central parts of the patches (Fig. 6.11). In these positions mean sediment accumulation reached 11 cm for *C. cophocarpa*, 6 cm for *R. peltatus* and 5 cm for *E. canadensis* relative to the sediment level upstream of the patches (Fig. 6.12). These mean values were found in

Figure 6.11. Mean and standard deviation (SD) of velocity 1 cm above the sediment surface in longitudinal transects through the mid-axis of patches of four plant species. Each measurement was based on 1000 registrations of velocity over 50 seconds, yielding a mean value and a standard variation. Patterns are averages for six patches of each species. Thus, the standard deviation shown is the average of the standard deviations from the six patches (Sand-Jensen orig.).

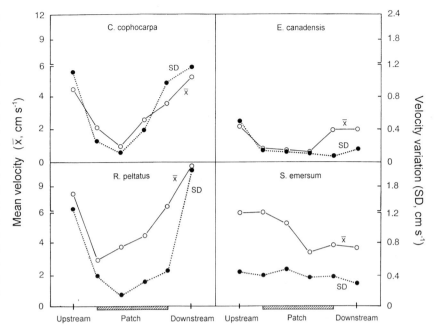

studies during summer including at least 6 patches of each species in several Danish streams. The upstream and central positions within patches were also strongly enriched with fine-grained particles, organic matter and nutrients (N and P, Fig. 6.13). In the

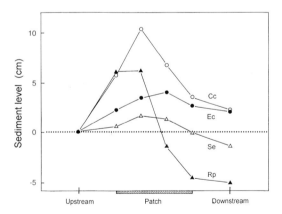

Figure 6.12. Sediment surface level measured along a longitudinal transect through the mid-axis of the plant patches. Patterns are averages for at least six patches of each species. Each measurement was normalised to the sediment level upstream of the patch (Sand-Jensen orig.).

downstream part of the patches, sedimentation was much smaller, and erosion occurred behind the large patches of *R. peltatus*. In these downstream positions the sediment was more coarse-grained and nutrient-poor.

The median grain size and the size variability (d_{16}-d_{84} percentiles) were both significantly smaller in the surface sediment (0-1 cm) located in the upstream and central part of the patches of *C. cophocarpa*, *E. canadensis* and *R. peltatus* compared with the sediment outside of the patches (Table 6.3). These differences were still present at 1-5 cm depth in the sediment. In deeper sediments between 5 and 15 cm, however, the differences between inside and outside positions vanished, because this depth stratum has not been influenced by sedimentation and has a composition similar to the sediments during winter when the flow is high and few plants are present. The accumulated sediment has, therefore, more fine and nutrient-rich particles than the general stream sediment, either because these particles are trapped selectively in the upstream positions of the

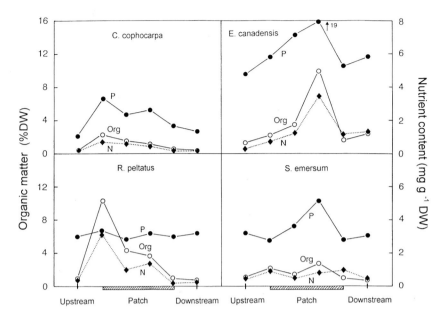

Figure 6.13. Content of organic matter, nitrogen and phosphorus in the surface sediment (0-1 cm) measured along a longitudinal transect through the mid-axis of the plant patches. Values are presented as the mean with standard error for six patches of each species (Sand-Jensen orig.).

patches, or because fine-grained particles dominate the suspended and bed-load transport at the relatively low velocities between spring and summer when the patches expand.

Patterns of sediment topography and composition were radically different within patches of *S. emersum*. Near-bed velocity, sediment level, median grain size and content of organic matter and nutrients were not significantly elevated inside the patches of *S. emersum* (Figs. 6.11-6.13, Table 6.3). This result could lead to the conclusion that *S. emersum* does not influence the sedimentation along stream stretches. But this conclu-

sion is premature because *S. emersum*, by its presence, can reduce the mean velocity along stretches and thereby promote overall sedimentation, though the local enrichment inside relative to outside of the patches is not significant. The strong local enrichment within dense patches of other species may also, in part, reflect redistribution of the fine-grained particles from the open flow channels between the patches to the protected sediments inside the patches.

Table 6.3. Grain size (μm) of mineral particles measured in surface sediments (0-1cm) upstream and inside the patches of four common macrophytes in several Danish streams during summer. Medians are given with $d_{16} - d_{84}$ percentiles in parentheses for at least 6 patches of each species. (Sand-Jensen, orig.).

| | *C. cophocorpa* | | *E. canadensis* | | *R. peltatus* | | *S. emersum* | |
	Upst.	Ins.	Upst.	Ins.	Upst.	Ins.	Upst.	Ins.
	385	215	300	180	300	225	180	195
	(215-780)	(130-430)	(140-750)	(100-500)	(200-535)	(130-380)	(140-580)	(170-610)

Sediment enrichment and plant growth

The mean pool of organic material and nutrients in the accumulated surface sediment inside the plant patches ranged from 900 g organic matter, 30 g N and 25 g P per m² for *C. cophocarpa* to only 190 g organic mater, 6.6 g N and 3.3 g P per m² for *S. emersum*. The plant biomass typically contains 150-250 g organic matter, 5-8 g N and 0.5-0.8 g P per m² within dense patches. Hence, the content of organic matter and nitrogen in the accumulated sediment within the patches was 4- to 5-fold higher than the content in the biomass for *C. cophocarpa* and of similar magnitude for *S. emersum*. The retention of phosphorus was 35-fold higher in the sediment than in the plant biomass for *C. cophocarpa* and 4-fold higher for *S. emersum*. The sediment retention of organic matter and nutrients in patches of *E. canadensis* and *R. peltatus* was close to the values for *C. cophocarpa*.

These results illustrate that nutrient retention in sediments within plant patches in Danish streams usually exceeds the pools contained in the plant biomass particularly for phosphorus. The self-enrichment of the sediment generated by the presence of the plants, therefore, has the potential to alleviate nutrient limitation of plant growth. This effect on plant growth should be most significant in nutrient-poor streams with coarse-textured sediments and low concentrations of inorganic nutrients (e.g. < 100 µg N l⁻¹ and 5 µg P l⁻¹), which are exceedingly rare in Denmark but common elsewhere. Nutrient limitation of plant growth has been demonstrated, for example, in Canadian streams (Chambers et al. 1991). The majority of Danish streams and of lowland streams in other countries are influenced by cultivation and sewage effluents and have fine-textured sediments and nutrient-rich waters with high concentrations of dissolved inorganic nutrients; e.g. typically at 1-10 mg N l⁻¹ and 0.05-0.30 mg P l⁻¹ in Danish streams (Kristensen et al. 1990). Danish stream macrophytes contain high nutrient concentrations (Kern-Hansen and Dawson 1978, Jacobsen 1993a) and are not very likely to experience nutrient limitation and self-enrichment of growth by enhanced sedimentation, unlike the macrophytes in oligotrophic streams with coarse sediments.

Retention and budgets of organic matter and nutrients in stream systems

Particle retention within macrophyte patches during summer can significantly influence

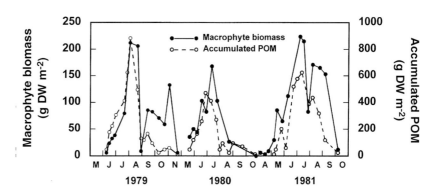

Figure 6.14. Seasonal changes in above-ground biomass of macrophytes and content of accumulated particulate organic matter (POM) in the surface sediment (0-1 cm) at a reach in the River Suså over three years. All values are means of 20 measurements randomly distributed along the reach at each sampling date. Redrawn from Sand-Jensen et al. (1989b) with permission from Blackwell Science Ltd.

the sediment environment and the transport of material to lakes and estuaries further downstream. Detailed studies over three years along stretches of the River Suså showed that the content of organic material and nitrogen in the surface sediment was close to zero during winter, when the macrophytes disappeared, the discharge and velocity were high and unconsolidated fine particles were washed out (Fig. 6.14). During summer a cm-thick surface layer rich in organic matter accumulated in association with the dense plant cover and the low velocities. Peak accumulations during summer amounted to 870-1300 g organic matter and 29-43 g N per m^2 averaged over stream stretches, and these accumulations exceeded the maximum plant biomasses from 3- to 9-fold (Jeppesen et al. 1984).

In the River Suså system, approximately 80% of the particle transport was retained in the sediment during summer (Jeppesen et al. 1984). This retention constituted only 10-20% of the total transport of nitrogen because the main nitrogen form, dissolved nitrate, remains unaffected. The retention of organic particles is also relatively small, compared with the total transport of organic material, because dissolved organic matter is the main organic form which typically exceeds the particle concentrations 5-10-fold. Particle retention is, on the other hand, much more important for the downstream transport of phosphorus, because concentrations of phosphate dissolved in the water are smaller than phosphorus bound in organic and mineral particles and adsorbed to mixed flocculates of aluminium-iron oxyhydroxides and organic material (Kronvang 1992). Phosphorus retention in the plant beds during summer amounted to 25-60% of the transport in some stream systems (Svendsen and Kronvang 1993, Svendsen et al. 1996).

Nutrient retention in the stream bed during summer will significantly reduce the input to lakes and estuaries located further downstream. Because these ecosystems are often dominated by phytoplankton communities, which are nutrient-limited during summer, the retention of nutrients in the streams will reduce phytoplankton blooms in lakes and estuaries. The particle retention is, however, only temporary in most streams, and the particles are eroded and transported downstream during autumn and winter when discharge is high and plant biomass is low (Svendsen and Kronvang 1993). Nevertheless, the transport will take place at a time of the year when phytoplankton growth in lakes and estuaries is mostly light limited and nutrients may be flushed through to open marine waters. The retention of organic material in the stream bed during summer will also result in substantial degradation. The material resuspended later will be more refractory, and the reduced rate of decomposition will reduce the risk of oxygen depletion in the downstream ecosystems relative to the situation where organic material is transported directly through the streams.

The accumulation of organic material in the plant beds has several additional effects on plants, animals and microbial processes in the sediments (Fig. 6.1). Microalgae on the sediment will be buried below the settled particles, which will restrict their photosynthesis and growth unless they can move to the surface (Sand-Jensen et al. 1989b). Sedimentation and enrichment of the sediment with respect to fine particles and organic matter will increase the rate of sediment oxygen consumption, reduce water percolation through the sediment and, thereby, increase the extent of sediment anoxia. Because photosynthesis of sediment microalgae declines markedly, enhanced by shading of the macrophyte canopy, sediment metabolism will change from mixed photoautotrophic-heterotrophic to predominantly heterotrophic (Jeppesen et al. 1984). The increased content of organic material and the enlarged anoxic zones may also account for the enhanced rates of denitrification in vegetated sediments relative to bare sediments (Christensen and Sørensen 1988, Christensen et al. 1990). Denitrification leads to a permanent loss of nitrogen which can amount to about 10% of the downstream nitrogen transport at low discharge during summer, but on an annual basis the loss by denitrification is only a few per cent of the nitrogen transport (Thyssen et al. 1990).

Influence of macrophytes on epiphytic and benthic microalgae

Rooted macrophytes regulate microalgal communities in the streams by forming surfaces for attachment of epiphytic algae and restricting the development of algae on the sediments (Fig. 6.1). A main bio-engineering effect of the macrophytes is to form a large substratum for mixed communities of microalgae, bacteria and invertebrates. The mixed metabolism of the epiphytic community has been observed to change from mainly photoautotrophic in spring and early summer to mainly heterotrophic during mid-summer in the River Suså (Sand-Jensen et al. 1989a). In spring and early summer, macrophyte density is low, and much light reaches the plant surfaces, stimulating growth of epiphytic microalgae. During mid-summer macrophyte density, self-shading and temperature are high, and less light reaches the plant surfaces, thereby impeding growth of microalgae and promoting growth of bacteria. Thus, epiphytic microalgae are usually much less important primary producers than macrophytes and benthic microalgae. In the River Suså, for example, the epiphytic microalgae contributed just a few per cent to the annual primary production. The degradation of dissolved organic matter by the bacterial community on the macrophyte surface was more significant because it constituted about 10% of the total annual oxygen consumption in the stream (Sand-Jensen et al. 1989a).

The spatial and temporal variability in bio-

mass and photosynthesis of microalgae on the stream bottom is mainly regulated by sediment stability, light availability and grazing loss (e.g. Sand-Jensen et al. 1988, Biggs 1996). The dissolved nutrients in most Danish lowland streams are probably present at concentrations saturating to microalgal growth (Iversen et al. 1990, 1991). The macrophytes, therefore, play a key-role in the development of benthic microalgae, due to their strong influence on light penetration to the sediment surface (Sand-Jensen et al. 1988), but they can also influence the microalgae by modifying sediment stability and sedimentation (Iversen et al. 1990, 1991, Thyssen et al. 1990).

The biomass of microalgae on fine-grained sediments of streams in East Denmark follows a regular seasonal course which is a direct function of the irradiance reaching the sediment surface (Fig. 6.15). The extensive biomass of benthic microalgae is restricted to a short spring period of low water level and macrophyte cover. The rapid biomass

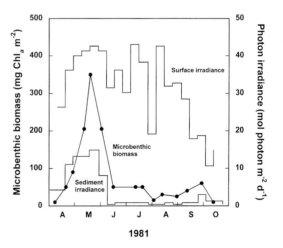

Figure 6.15. Seasonal changes in biomass of benthic microalgae and daily irradiance (400-700 nm) reaching the stream surface and the sediment surface at a reach in the River Suså. Biomass values are means of 20 measurements randomly distributed along the reach at each sampling date. Redrawn from Sand-Jensen et al. (1989b) with permission from Blackwell Science Ltd.

increase of benthic microalgae accompanies the coupled reduction of discharge, velocity and water level in spring. Thus, increased light availability due to increasing surface irradiance and low water level coincides with increased sediment stability. The amount of microalgae rapidly declines as a dense macrophyte cover is established and shades the microalgae (Fig. 6.15). Particle sedimentation may further impede the microalgae and contribute to the presence of high proportions of pigment degradation products.

Because dominance of benthic microalgae and macrophytes often replace each other seasonally, macrophyte cutting during summer may induce secondary maxima of microalgae, though usually of more modest magnitude than the primary maximum in spring (Iversen et al. 1984, Sand-Jensen et al. 1988).

Macrophytes and benthic microalgae are usually the main contributors to the annual primary production, and rates of gross photosynthesis can be very high (6-20 g O_2 m^{-2} d^{-1}) during peak biomasses in eutrophic streams (Jeppesen et al. 1984, Bijl et al. 1989). Extensive amplitudes of oxygen concentrations between night and day are, therefore, found during the spring bloom of benthic microalgae and the summer bloom of macrophytes (Fig. 6.5).

Influence of macrophytes on invertebrates and fish

Macrophytes are important for the composition and abundance of macroinvertebrates and fish in lowland streams. The macrophytes influence the animals by providing food, substrate and shelter and by modifying

Table 6.4. Number within dominant groups, or species, of macroinvertebrates in the vegetation and in the sediment at a reach in the River Suså in July 1979. The macrophyte biomass was 51 g DW m^{-2}. Mean numbers are calculated per m^2 of the stream bottom. From Iversen et al. (1985). Main feeding modes are indicated: a – filter feeder, b – grazer, c – deposit feeder, d – carnivore.

Invertebrate group/species	Vegetation	Sediment
Chironomidae		
Orthocladinae[b]	205,000	4,990
Rheotanytarsus sp.[a]	96,800	1,000
Poratanytarsus sp.[b]	42,900	21,200
Cladytanytarsus sp.	10	2,770
Chironomini[c]	310	4,920
Simulidae		
Boophthora erythrocephala[a]	76,200	220
Hydra sp.[a]	3,120	300
Baetis sp.[b]	670	220
Oligochaeta[c]	240	9,350
Hirudinae[d]	110	650
Asellus aquaticus [c]	2	2,270
Gammerus pulex [b,c]	0	275
Sialis lutaria [d]	0	118

physical and chemical conditions and food sources (Mortensen 1977, Kern-Hansen et al. 1980, Iversen et al. 1985, Iversen and Thorup 1988).

Several feeding mechanisms are represented among macroinvertebrates in the vegetation. Filtration of particles in flowing water, grazing on the epiphyte community, consumption of the host plant and carnivory are the most important feeding habits (Iversen et al. 1984, Table 6.4). Filtration and grazing are obvious feeding modes in the vegetation. Because flowing water continuously passes across the plant surface with high concentrations of organic particles, passive filtration is an attractive feeding mode. Important passive filtrators include larvae of simulids and certain chironomids (e.g. *Rheotanytarsus sp.*) which can develop very large populations (> 100,000 ind. m^{-2} in the River Suså, Table 6.4) with short generation time and high production (Iversen et al. 1984, 1985). The short generation time is important for the attached filtrators because the submerged leaves only live for one to two months before they decay and are replaced by new leaves (Nielsen et al. 1985, Jacobsen and Sand-Jensen 1994a). The animals have to finish their entire development during the lifetime of the leaf to avoid changing position. The suitability of the leaves for invertebrates may depend both on the lifetime and the flow pattern surrounding the leaf. The ribbon-like leaves of *Sparganium emersum* are apparently suitable for simulid larvae which are abundant on the leaves. The efficiency of filtration of simulids depends on the flow velocity and the thickness of the viscous sublayer in contact with the leaf surface (Chance and Craig 1986). It is plausible that the high flow through the open patches and around the leaves of *S. emersum* are advantageous to the filtration, compared with the situation for other plant species such as *Callitriche cophocarpa*, where the main velocity gradient is confined to the outer surface of the patches and the

flow is strongly restricted inside the patches. For macroinvertebrates with different feeding modes and flow tolerances, however, other plant species and locations within the patches may be preferred due to the different architecture and flow pattern within the patches (Gregg and Rose 1985, Tokeshi 1986).

Grazers constitute another abundant feeding group in the vegetation. Snails and certain species of mayflies (e.g. *Baetis rhodanii*) and chironomids are important grazers (Iversen et al. 1984, 1985). *Gammarus pulex* is also abundant within plant patches, probably because of reduced velocity and protection against fish predation. Moreover, *G. pulex* consumes the epiphyte community and, in particular, the fine organic particles retained within the vegetation. In some cases it also consumes the macrophyte tissue (Sand-Jensen and Madsen 1989).

Historically, macrophyte tissue was thought to have a poor nutritional value and, for that reason, to be consumed only very little by macroinvertebrates (Sand-Jensen and Madsen 1989, Jacobsen and Sand-Jensen 1992). Most stream plants have, however, a high nutritional value in terms of the content of nitrogen (protein) and phosphorus (Jacobsen 1993a). The nutrient content exceeds that of most terrestrial leaves (Newman 1991), including beech leaves which are used extensively by invertebrate shredders in many Danish forest streams, and the content resembles that of alder leaves which are highly preferred (Jacobsen 1994). Several investigations have shown that large shredders, for example the caddis larva (*Anabolia nervosa*), consume fresh macrophyte leaves in streams and can metabolise and grow well on this food source (Jacobsen and Sand-Jensen 1994b). Particularly during periods of relative food shortage in the streams, grazing of fresh macrophyte leaves can become intense, though less than 10% of the annual macrophyte production is usually consumed (Jacobsen 1993a). The consumption is,

Table 6.5. Primary consumer production (P) and mean P: B ratios of total invertebrate communities in low order streams in forests, mixed forest- open landscape and open landscape. Production is shown as g ash-free dry matter per m^2 and year. Compiled from various sources in Iversen et al. (1984) and Iversen (1988).

Stream type	Production (g m^{-2} y^{-1})	P : B (y^{-1})
Forest streams		
Caribau River, Minn.	4.1	4.3
Blackhoof River, Minn.	5.6	4.2
Hinau, N. Z.	6.8	7.9
Upper Ball Creek, N. Cor.	7.2	2.7
Rold Kilde, DK	8.7	2.7
Mixed forest-open streams		
Horokiwi, N. Z.	17	6.0
Nath Branch Creek, Minn	18	5.1
Bisballe Bæk, DK	21	3.8
Hinqau, N. Z.	28	7.5
Open streams – pasture/field		
Horokiwi, N. Z.	30	6.4
Hinau, N. Z.	65	7.3
River Suså, DK[a]	90	
Cone Spring, Iowa[b]	129	10.7

[a] High production of benthic microalgae and macrophytes.

[b] Dense biomass of emergent plants.

nevertheless, substantial when macrophyte production is high. Fresh macrophytes are mainly utilized as an alternative food source among macroinvertebrate shredders-collectors which feed on a wide spectrum of food types. Only few grazers have apparently specialised on consuming freshwater macrophytes. Whether the presence of only a few specialised grazers is due to some unidentified toxic substances in freshwater macrophytes, a postulated unpredictable occurrence of macrophytes, insufficient time to have allowed the evolution of specialised grazers, or some other mechanism, is open to discussion (Jacobsen 1993a,b).

The presence of macrophytes in unshaded lowland streams usually stimulates a much greater production of macroinvertebrates than in the plant-free situation (Iversen 1988, Table 6.5). The elevated invertebrate production probably results from the formation of the dense communities associated with the plant patches (Iversen et al. 1984, 1985). The enhanced invertebrate production can in turn support denser and more productive fish populations (Mortensen 1977, Madsen 1995). Brown trout, which is the most important fish in Danish lowland streams, both ecologically and for sport fishing, is also more abundant in the presence of macrophytes because they increase the physical variability and the number of fish territories (Mortensen 1977, Madsen 1995). Fish production is very high in Danish streams when

Figure 6.16. The small Danish streams have a rich vegetation of sub-merged and amphibious species. Photograph by Klaus Brodersen.

food availability and physical variability are high and environmental conditions suitable. The high food quality of the autochthonous production of macrophytes and microalgae in small, unshaded streams can also account for the higher production of invertebrates and fish as compared to shaded forest streams, though the allochthonous input of detritus from the forest can match or exceed the rates of primary production in the unshaded streams (Jacobsen and Friberg 1997).

Critical conditions for the development of certain animals may, however, arise in streams with very dense vegetation and low discharge during summer. If the stream bottom is covered by dense homogeneous vegetation, critically low oxygen concentrations may arise in the sediment and in the water column at night because flow velocity, turbulence and reaeration are low, while organic sedimentation, oxygen consumption and temperature are high (Larsen 1975, Iversen et al. 1984). These conditions will restrict salmonid fishes and invertebrates with high oxygen demands, and large burrowing species in the sediment will disappear. At high macrophyte density and critical oxygen conditions close to the sediment, benthic species may move up into the vegetation where the velocity and oxygen fluxes are higher (Iversen et al. 1984). An optimal strategy in the management of streams would, therefore, be to maintain an open flow channel and a mosaic structure of plant patches to ensure a sufficient velocity and reaeration capacity as well as a great variability of habitats and food sources (Jensen et al. 1994, Wiberg-Larsen et al. 1994).

Concluding remarks

It is difficult to study stream ecology in Denmark because of the intense management of the streams. All Danish stream research faces the risk of extensive disturbance by unscheduled macrophyte cutting, which is potentially disastrous to any study. The evaluation of the key-role of macrophytes in lowland streams presented here has, therefore, been based mainly on comparisons of selected processes and environmental and biological conditions in situations with or without plants; investigations have been performed at spatial scales of short reaches, or individual patches, and temporal scales of one year or less. Because many ecological and biological processes also respond to larger scales, it would be preferable to establish future comparisons of plant-dominated *versus* plant-free situations over long reaches (< 1 km), entire stream systems and several years.

The review demonstrates that macrophytes have a strong bio-engineering influence in the shallow lowland streams and that they regulate or modify most physical, chemical and biological features. Macrophytes probably have a stronger influence on the ecology in lowland streams than in any other aquatic ecosystem because the areal cover and the influence on flow and sediment stability are so profound. A similar key-role of macrophytes is observed in the shallow regions of transparent lakes and marine waters (e.g. Sand-Jensen and Borum 1991), as in most terrestrial ecosystems (Jones et al. 1994), and it deserves further quantitative analysis in the future.

Literature cited

Andersen, T. L., and K. Andersen. 1991. Ø-dynamik hos *Callitriche cophocarpa*. MS-Thesis. Botanical Institute, University of Aarhus, Denmark.

Biggs, B. J. F. 1996. Patterns in benthic algae of streams. Pages 31-56 *in* R. J. Stevenson, M. L. Bothwell and R. L. Lowe. Algal ecology. Freshwater benthic ecosystems. Academic Press, San Diego, USA.

Bijl, L. van der, K. Sand-Jensen, and A. L. Hjermind. 1989. Photosynthesis and canopy structure of a submerged plant, *Potamogeton pectinatus,* in a Danish lowland stream. Journal of Ecology 71: 161-175.

Butcher, R. W. 1933. Studies on the ecology of rivers. I. On the distribution of macrophytic vegetation in the rivers of Britain. Journal of Ecology 21: 58-91.

Chambers, P. et al. 1991. Current velocity and its effect on aquatic macrophytes in flowing rivers. Ecological Applications 1: 249-257.

Chow, V. T. 1959. Open channel hydraulics. McGraw-Hill, New York, USA.

Christensen, P. B., L. P. Nielsen, J. Sørensen, and N. P. Revsbech. 1990. Denitrification in nitrate-rich streams; diurnal and seasonal variation related to benthic oxygen metabolism. Limnology and Oceanography 35: 336-347.

Christensen, P. B., and J. Sørensen. 1988. Denitrification in sediment of lowland streams: Regional and seasonal variation in Gelbæk and Rabis Bæk,

Denmark. FEMS Microbial Ecology 53: 335-344.

Dawson, F. H. 1978a. The annual production of the aquatic macrophyte *Ranunculus penicillatus* var. *calcareus* (R. W. Butcher) C. D. Cook. Aquatic Botany 2: 51-73.

Dawson, F. H. 1978b. The seasonal effects of aquatic plant growth on the flow of water in a stream. Pages 71-78 *in* Proceedings of the European Weed Research Society Fifth Symposium on Aquatic Weeds. Amsterdam, The Netherlands.

Dawson, F. H., E. Castellano, and M. Ladle. 1978. Concept of species succession in relation to river vegetation and management. Verhandlungen Internationale Vereinigung für Theoretische und Angewandte Limnologie 20: 1439-1444.

Dawson, F. H., and U. Kern-Hansen. 1979. The effects of natural and artificial shade on the macrophytes of lowland streams and the use of shade as a management technique. Internationale Revue der gesamten Hydrobiologie 64: 437-455.

Gregg, W. W., and F. Rose. 1985. Influences of macrophytes on invertebrate community structure, guild structure and microdistribution in streams. Hydrobiologia 128: 45-56.

Henry, C. P., and C. Amoros. 1996. Are the banks a source of recolonization after disturbance: an experiment on aquatic vegetation in a former channel of Rhone River. Hydrobiologia 330: 151-162.

Holmes, N. T. H., and C. Newbold. 1984. River plant communities – Reflectors of water and substrate chemistry. No. 9. Nature Conservancy Council, Peterborough. England.

Iversen, T. M. 1988. Secondary production and trophic relationships in a spring invertebrate community. Limnology and Oceanography 33: 582-592.

Iversen, T. M., E. Jeppesen, K. Sand-Jensen, and J. Thorup. 1984. Økologiske konsekvenser af reduceret vandføring i Susåen. Bind I. Den biologiske struktur. Freshwater Biological Laboratory, University of Copenhagen, Denmark.

Iversen, T. M., and J. Thorup. 1988. A three years' study of life cycle, population dynamics and production of *Asellus aquaticus* L. in a macrophyte rich stream. Internationale Revue der gesamten Hydrobiologie 73: 73-94.

Iversen, T. M., J. Thorup, T. Hansen, J. Lodal., and J. Olsen. 1985. Quantitative estimates and community structure of invertebrates in a macrophyte rich stream. Archiv für Hydrobiologie 102: 291-301.

Iversen, T. M., J. Thorup, N. Thyssen, K. Kjeldsen, L. P. Nielsen, P. Lund-Thomsen, N. B. Jensen, C. Pedersen, and T. Winding. 1990. Bundlevende algers regulering i små vandløb. NPO-forskning fra Miljøstyrelsen, nr. C7. København, Denmark.

Iversen, T. M., J. Thorup, K. Kjeldsen, and N. Thyssen. 1991. Spring bloom development of microbenthic algae and associated invertebrates in two reaches of a small stream with contrasting sediment stability. Freshwater Biology 26: 189-198.

Jacobsen, D. 1993a. Herbivory of invertebrates on submerged macrophytes in freshwaters. Ph.D.-Thesis. Freshwater Biological Laboratory, University of Copenhagen, Denmark.

Jacobsen, D. 1993b. Trichopteran larvae as consumers of submerged angiosperms in running waters. Oikos 67: 379-383.

Jacobsen, D. 1994. Food preference of the caddis larva *Anabolia nervosa* feeding on aquatic macrophytes. Verhandlungen Internationale Vereinigung für Theoretische und Angewandte Limnologie 25: 2478-2481.

Jacobsen, D., and N. Friberg. 1997. Macroinvertebrates communities in Danish streams; the effect of riparian forest cover. Chapter 13 *in* K. Sand-Jensen and O. Pedersen, editors. Freshwater Biology. Priorities and Development in Danish Research. Gad, Copenhagen, Denmark.

Jacobsen, D., and K. Sand-Jensen. 1992. Herbivory of invertebrates on submerged macrophytes from Danish freshwaters. Freshwater Biology 28: 301-308.

Jacobsen, D., and K. Sand-Jensen. 1994a. Invertebrate herbivory on the submerged macrophyte *Potamogeton perfoliatus* in a Danish stream. Freshwater Biology 31: 43-52.

Jacobsen, D., and K. Sand-Jensen. 1994b. Growth and energetics of a trichopteran larva feeding on fresh submerged and terrestrial plants. Oecologia 97: 412-418.

Jenkins, J. T., and M. C. F. Proctor. 1985. Water velocity, growth-form and diffusion resistance to photosynthetic CO_2 uptake in aquatic bryophytes. Plant, Cell and Environment 8: 317-323.

Jensen, J. J., J. Skriver, and L. Skjødsholm. 1994. Effekter af miljøvenlig vedligeholdelse i amtsvandløb, Aarhus Amt 1987-1992. Section for Nature and Environment, Aarhus County, Denmark.

Jeppesen, E., T. M. Iversen, K. Sand-Jensen, and C. P. Jørgensen. 1984. Økologiske konsekvenser af reduceret vandføring i Susåen. Bind II. Ilt- og vandkvalitetsforhold. Freshwater Biological Laboratory, University of Copenhagen, Denmark.

Johansson, M. E., and C. Nilsson. 1993. Hydrochory, population dynamics and distribution of the clonal aquatic plant *Ranunculus lingua*. Journal of Ecology 8: 81-91.

Jones, C. G., J. H. Lawton, and M. Shachak. 1994. Organisms as ecosystem engineers. Oikos 69: 373-386.

Jones, H. 1955. Studies on the ecology of the river Rheidol. Journal of Ecology 43: 462-476.

Jørgensen, T. B., and H. O. Enevoldsen. 1989. Strømhastighedens betydning for fotosyntese, respiration og vækst af submerse akvatiske makrofyter. MS.-Thesis, Botanical Institute, University of Aarhus, Denmark.

Kelly, M. G., N. Thyssen, and B. Moeslund. 1983. Light and the annual variation of oxygen- and carbon-based productivity in a macrophyte-dominated river. Limnology and Oceanography 28: 503-515.

Kern-Hansen, U., and F. H. Dawson. 1978. The standing crop of aquatic plants of lowland streams in Denmark and the inter-relationship of nutrients in plant, sediment and water. Pages 143-150 *in* Proceedings of European Weed Research Society Fifth Symposium on Aquatic Weeds. Amsterdam, The Netherlands.

Kern-Hansen, U., T. F. Holm, B. L. Madsen, N. Thyssen, and J. Mikkelsen. 1980. Vedligeholdelse af vandløb. Miljøprojekter 30. Ministry of Environmental Protection, Copenhagen, Denmark.

Kristensen, P., B. Kronvang, E. Jeppesen, P. Græsbøll, M. Erlandsen, Aa. Rebsdorf, A. Bruhn, and M. Søndergaard. 1990. Ferske vandområder – vandløb, kilder og søer. Vandmiljøplanens Overvågningsprogram. Danmarks Miljøundersøgelser, Silkeborg, Denmark.

Kronvang, B. 1992. The export of particulate matter, particulate phosphorus and dissolved phosphorus from two agricultural river basins: Implications on estimating the non-point phosphorus load. Water Research 26: 1347-1358.

Larsen, V. 1975. Grødeskæringens betydning for iltforhold og plantevækst i vandløb. Hedeselskabets Tidsskrift 96: 14-21.

Larsen, T., J.-O. Frier, and K. Vestergaard. 1990. Discharge/stage relations in vegetated Danish streams. Pages *55-62 in* W. R. White. Proceedings of International Conference on River Flood Hydraulics. 17-20 September 1990, Wallingford, England.

Larsen, T., K. Vestergaard, and J.-O. Frier. 1991. Hydraulic aspects of vegetation maintenance in streams. Pages 206-211 *in* Proceedings of XXIV Congress of the International Association on Hydraulic Research, 9-13 September. Madrid, Spain.

Madsen, B. L. 1995. Danish watercourses. Miljønyt nr. 11. Ministry of Environmental Protection, Copenhagen, Denmark.

Madsen, T. V., H. O. Enevoldsen, and T. B. Jørgensen. 1993. Effects of water velocity on photosynthesis and dark respiration in submerged stream macrophytes. Plant, Cell and Environment 16: 317-327.

Madsen, T. V., and S. C. Maberly. 1991. Diurnal variation in light and carbon limitation of photosynthesis by two species of submerged freshwater macrophytes with a differential ability to use bicarbonate. Freshwater Biology 26: 175-187.

Madsen, T. V., and E. Warncke. 1983. Velocities of currents around and within submerged aquatic vegetation. Archiv für Hydrobiologie 97: 389-394.

Moeslund, B. 1995. Grøde og vandløbenes form. Vand og Jord 2: 69-71.

Moeslund, B., B. Løjtnant, H. Mathiessen, A. Pedersen, N. Thyssen, and J. C. Schou. 1990. Danske vandplanter. Miljønyt nr. 2. Danmarks Miljøundersøgelser. Ministry of Environmental Protection, Copenhagen, Denmark.

Mortensen, E. 1977. Density-dependent mortality of trout fry (*Salmo trutta*) and its relationship to the management of small streams. Journal of Fish Biology 11: 613-617.

Newman, R. M. 1991. Herbivory and detrivory on freshwater macrophytes by invertebrates: a review. Journal of the North American Benthological Society 10: 89-114.

Nielsen, G. 1986. Dispersion of brown trout (*Salmo trutta* L.) in relation to stream cover and water depth. Polische Archiv für Hydrobiologie 33: 475-488.

Nielsen, L. W., K. Nielsen, and K. Sand-Jensen. 1985. High rates of production and mortality of submerged *Sparganium emersum* Rehman during its short growth season in a eutrophic Danish stream. Aquatic Botany 22: 325-334.

Nørgaard, C. 1992. Genvækst af *Potamogeton perfoliatus* L. (Hjertebladet Vandaks) efter grødeskæring. MS.-Thesis, University of Odense, Denmark.

Rebsdorf, Aa, N. Thyssen, and M. Erlandsen. 1991. Regional and temporal variation in pH, alkalinity and carbon dioxide in Danish streams related to soil type and land use. Freshwater Biology 27: 15-32.

Rostrup, E., and C. A. Jørgensen 1961. Den danske flora. Gyldendal, Copenhagen, Denmark.

Sand-Jensen, K. 1983. Photosynthetic carbon sources of stream macrophytes. Journal of Experimental Botany 34: 198-210.

Sand-Jensen, K. 1995. Nøgleorganismer og biologiske ingeniører. Vand og Jord 201-202.

Sand-Jensen, K., and J. Borum. 1991. Interactions among phytoplankton, periphyton, and macrophytes in temperate freshwaters and estuaries. Aquatic Botany 41: 137-175.

Sand-Jensen, K., D. Borg, and E. Jeppesen. 1989a. Biomass and oxygen dynamics in the epiphyte community in a Danish lowland stream. Freshwater Biology 22: 431-443.

Sand-Jensen, K., P. A. Brodersen, T. V. Madsen, T. Sand-Jespersen, and B. Kjøller. 1995. Planter og CO$_2$-overmætning i vandløb. Vand og Jord 2: 72-77.

Sand-Jensen, K. E. Jeppesen, K. Nielsen, L. van der Bijl, A.-L. Hjermind, L. W. Wiggers, and T. M. Iversen. 1989b. Growth of macrophytes and ecosystem consequences in a Danish lowland stream. Freshwater Biology 22: 15-32.

Sand-Jensen, K., and C. Lindegaard. 1996. Økologi i søer og vandløb. Gad, Copenhagen, Denmark.

Sand-Jensen, K., and T. V. Madsen. 1989. Invertebrates graze submerged rooted macrophytes in lowland streams. Oikos 55: 420-423.

Sand-Jensen, K., and T. V. Madsen. 1992. Patch dynamics of the stream macrophyte, *Callitriche cophocarpa*. Freshwater Biology 27: 277-282.

Sand-Jensen, K., and J. R. Mebus. 1996. Fine-scale patterns of water velocity within macrophyte patches in streams. Oikos 76: 169-180.

Sand-Jensen, K., J. R. Mebus, and T. V. Madsen. 1996. Vandplanter påvirker strøm og sediment. Vand og Jord 5: 191-195.

Sand-Jensen, K., J. Møller, and B. H. Olesen. 1988. Biomass regulation of microbenthic algae in Danish lowland streams. Oikos 53: 332-340.

Sand-Jensen, K., M. F. Pedersen, and S. L. Nielsen. 1992. Photosynthetic use of inorganic carbon among primary and secondary water plants in streams. Freshwater Biology 27: 283-293.

Svendsen, L. M., and B. Kronvang. 1993. Retention of nitrogen and phosphorus in a Danish lowland river system: Implications for the export from the watershed. Hydrobiologia 251: 123-135.

Sode, A. 1996. Effekt af strømrendeskæring på vand-føringsevne og sammensætning og mængde af grøde. Section for Nature and Aquatic Environment, Fyns County, Denmark.

Svendsen, L. M., B. Kronvang, A. R. Laubel, S. E. Larsen, and B. Andersen. 1996. Phosphorus retention in a Danish lowland river system. Verhandlungen Internationale Vereinigung für Theoretische und Angewandte Limnologie, in press.

Thyssen, N. 1981. Genluftning og fotosyntese i vandløb. Miljøprojekter 40. Ministry of Environmental Protection, Copenhagen, Denmark.

Thyssen, N., and M. Erlandsen 1987. Reaeration of oxygen in shallow, macrophyte rich streams: II. Relationship between the reaeration coefficient and hydraulic properties. Internationale Revue der gesamten Hydrobiologie 72: 575-594.

Thyssen, N., M. Erlandsen, B. Kronvang, and L. M. Svendsen. 1990. Vandløbsmodeller- biologisk struktur og stofomsætning. NPO-forskning. Nr. C 10. Ministry of Environmental Protection, Copenhagen, Denmark.

Tokeshi, M. 1986. Population dynamics, life histories and species richness in an epiphytic chironomid community. Freshwater Biology 16: 431-442.

Westlake, D. F. 1967. Some effects of low-velocity currents on the metabolism of aquatic macrophytes. Journal of Experimental Botany 18: 187-205.

Westlake, D. F. 1975. Macrophytes. Pages 106-128 *in* B. A. Whitton, editor. River Ecology. Blackwell, Oxford, England.

Wiberg-Larsen, P., P. S. Pedersen, T. Rugaard, and P. Geertz-Hansen. 1994. Bedre vandløbspleje giver flere fisk. Vand og Jord 1: 263-265.

7 Ecology of coastal waters and their response to eutrophication

By Jens Borum

Shallow coastal waters are highly dynamic ecosystems. Rates of primary production and mineralization are high, and pelagic and sediment processes are tightly coupled. Shallow coastal waters are naturally eutrophic ecosystems, and Danish areas receive, in addition, high nutrient input from land. Coastal waters function as filters by retaining a substantial proportion of the nutrients transported from land to the open sea. Coastal eutrophication increases the importance of pelagic processes relative to benthic processes in primary production and mineralization of organic matter, and eutrophication increases the temporal and spatial separation of oxygen evolution and consumption resulting in larger fluctuations in oxygen concentrations.

The Freshwater-Biological Laboratory has a long tradition for including research in coastal marine areas among its activities, but during the last 10-15 years our efforts within coastal research have increased markedly. While concern with freshwater ecosystems had a prime environmental priority in Denmark during the 1960's and 1970's, eutrophication problems in coastal ecosystems first became acknowledged and began to draw the attention of media, and thus the politicians, in the 1980's. In addition, the traditional aquatic research had a strong focus on autecology of organisms or systems ecology in either large lakes or open marine waters, leaving the brackish coastal waters in a sort of scientific vacuum. With a gradual change of research interests towards more process-oriented and mechanistic studies, the interfaces between land and aquatic ecosystems became more important areas of research interest. At present, wetlands and estuaries are very much in the focus of international science, and the traditional limnology and oceanography profit from the new information and concepts gained.

The physical dynamics and the ecology of shallow coastal waters are strongly influenced by the runoff of freshwater from land and the exchange of water with adjacent open marine waters. Freshwater influences coastal hydrography by creating salinity gradients and vertical stratification, and rivers and streams carry large amounts of silt, organic material and inorganic nutrients to the coastal areas. The open marine areas, on the other hand, impose large scale physical forcing on the coastal ecosystems with tide and wind-generated water exchange. The massive water exchange with the open sea also carries large amounts of organic material and nutrients subsidizing coastal primary and secondary production.

The shallow water column is the most characteristic and important feature of the coastal waters. The water column is often well mixed, resulting in tight coupling between benthic and pelagic processes. Organic matter is produced both by phytoplankton in the pelagic and by a large variety of attached primary producers such as benthic microalgae, macroalgae and rooted macrophytes. Efficient water column mixing and frequent resuspension events ensure fast vertical transport of inorganic and organic matter integrating, to a large extent, the pelagic and benthic food webs and the biogeochemical processes. Consequently, the dynamics of nutrients, carbon and oxygen in shallow coastal waters differ markedly from those in deep, stratified aquatic ecosystems.

Our research has been performed in a variety of Danish marine areas, in coastal areas of the Mediterranean and in seagrass beds and mangroves of South East Asia. However, most of the work has been done in the shallow estuary, Roskilde Fjord (Rasmussen 1973), which is representative of a typical Danish coastal ecosystem. Danish coastal areas have sandy or silty sediments, and most of the numerous estuaries and embayments are small and well protected from strong wave exposure. Mean depths are often less than 5 m, and all of the areas are subject to heavy nutrient loading from land.

This chapter focuses on characteristic differences between the ecology of shallow coastal marine waters and the ecology of deep lakes and open marine areas based, primarily, on results of studies involving the Freshwater-Biological Laboratory. The paper discusses how nutrient loading from land is buffered in the coastal zone, and how eutrophication influences plant communities, ecosystem metabolism, and oxygen dynamics.

Nutrient dynamics

Coastal waters are by nature eutrophic ecosystems due to the continuous water exchange with the open sea. In the shallow waters, sessile organisms can exploit the con-

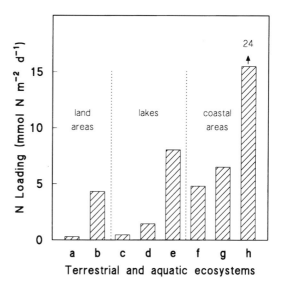

Figure 7.1. Daily nitrogen loading of a variety of Danish terrestrial (a-b), freshwater (c-e) and coastal marine (f-h) ecosystems: a) unfertilised land areas, b) agricultural areas, c) Lake Esrom, d) Lake Furesø, e) Lake Glumsø, f) Limfjord, g) Roskilde Fjord, and h) Norsminde Fjord. The coastal areas are among the most heavily nitrogen-loaded ecosystems.

tinuous supplies of inorganic carbon and nutrients and the organic material imported from the open sea. In addition, the massive nutrient runoff from the urbanized and extensively cultivated land areas makes the Danish coastal waters, like other north-temperate coastal waters (Nixon et al. 1986), some of the most strongly nutrient-subsidized ecosystems (Fig.7.1).

High biological activity and the close spatial and temporal coupling between processes in the water column and in the sediment imply that coastal ecosystems have large potentials as transformers and sinks of organic and inorganic substances during their transport from land to open sea. Thereby, coastal waters are filters that, to some extent, buffer anthropogenic loadings of organic material, toxic substances and nutrients (Sharp et al. 1984, Kamp-Nielsen 1992, Borum 1996). Their efficiency as filters most

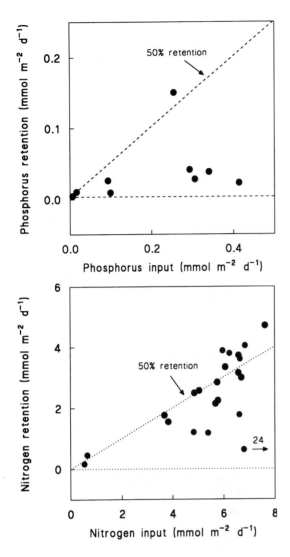

Figure 7.2. Nitrogen and phosphorus retention as functions of nutrient loading for a variety of north-temperate coastal areas (redrawn from Borum (1996)). Coastal waters retain variable amounts of imported nutrients, depending on flushing rate and loading.

likely depends on the biological, morphometric and hydrographic characteristics of the different types of coastal ecosystems. Fjords and semi-enclosed bays are mostly sinks with respect to inorganic nutrients (Fig. 7.2). From a few to almost 80% of the total phosphorus input to the coastal areas (runoff from land plus import from open sea) are

retained by sediment burial (Borum 1996). Nitrogen losses via permanent sediment burial and denitrification are generally higher but also range from 2 to 75% of the total nitrogen input. During periods of sudden reductions in nutrient loading due to climatic changes or remedial actions, coastal ecosystems may temporally act as net sources to nutrients, with export exceeding import. Such transition periods are likely to be short (a few years) because of the high rates of water exchange in the coastal areas, and the net export of nutrients stops as soon as new balances are established between water column and sediment nutrient pools (Jørgensen 1995, Ringkøbing Amtskommune 1992).

For predictive management purposes it is of interest to know, how the coastal filtration capacity changes with nutrient loading. The absolute rate of nutrient retention tends to be positively related to the rate of nutrient loading (Fig. 7.2), but the relative retention shows no clear trends with loading (Borum 1996). A recently published analysis, based on nutrient mass balances of 17 different coastal ecosystems, showed that changes in the relative retention of nitrogen could best be explained by differences in flushing rate among systems (Kaas et al. 1996). It has been suggested that coastal filtration should be more efficient under modest nutrient loadings than under high. Reduced eutrophication is expected to result in higher abundance of benthic infauna, which would increase bioturbation and the construction of ventilated burrows. Reduced eutrophication would also increase the abundance of rooted benthic plants, and both these changes should stimulate the loss of inorganic nitrogen via denitrification (Seitzinger 1988, Aller 1988). On the other hand, the organisms also tend to mobilize sediment nutrient pools that otherwise might have remained permanently buried in the sediment. A fair hypothesis could be based on the simple argument that ecosystems with low nutrient inputs have, in general, more conservative elemental cycles with low drains and sinks (Odum 1971), while nutrient-rich systems are likely to have less efficient exploitation of resources. A tempting analogy could be made in regards to the mechanisms controlling personal economy. Drains are obviously much more prevalent and economic investments less sensible during periods of high resource availability compared to times of economic low tide.

Plant-nutrient interactions

The total primary production of shallow coastal waters is not limited by nutrients to the same extent as phytoplankton production in large, deep lakes or oceans (Borum and Sand-Jensen 1996). The perception of the effects of nutrient loading on primary production in shallow waters seems to be biased by the traditional focus on phytoplankton production in deep stratified waters. Nutrient loading, expressed on an areal or volumetric basis, is a complicated and dubious term to use when the primary production of attached organisms is evaluated. Even under conditions with low nutrient concentrations in the water column, attached plants are exposed to a continuous flow of nutrients which far exceeds their requirements for growth. This situation resembles the nutrient environment of attached macrophytes in streams. Individual plants may temporarily experience a scarcity of nutrients, or inorganic carbon, in coastal marine waters, but given a modest growth rate, a long generation time and a solid anchor to the substratum, all attached autotrophs will be able to steadily exploit the water column or sediment nutrient resources, and they can also build up internal stores of nutrients during periods of excess availability.

We have attempted to establish the coupling between growth and nutrient dynamics of marine plants of different life form and growth strategy (Pedersen 1993, Hein et al. 1995, Pedersen and Borum 1996).

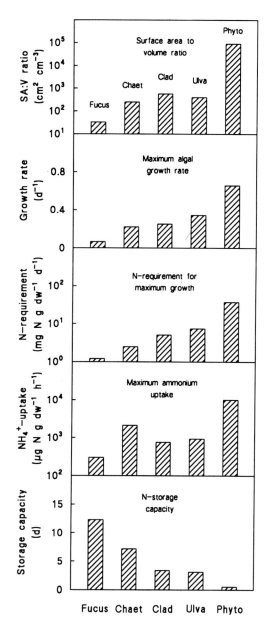

Figure 7.3. Relative surface area (SA:V), maximum specific growth rate, nitrogen requirement at maximum growth rate, maximum ammonium uptake rate, and nitrogen storage capacity of phytoplankton (Phyto) and the macroalgae *Chaetomorpha linum* (Chaet), *Cladophora serica* (Clad), *Ulva lactuca* (Ulva), and *Fucus vesiculosus* (Fucus). Plants with high SA:V ratios have high growth rates, nutrient requirements and uptake rates, but internally stored nutrients can only sustain maximum growth (storage capacity) for short periods without external nutrient supplies (data from Pedersen 1993, Pedersen and Borum 1996).

Small, fast-growing planktonic algae with large surface areas relative to biomasses exhibit fast nutrient uptake from the water column compared to slow-growing macro-algae (Fig. 7.3). The general allometric relationship between size and nutrient uptake has been used to conclude that small plant size is competitively advantageous at low nutrient availability (e.g. Schlesinger et al. 1981, Kiørboe 1993). Other studies on phytoplankton competition have, for various reasons, come to the opposite conclusion (e.g. Smith and Kalff 1982, Turpin 1988), and comparative studies on nitrogen dynamics encompassing a broader scale of algal size classes have shown that large, slow-growing plants are likely to be more successful in nutrient-poor environments (Pedersen 1993, Hein et al. 1995, Pedersen and Borum 1996).

Competitive advantages related to nutrient dynamics are not solely based on the absolute rates of nutrient uptake but also on differences in nutrient requirements, storage capacity and recycling within individuals or populations. Rates of biomass losses, for example due to grazing, are positively related to algal size, as are growth rates (Cebrián and Duarte 1994), and the loss rates tend to balance the differences in growth rates. The organisms having the greatest competitive advantage may, therefore, be those which are best able to cover their nutrient requirements independently of the absolute rates of growth or nutrient uptake. A comparative study showed that the large, thick, slow-growing plants can satisfy their nutrient requirements for growth at the same, or even lower, water column nutrient concentrations as planktonic algae or fast-growing macrophytes with thin thalli (Hein et al. 1995). Furthermore, the larger plants can survive during long periods of low external nutrient availability by exploiting internal stores of nutrients which are only slowly depleted due to low cell quota and low growth rates (Fig. 7.3). Finally, plants with highly differentiated tissues and continuous renewal of tissues, such as seagrasses and some kelps, can recycle nutrients internally and thus reduce the demand for externally derived nutrients (Borum et al. 1989, Pedersen and Borum 1992). Consequently, large, slow-growing plants are well adapted to environments with permanently low nutrient concentrations, or with nutrient resources fluctuating over a seasonal cycle, while fast-growing organisms, such as phytoplankton and ephemeral macroalgae, are more opportunistic and require continuous nutrient supplies from the water column to maintain maximum growth.

Plant community responses to eutrophication

Benthic plant communities in temperate coastal waters with low anthropogenic influence are dominated by seagrasses or large brown algae. The only true seagrass species in Danish waters, eelgrass (*Zostera marina* L.), is perennial and forms dense stands down to depths of 5-10 m. Annual primary production of closed eelgrass stands is between 500 and 1000 g C m^{-2} (Sand-Jensen 1975, Wium-Andersen and Borum 1984, Pedersen and Borum 1993), or about the same as the annual production of kelps on rocky shores (Mann 1972). Eelgrass beds appear to be relatively stable with time but are actually highly dynamic, with continuous losses of individual plants and patches while continuously forming new plants and patches (Olesen and Sand-Jensen 1994). Unperturbed waters also contain a large number of more opportunistic, fast-growing macroalgae, but their contribution to total plant biomass is low, probably because they are more heavily grazed than slow-growing species (Geertz-Hansen et al. 1993, Christiansen and Hansen 1989). Phytoplankton biomasses are also low due to moderate nutrient availability (Sand-Jensen et al. 1994a) and efficient herbivore grazing (Horsted et al. 1988, Riemann et al. 1988, Sand-Jensen et al. 1994b).

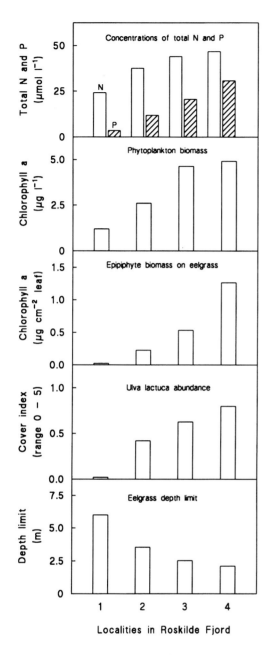

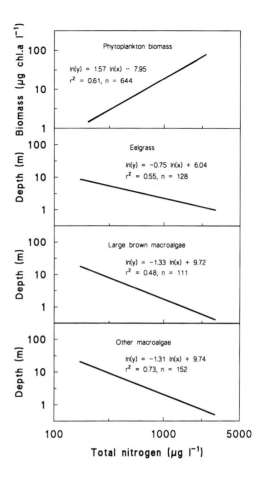

Figure 7.5. Relationships between the mean summer concentration of total nitrogen and phytoplankton biomass and between total nitrogen and the depth penetration of eelgrass, large brown algae and other macroalgae from different Danish coastal areas (regressions from Sand-Jensen et al. 1994a). Increasing nitrogen richness leads to increased phytoplankton abundance and a systematic reduction in macrophyte depth penetration. Plants with thin and morphologically simple thalli penetrate to greater depths because they have lower light requirements (Markager and Sand-Jensen 1992).

Figure 7.4. Changes in plant communities along a nutrient gradient in Roskilde Fjord. Mean concentrations of total nitrogen and phosphorus, mean phytoplankton biomass and epiphyte biomass on eelgrass during summer, abundance of *Ulva lactuca*, and depth penetration of eelgrass (data from Borum 1983, 1985 and Hansen 1984). With increasing nutrient richness, phytoplankton and, especially, epiphyte biomass increases. The abundance of the fast-growing macroalgae, *Ulva lactuca*, also increases while the depth penetration of eelgrass decreases due to a reduced light climate.

With increasing nutrient availability in the water column, plants with higher nutrient requirements, such as planktonic microalgae, periphytic microalgae and opportunistic macroalgae, become more abundant (Fig. 7.4). Well-defined relationships have been established among nitrogen richness, phytoplankton biomass, light attenuation in the water column and depth penetration of eelgrass and large brown algae for Danish coastal waters (Fig. 7.5). The integrated response of primary producers to increased nutrient richness, therefore, is that the composition of the plant communities change from dominance of slow-growing benthic species penetrating to large depths to fast-growing species living in the water column or confined to very shallow water (Sand-Jensen and Borum 1991). Total primary production, however, seems to be largely unaffected (Borum and Sand-Jensen 1996).

Herbivore control of phytoplankton
Although 60 % of the overall variation in phytoplankton biomass in the shallow Danish areas can be explained by variations in nutrient richness (Sand-Jensen et al. 1994a), a large residual variation remains unaccounted for and is probably due to site specific differences in biological structure, morphometry and hydrography. In addition, these characteristics themselves vary with nutrient richness and, therefore, a more comprehensive understanding of phytoplankton dynamics can only be achieved by evaluating the complex interactions among nutrients, light conditions, turbulence and herbivore dynamics (Riemann et al. 1990).

Control of the phytoplankton biomass by pelagic herbivores seems to be much less prominent in coastal waters than in the stratified water column of deeper, open waters (Kiørboe 1993, 1996). The mesozooplankton of shallow Danish estuaries is frequently dominated by the copepod, *Acartia tonsa* (Dana), but the species seldom occurs in large numbers. Estimates of *A. tonsa* clearance rates in Roskilde Fjord were consistently below 5% of the water column per day (Horsted et al. 1988). Microzooplankton, such as tintinnids and rotifers, were more abundant, but total daily zooplankton clearance rates still remained below 50% of the

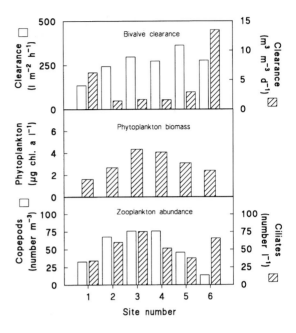

Figure 7.6. Clearance rates of benthic suspension feeding *bivalves (Mya arenaria, Mytilus edulis*, and *Cardium edulis)* along a coast to coast transect in one of the outer basins of Roskilde Fjord. Clearance rates are estimated from faunal biomass and maximum clearance rates determined experimentally in the laboratory. A) Estimated clearance by bivalves expressed per m² and per m³, B) phytoplankton biomass along the transect, and C) abundance of pelagic copepodes and ciliates (data from Sand-Jensen et al. 1994b). The theoretical benthic filtration capacity is large, and the spatial variation in the abundance of phytoplankton and zooplankton suggests that bivalves control the biomass of these pelagic components.

water column. For a mid-summer period, Riemann et al. (1988) estimated that mesozooplankton consumed 5% of daily phytoplankton production in the Limfjord and microzooplankton consumed another 7%. Consequently, the zooplankton community was unlikely to exert efficient herbivore control over the phytoplankton community in these shallow estuaries.

The reasons for the relatively low zooplankton biomass in shallow estuaries are not totally clear, but the mesozooplankton biomass is probably constrained by high abundance of planktivorous fish (Horsted et al. 1988, Riemann et al. 1990) or the jellyfish, *Aurelia aurita* L., (Olesen et al. 1994). Low abundances of microzooplankton may also result from filtration by benthic, suspension-feeding bivalves like *Mytilus edulis* L. (Horsted et al. 1988). *M. edulis* is able to catch and consume all stages of *A. tonsa*, although large copepodits and adults can escape the siphonal flow, at least when turbulence in the water column is low (Sand-Jensen et al. 1994b).

Benthic suspension feeding herbivores, such as bivalves and acidians, seem to have a much larger potential for controlling the phytoplankton community in shallow coastal areas than pelagic herbivores. In the northern part of Roskilde Fjord potential clearance rates of populations of *Mya arenaria* L., *Mytilus edulis* and *Cardium edule* L. ranged from 140 to 360 l m^{-2} h^{-1} at different depths (Sand-Jensen et al. 1994b, Fig. 7.6). These clearance capacities corresponded to filtration of the water column from 1 to more than 10 times per day with the highest filtration rates in the shallow littoral zone. Enclosure experiments with caged mussels deployed at realistic densities at different depths in the water column, showed that the benthic suspension feeders could control phytoplankton development throughout summer and autumn (Riemann et al. 1990). However, estimates of total clearance from experimentally determined filtration rates, or from clearance by caged mussels located in the water column, cannot be directly translated to *in situ* clearance rates. Stratification of the water column during calm and warm periods constrains the replenishment of food sources in the bottom-near water layers to which the benthic suspension feeders have access (Sand-Jensen et al. 1994b). Furthermore, the bivalves living in the sediment are periodically subject to low oxygen concentrations hampering filtration activity (Kaas et al. 1996). Accordingly,

the filtration capacity of the benthic fauna might be fully expressed only during periods of high turbulence and total water column mixing, and not during the frequent periods of stratification or low oxygen concentrations in the bottom water. Nevertheless, the benthic suspension feeders seem to have a strong control over the phytoplankton community in shallow coastal waters.

Herbivore control of epiphytes and macroalgae

The abundance of attached plants is also highly dependent on grazing pressure. Herbivory changes systematically with plant size and plant growth rate, resulting in a larger fraction of biomass and production being consumed daily for fast-growing plants compared to slow-growing (Cebrián and Duarte 1994). Periphytic microalgae have high growth rates similar to phytoplankton and are subject to heavy herbivore grazing (Borum 1987), while the slow-growing brown alga *Fucus vesiculosus* L. is much less heavily grazed (Christiansen and Hansen 1989). Epiphytic microalgae on eelgrass leaves have specific growth rates of up to 0.4 d^{-1} (calculated from an exponential growth model) and can form dense communities impeding leaf growth of the host plants, if grazing by isopods, amphipods and snails does not constrain biomass development (Borum 1987). This delicate balance between epiphyte growth and grazing may be disturbed by eutrophication of the coastal areas due to changes in herbivore abundance. While phytoplankton biomass increased 5-fold, the biomass of epiphytic algae on eelgrass increased up to 50-fold during summer when moving from nutrient-poor to nutrient-rich sites in Roskilde Fjord (Borum 1985, Fig. 7.4). This increase could result from either increased algal growth rates, due to higher nutrient availability, or from lower herbivore grazing.

The high abundance of ephemeral macroalgae under nutrient-rich conditions in shallow coastal areas seems, primarily, to reflect an induced imbalance between algal growth and grazing. The free-floating thalli of the macroalgae, *Ulva lactuca* L., form mass accumulations in the inner, and most eutrophic, parts of Roskilde Fjord throughout spring and summer while in the outer parts *Ulva* blooms during spring only (Hansen 1984, Geertz-Hansen et al. 1993). A study on the seasonal regulation of *U. lactuca* growth and grazing rates showed that growth rates were about the same along the estuary whereas grazing rates were much higher in the outer part, suggesting that the difference in grazing pressure was most important for regulating the accumulation of algal biomass (Geertz-Hansen et al. 1993). Similar patterns of growth and grazing were recorded for *Fucus vesiculosus*, but no mass accumulation was observed in the most eutrophic areas, probably because growth and loss processes are more balanced for this species in general (Christiansen and Hansen 1989).

Pelagic metabolism

Pelagic metabolism contributes substantially to total systems metabolism of shallow coastal waters, though benthic respiration plays a more important role in shallow waters than in deep open waters (Hargrave 1973). Outside the vegetated littoral zone of Roskilde Fjord, pelagic respiration exceeded benthic respiration (Fig. 7.7) and increased relatively more than benthic respiration with increasing eutrophication (Jensen et al. 1990). This pattern is consistent with the change in dominance from benthic to pelagic plants with increasing nutrient richness. Along the nutrient gradient in Roskilde Fjord, phytoplankton algae were responsible for between 26 and 41 % of total pelagic respiration and bacteria for between 38 and 43 % (Jensen et al. 1990). The remaining pelagic respiration due to heterotrophic flagellates, ciliates, rotifers and crustacean zooplankton ranged from 17 to 36 % and decreased systematically with

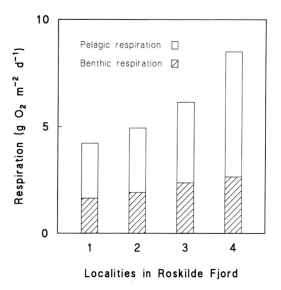

Figure 7.7. Average pelagic respiration (corrected for differences in water depth) and benthic respiration during the period from April through September in the deeper areas of four basins in Roskilde Fjord (data from Jensen et al. 1990). Stations 1 to 4 represent a gradient of increasing nutrient richness. Nutrient enrichment stimulated both pelagic and benthic respiration, but the benthic proportion of total respiration declined with increasing nutrient richness.

increasing eutrophication. Accordingly, the data suggest that with increasing eutrophication and light attenuation in the water column, an increasing proportion of pelagic primary production is lost through respiration by the planktonic algae themselves and by bacteria resulting in a less efficient transfer of organic matter to higher trophic levels.

Bacterial biomass is a function of production and losses through grazing. Heterotrophic nanoflagellates probably control bacterial abundance and, at least during summer, tightly coupled oscillations in the biomass of bacteria and nanoflagellates have been observed (Bjørnsen et al. 1988). In a study on temporal and spatial variability of the plankton community of Roskilde Fjord, we found that bacterial production covaried only with phytoplankton production indicating that pelagic primary production, provided the main substrate for bacterial metabolism in the water column (Sand-Jensen et al. 1994b). Stochastic events such as strong winds or currents can resuspend settled organic material from the sediment of shallow waters and thereby enrich the pelagic pool of organic substrates for bacterial metabolism (Ritzrau and Graf 1992). However, neither bacterial biomass nor production responded significantly to resuspension events in the outer part of Roskilde Fjord (Sand-Jensen et al. 1994b), suggesting that material originally produced in the water column and later settled on the sediment was essentially lost from pelagic metabolism.

Oxygen dynamics

Environmental problems with anthropogenic nutrient loading leading to oxygen depletion have most often been discussed for the

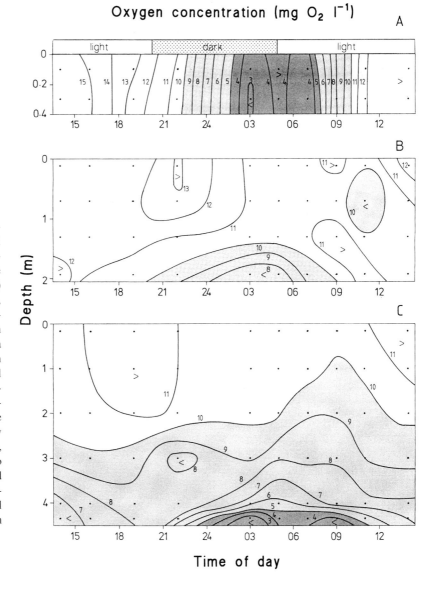

Figure 7.8. Isopleth diagrams showing diel changes of oxygen concentrations in A) the shallow littoral zone, B) at an intermediate depth, and C) in the deeper areas of a nutrient-rich basin in Roskilde Fjord (data 9-10 June 1988, Borum et al. 1990). The diel changes of oxygen concentrations are substantial, particularly in the most shallow areas. Low oxygen concentrations, potentially detrimental to flora and fauna, occurred during the night and early morning in the littoral zone and in the bottom waters of the deep area.

bottom waters of deep lakes and open marine areas. It is true that oxygen depletion events are more prolonged and easier to detect in deep ecosystems. However, oxygen depletion, with potentially detrimental effects on the flora and fauna, occurs much more frequently in the shallow coastal waters because of the high volumetric rates of biological activity.

Rates of total pelagic respiration are very high in coastal waters because of the high autotrophic production and resemble rates measured in eutrophic lakes (Jensen et al. 1990). In Roskilde Fjord, pelagic oxygen consumption ranged between 0.5 and 5.0 mg O_2 l^{-1} d^{-1} during the biologically most active period from March to September (Borum et al. 1990, Jensen et al. 1990, Sand-Jensen et al. 1994b). On an areal basis, integrated oxygen evolution and consumption were about 2-fold higher than in the open waters of the Kattegat, while volumetric rates were more than 10-fold higher (Kruse 1993). Hence, marked diel changes in oxygen concentrations were generated in the water column of both shallow and deeper areas of Roskilde Fjord (Fig. 7.8). In the littoral zone, at less than 0.5 m water depth, the combined sediment and water column oxygen consumption during the night and oxygen evolution during the day changed the oxygen concentration from less than 2 mg O_2 l^{-1} to more than 15 mg O_2 l^{-1} during the diel cycle. Despite the high gas exchange rate between the shallow water column and the atmosphere, oxygen concentrations could nonetheless reach minimum levels at night, which may be detrimental to flora and fauna and force mobile invertebrates to escape to deeper waters with less extreme fluctuations in oxygen concentrations. In deeper waters, diel oxygen changes were more moderate, but low oxygen concentrations still occurred during the night in the bottom water (Fig. 7.8). The diel changes in oxygen concentration in shallow water were larger in the most eutrophic part of Ros-

kilde Fjord than in the nutrient-poor area, and concentrations at night tended to be lower in the deep bottom waters (Borum et al. 1990).

Oxygen conditions in coastal waters are highly dependent on site-specific differences in water column mixing and accumulation of organic material and, in addition, depend on climatic conditions (Kaas et al. 1996). It is, therefore, difficult to establish clear causal relationships between environmental conditions and oxygen dynamics in the coastal zone. Kaas et al. (1996) analysed a large data-base on oxygen concentrations measured in Danish inlets and found that oxygen concentrations in the coastal bottom waters were primarily regulated by the duration of vertical water column stratification during summer. Despite the shallow water depths, vertical stratification occurs for shorter or longer periods due to thermal or salinity-derived differences in water density, allowing the pool of oxygen in the bottom water to be depleted within a few days, given the high combined respiratory rates in water column and sediment. The concentration of oxygen in the bottom waters was also negatively correlated with nitrogen loading from land (Kaas et al. 1996), strongly suggesting that eutrophication contributes to create oxygen conditions which may be critical to the benthic fauna. While their analysis showed that the benthic macroinvertebrates primarily seemed to profit from increased phytoplankton production in eutrophic shallow waters (Kaas et al. 1996), sustained oxygen deficiency may be fatal to the populations of suspension-feeding bivalves, which are so important for the overall ecological conditions of the coastal areas.

Concluding remarks
Eutrophication of shallow coastal waters represents an unintended, well replicated, large-scale manipulation experiment. The analysis of coastal ecosystem responses to eutrophication has provided a more compre-

hensive understanding of interactions among different plant types, the tight coupling of benthic and pelagic processes, and the highly dynamic oxygen conditions in shallow waters. Coastal areas have a wide range of plant forms with different growth strategies which respond predictably to eutrophication in the competition for light and nutrients. Benthic suspension feeding bivalves play a key-role in coastal waters by regulating the pelagic communities and processes depending on the intensity of vertical mixing. Our

analyses have shown that oxygen concentrations change dramatically on a diel basis, reflecting the high areal and volumetric rates of ecosystem metabolism, and that eutrophication increases the role of pelagic metabolism relative to sediment metabolism. Overall, the research has provided new insights with respect to the functioning of shallow coastal ecosystems, and this understanding of processes and regulatory mechanisms may be profitably used in comparative analyses of other aquatic ecosystems.

Literature cited.

Aller, R. C. 1988. Benthic fauna and biogeochemical processes in marine sediments. Pages 301-338 *in* T. H. Blackburn and J. Sørensen, editors. Nitrogen Cycling in Coastal Marine Environments. John Wiley and Sons, Chichester, England.

Bjørnsen, P. K., B. Riemann, S. J. Horsted, T. G. Nielsen, and J. Pock-Steen. 1988. Trophic interactions between heterotrophic nanoflagellates and bacterioplankton in manipulated seawater enclosures. Limnology and Oceanography 33: 409-420.

Borum. J. 1983. The quantitative role of macrophytes, epiphytes, and phytoplankton under different nutrient conditions in Roskilde Fjord, Denmark. Pages 35-40 *in* Proceedings of the International Symposium on Aquatic Macrophytes. Faculty of Science, Nijmegen, The Netherlands.

Borum, J. 1985. Development of epiphytic communities on eelgrass (*Zostera marina*) along a nutrient gradient in a Danish estuary. Marine Biology 87: 211-218.

Borum, J. 1987. Dynamics of epiphyton on eelgrass (*Zostera marina* L.) leaves: Relative roles of algal growth, herbivory, and substratum turnover. Limnology and Oceanography 32: 986-992.

Borum, J. 1996. Shallow waters and land/sea boundaries. Pages 179-203 *in* B. B. Jørgensen and K. Richardson, editors. Eutrophication in Coastal Marine Ecosystems. American Geophysical Union, Washington DC, USA.

Borum, J., and K. Sand-Jensen. 1996. Is total primary production in shallow coastal marine waters stimulated by nitrogen loading? Oikos 76: 406-410.

Borum, J., L. Murray, and W. M. Kemp. 1989. Aspects of nitrogen acquisition and conservation in eelgrass plants. Aquatic Botany 35: 289-300.

Borum, J., O. Geertz-Hansen, K. Sand-Jensen, and S.

Wium-Andersen. 1990. Eutrofiering – effekter på marine primærproducenter. NPo-forskning fra Miljøstyrelsen. Miljøstyrelsen, Copenhagen, Denmark.

Borum, J., M. F. Pedersen, L. Kær, and P. M. Pedersen. 1994. Vækst- og næringsstofdynamik hos marine planter. Havforskning fra Miljøstyrelsen. Miljøstyrelsen, Copenhagen, Denmark.

Cebrián, J., and C.M. Duarte. 1994. The dependence of herbivory on growth rate in natural plant communities. Functional Ecology 8: 518-525.

Christiansen, A., and D. F. Hansen. 1989. Invertebratgræsning på makroalger. MS Thesis. Freshwater-Biological Laboratory, University of Copenhagen, Denmark.

Geertz-Hansen, O., K. Sand-Jensen, D. F. Hansen, and A. Christiansen. 1993. Growth and grazing control of abundance of the marine macroalga, *Ulva lactuca* L. in a eutrophic Danish estuary. Aquatic Botany 46: 101-109.

Hansen, F. G. 1984. Fordelingen af makroalger samt vækst af *Ulva lactuca og Fucus vesiculosus* langs en næringsstofgradient i Roskilde Fjord. MS Thesis. Freshwater-Biological Laboratory, University of Copenhagen, Denmark.

Hargrave, B. T. 1973. Coupling carbon flow through some pelagic and benthic communities. Journal of the Fisheries Research Board of Canada 30: 1317-1326.

Hein, M., M. F. Pedersen, and K. Sand-Jensen. 1995. Size-dependent nitrogen uptake in micro- and macroalgae. Marine Ecology Progress Series 118: 247-253.

Horsted, S. J., T. G. Nielsen, B. Riemann, J. Pock-Steen, and P. K. Bjørnsen. 1988. Regulation of zooplankton by suspension-feeding bivalves and fish in estuarine enclosures. Marine Ecology Progress Series 48: 217-224.

Jensen, L. M., K. Sand-Jensen, S. Marcher, and M. Hansen. 1990. Plankton community respiration along a nutrient gradient in a shallow Danish estuary. Marine Ecology Progress Series 61: 75-85.

Jørgensen, C. 1995. Modelling of nutrient release from the sediment in a tidal inlet, Kertinge Nor, Funen, Denmark. Ophelia 42: 163-178.

Kaas, H., F. Møhlenberg, A. Josefson, B. Rasmussen, D. Krause-Jensen, H. S. Jensen, L. M. Svendsen, J. Windolf, A. L. Middelboe, K. Sand-Jensen, and M. F. Pedersen. 1996. Marine områder. Danske fjorde – status over miljøtilstand, årsagssammenhænge og udvikling. Vandmiljøplanens Overvågningsprogram 1995. Faglig rapport fra DMU nr. 179. Danmarks Miljøundersøgelser, Copenhagen, Denmark.

Kamp-Nielsen, L. 1992. Benthic-pelagic coupling of nutrient metabolism along an estuarine eutrophication gradient. Hydrobiologia 235/236: 457-470.

Kiørboe, T. 1993. Turbulence, phytoplankton cell size, and the structure of pelagic food webs. Advances in Marine Biology 29: 1-72.

Kiørboe, T. 1996. Material flux in the water column. Pages 67-94 *in* B. B. Jørgensen and K. Richardson, editors. Eutrophication in coastal marine ecosystems. American Geophysical Union, Washington DC, USA.

Kruse, B. 1993. Measurement of plankton O_2 respiration in gas-tight plastic bags. Marine Ecology Progress Series 94: 155-163.

Mann, K. H. 1972. Ecological energetics of the seaweed zone in a marine bay on the Atlantic coast of Canada.. II: Productivity of the seaweeds. Marine Biology 14: 199-209.

Markager, S., and K. Sand-Jensen. 1992. Light requirements and depth zonation of marine macroalgae. Marine Ecology Progress Series 88: 83-92

Nixon, S. W., C. A. Oviatt, J. Frithsen, and B. Sullivan, 1986. Nutrients and the productivity of estuarine and coastal ecosystems. Journal of the Limnological Society of Southern Africa 12: 43-71.

Odum, E.P. 1971. Fundamentals of Ecology. W.B. Saunders, Philadelphia, USA.

Olesen, B., and K. Sand-Jensen. 1994. Patch dynamics of eelgrass, *Zostera marina*. Marine Ecology Progress Series 166: 147-156.

Olesen, N.J., K. Frandsen, and H. U. Riisgård. 1994. Population dynamics, growth and energetics of jellyfish (*Aurelia aurita*) in a shallow fjord. Marine Ecology Progress Series 105: 9-18.

Pedersen, M.F. 1993. Vækst og næringsstofdynamik hos marine planter. Ph.D. thesis. Freshwater-Biological Laboratory, University of Copenhagen, Denmark.

Pedersen, M. F., and J. Borum. 1992. Nitrogen dynamics of eelgrass *Zostera marina* during a late summer period of high growth and low nutrient availability. Marine Ecology Progress Series 80: 65-73.

Pedersen, M. F., and J. Borum. 1993. An annual nitrogen budget for a seagrass *Zostera marina* population. Marine Ecology Progress Series 80: 65-73.

Pedersen, M. F., and J. Borum. 1996. Nutrient control of algal growth in estuarine waters. Nutrient limitation and the importance of nitrogen requirements and nitrogen storage among phytoplankton and species of macroalgae. Marine Ecology Progress Series 142: 261-272.

Rasmussen, E. 1973. Systematics and ecology of the Isefjord marine fauna (Denmark) with a survey of the eelgrass (*Zostera*) vegetation and its communities. Ophelia 11: 1-495.

Riemann, B., T. G. Nielsen, S. J. Horsted, P. K. Bjørnsen, and J. Pock-Steen. 1988. Regulation of phytoplankton biomass in estuarine enclosures. Marine Ecology Progress Series 48: 205-215.

Riemann, B., H. M. Sørensen, P. K. Bjørnsen, S. J. Horsted, L. M. Jensen, T. G. Nielsen, and M. Søndergaard. 1990. Carbon budgets of the microbial food web in estuarine enclosures. Marine Ecology Progress Series 65: 159-170.

Ringkøbing Amtskommune. 1992. Rapport om Næringssalte og Vandskifte i Ringkøbing Fjord. Torben Larsen Hydraulics Aps, Denmark.

Ritzrau, W., and G. Graf. 1992. Increase of microbial biomass in the benthic turbidity zone of Kiel Bight after resuspension by a storm event. Limnology and Oceanography 37: 1081-1086.

Sand-Jensen, K. 1975. Biomass, net production and growth dynamics in an eelgrass (*Zostera marina* L.) population in Vellerup Vig, Denmark. Ophelia 14: 185-201.

Sand-Jensen, K., and J. Borum. 1991. Interactions among phytoplankton, periphyton, and macrophytes in temperate freshwaters and estuaries. Aquatic Botany 41: 137-175.

Sand-Jensen, K., S. L. Nielsen, J. Borum, and O. Geertz-Hansen. 1994a. Fytoplankton- og makrofytudvikling i danske kystområder. Havforskning fra Miljøstyrelsen nr. 30. Miljøstyrelsen, Copenhagen, Denmark.

Sand-Jensen, K., J. Borum, O. Geertz-Hansen, J. N. Jensen, A. B. Josefson, F. Møhlenberg, and B. Riemann. 1994b. Resuspension og stofomsætning i Roskilde Fjord. Havforskning fra Miljøstyrelsen nr. 51. Miljøstyrelsen, Copenhagen, Denmark.

Seitzinger, S. P. 1988. Denitrification in freshwater and coastal marine ecosystems. Limnology and Oceanography 33: 702-724.

Sharp, J. H., J. R. Pennock, T. M. Church, J. M. Tramontano, and L. A. Cifuentes. 1984. The estuarine interaction of nutrients, organics, and metals: A case study in the Delaware Estuary. Pages 241-258 *in* V. S. Kennedy, editor. The Estuary as a Filter. Academic Press, New York, USA.

Schlesinger, D. A., L. A. Molot, and B. J. Shuter. 1981. Specific growth rates of freshwater phytoplankton in relation to cell size and light intensity. Canadian Journal of Fisheries and Aquatic Sciences 28: 1052-1058.

Smith, R. E. H., and J. Kalff. 1982. Size-dependent phosphorus uptake kinetics and cell quota in phytoplankton. Journal of Phycology 18: 275-284.

Turpin, D. H. 1988. Physiological mechanisms in phytoplankton resource competition. Pages 316-368 *in* C. D. Sandgren, editor. Growth and reproductive strategies of freshwater phytoplankton. Cambridge University Press, Cambridge, England.

Wium-Andersen, S., and J. Borum. 1984. Biomass variation and autotrophic production of an epiphyte-macrophyte community in a coastal Danish area: I. Eelgrass (*Zostera marina* L.) biomass and net production. Ophelia 23: 33-46.

8 Nutrient dynamics and modelling in lakes and coastal waters

By Lars Kamp-Nielsen

Nutrient dynamics and their role in determining the trophic state of aquatic ecosystems have been studied at the Freshwater Biological Laboratory (FBL) by field and experimental methods together with empirical and dynamic modelling. We always face the inherent conflict between analytical tractability, realism, and precision in prediction when we model complex ecosystems. At the FBL, we have addressed this conflict by separating the total lake models into simple sub-models with few state variables. These simple models have been verified and validated on mesocosms before they were implemented into total lake models. The changes in species composition which we observe during changing nutrient loading have been taken into account by structural, dynamic modelling in which the biological components are introduced or excluded from the model. The use of goal functions derived from ecological theories in this process, has both increased the predictive power of the models for tactical purposes and has provided us with a tool for more strategic investigations of ecosystem properties.

Lake typology and nutrients

From the earliest days of limnology both holistic and reductionistic approaches have been used (Rigler 1985). The lake typology proposed by Naumann (1919) and Thienemann (1921) classified lakes into eutrophic and oligotrophic types according to their nutritional status. Although the terminology referred to the nutrient content, most of the classification effort reflected the flora and fauna of lakes rather than their nutrient contents.

In the early lake typology studies, the nutrients were poorly defined and analysed and included both organic and inorganic compounds. Juday et al. (1928), Naumann (1930) and Gessner (1934) suggested that nitrogen and phosphorus could be the limiting factors for the trophic state of lakes, and Pearsall (1932) was the first to correlate the composition of phytoplankton species across lakes with the concentration of dissolved nutrients. Domogala et al. (1925) studied the nitrogen dynamics and stressed the importance of bacterial nitrification and nitrate reduction, and Einsele (1936, 1941), followed by Mortimer (1941, 1942), studied phosphorus and its relation to iron and oxygen dynamics.

In his extensive studies of phytoplankton abundance and water chemistry in Danish ponds and lakes, Nygaard (1938) found that phosphate concentrations were very low over a wide range of lake types. However, as for nitrate concentrations he did not find any correlations between phytoplankton abundance and the concentrations of dissolved nutrients. Nygaard (1938) concluded that analyses of inorganic nitrogen and phosphorus cannot be

Table 8.1. Early mass balances Lake Fure (1952-54, Berg et al 1958) and Lake Esrom (1969-70, Jónasson et al. 1974).

Lake	Inflow of total P (tons y^{-1})	Outflow of total P (tons y^{-1})	Inflow of total N (tons y^{-1})	Outflow of total N (tons y^{-1})
Lake Fure	4.031	0.087	24.210	3.408
Lake Esrom	10.800	2.900	85.200	2.900

used to classify lakes and ponds according to the Thienemann concept of productivity (Thienemann 1932). Comparisons between productivity and the averages of dissolved nutrients over the entire study period did not yield significant correlations either.

Although there was a growing understanding of the dynamic nature of the relationship between phytoplankton productivity and levels of inorganic nutrients, there were no attempts to quantify these relationships. The introduction of phosphate-containing detergents and increased use of fertilizers, however, made the eutrophication of lakes and coastal waters an obvious problem (Hasler 1947). The experimental approaches introduced by Einsele and Mortimer were supported by the introduction of radioactive tracers. Hutchinson and Bowen (1947, 1950), Hayes et al. (1952) and Rigler (1956) threw new light on the phosphorus kinetics between the different compartments in lakes and demonstrated very fast turnover rates of phosphorus in the water column and identified the inactivation in the sediments. In Switzerland, it was realized that the connection between nutrients and eutrophication of lakes lay in the areal loading, and the first mass balance appeared (Baldinger 1957).

Mass balance and sediment kinetics in Lake Fure

The eutrophication problem also increased in the densely populated Denmark. In 1948, the deteriorating water quality of Lake Fure had become a matter of public concern (Olsen and Larsen 1948), and an investigation programme was initiated (Berg et al. 1958). The initial scope was to describe the changes in the trophic state since the start of the century and to provide data for a future comparison with other lakes. This future-orientated project was strongly supported by collaboration with the Danish Association of Engineers who provided the first nutrient mass balance for a lake in Denmark. The initiative was continued by the local municipalities responsible for the lake and allowed us to reconstruct the loading history of the lake (Sand Jensen 1995, 1997).

The mass balance for 1952-1953 showed that 86% of the nitrogen load and 98% of the phosphorus load was retained in the lake (Table 8.1). The extraordinarily high loss of phosphorus was compared with the gross primary production measured by the newly introduced ^{14}C-technique, and from the Redfield ratio, the corresponding nitrogen and phosphorus uptake was calculated (Berg et al. 1958). Berg et al. (1958) realized, however, that the nitrogen and phosphorus atoms can be used several times, but they could not quantify the seasonal dynamics and to what degree nutrients were retained by other processes than sedimentation. Olsen (1958, 1964) considered the physico-chemical equilibrium between phosphate and the sediment as a possible explanation for the high reten-

Table 8.2. Release of phosphorus, nitrogen, iron, calcium and silicate from various lake sediments under aerobic (ae) and anaerobic (an) conditions (From Kamp-Nielsen 1974).

Lake	PO$_4$-P	NH$_4$-N	NO$_3$-N	Fe	Ca	Si
			mg m^{-2} d^{-1}			
Lake Fure (ae)	-2.0	-1.2	-10.8	-0.3	42	94.7
Lake Fure (an)	17.3	11.1	-17.7	3.9	75	95.7
Lake Esrom (ae)	-1.4	5.9	2.2	5.0		
Lake Esrom (an)	12.3	13.7	-12.1	7.4		
Lake Gribsø (ae)	0.2	6.7		2.6		
Lake Gribsø (an)	1.2	9.8		1.7		
Lake Grane (ae)	0.6	0.2	2.6	-0.2		4.7
Lake Grane (an)	0.8	3.4	-2.3	8.7		5.0

tion. After a visit to the FBA-laboratory in Windermere, he abandoned the use of undisturbed sediment cores as used by Mortimer (1941, 1942) because he did not find the theoretical background sufficient for the interpretation of such experiments. Inspired by soil sciences, he expected that sorption isotherms had a high explanatory value. He studied the sediment-phosphorus kinetics experimentally by the use of ^{32}P addition to slurries of minerals and sediments from different lakes kept at various temperatures and redox conditions. The resulting sorption isotherms showed that carbonate and calcareous and sandy sediments had low sorption capacities whereas iron hydroxide and deep water gyttja had high capacities. At ambient phosphate concentrations, the deep water sediments in Lake Fure adsorbed more than 1000 μg P g^{-1} dry sediment under aerobic conditions but only about 20 μg P g^{-1} dry sediment under anaerobic conditions. Unfortunately, he was not able to relate the experimental results to the phosphorus mass balance in a quantitative way.

Mass balances and nutrient dynamics in Danish lakes

In his 1955 – 1965 study of the relations between primary production and benthic production in Lake Esrom, Jonasson (1972) stressed the importance of nutrients as regulators of primary production. In 1969-1970 resources became available for nutrient dynamics and mass balance studies (Jonasson et al. 1974). The seasonal variations in nutrient concentrations showed that nitrogen was the limiting element from early March and the rest of the production season as phosphate was present at high concentrations throughout the year. The mass balance (Table 8.1) showed that 73% of the phosphorus and 97% of the nitrogen was retained in the lake. The difference between Lake Esrom and Lake Fure was only briefly addressed by Jonasson et al. (1974). Steemann Nielsen (1973) suggested that the difference between the two lakes, both of which have long water residence times of (13-18 years, was a phosphorus-rich inflow of groundwater in Lake Esrom. The correct explanation was a lower storage capacity in the Lake Esrom sediments. The mass balances for Lake Esrom and Lake Fure demonstrated that the capacity of the sediments to retain nutrients was essential to nutrient dynamics in lakes. Olsen (1958) described the retention as a sorption equilibrium, but he did not consider the influence of pH, the establishment of porewater concentrations from biological processes and the transport of nutrients to the sediment.

In my studies (Kamp-Nielsen 1974) with undisturbed sediment cores from the eutrophic-calcareous Lake Fure and Lake Esrom, the oligotrophic Lake Grane Langsø and the dystrophic Lake Gribsø, I showed high exchange rates in the eutrophic lakes and a strong influence of pH and redox conditions on nitrogen, phosphorus and iron exchange (Table 8.2). The sorption control of the aerobic phosphate exchange was demonstrated by the logarithmic dependence of bottom water concentration (Fig. 8.1). The upwards diffusion under anaerobic conditions was demonstrated by the linear relation of the concentration gradient across the sediment-water interphase (Fig. 8.2).

The stochiometric relations between exchanged phosphate, calcium and iron at different pH-values indicated that the anaero-

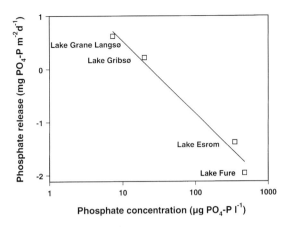

Figure 8.2. Aerobic phosphate release vs. the phosphate concentration in the bottom water in various lakes. Redrawn from Kamp-Nielsen (1974).

bic phosphate exchange in Lake Fure and Lake Esrom was controlled by the solubility of iron phosphate in the pH-interval 5.7 – 7.5 and by the formation of hydroxyapatite in the pH-interval 7.5 – 9 in Lake Fure. Poisoning the sediment surface with antibiotics showed that microbial mineralisation and nitrogen transformations were also important for the sediment-water exchange. Differential extraction of the sediments in Lake Esrom and Lake Grane Langsø demonstrated the existence of various forms of sediment-stored phosphorus and the pH and redox control of the relationships between these forms. In the surface layers, labile pools of organic phosphorus, iron-and aluminium-bound phosphorus and calcium-bound phosphorus decreased with depth and redox potential until 5-10 cm into the sediment. Below this depth, pH stabilised and the proportions between the inorganic phosphorus pools became constant.

The transfer of results from such mesocosm studies to mass balances for whole lakes cannot be done directly. If we consider phosphorus as a tracer for net deposition in simple mass balance studies, we can use the N/P ratio in the stabilized sediment to calculate the denitrification loss from a nitrogen

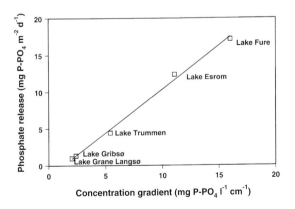

Figure 8.1. Anaerobic phosphate release vs. the concentration gradient across the surface 0.5 cm of the sediment in various lakes. Redrawn from Kamp-Nielsen (1974).

balance if we assume steady state and constant loading in the past period. Such assumptions and calculations were made by Jørgensen et al. (1973) for Lake Glumsø and by Andersen (1974) for six shallow lakes. Denitrification losses calculated from monthly mass balances varied from 18-47 g N m^{-2} y^{-1} for six of the lakes, corresponding to 20-50% of the external nitrogen loading. Losses were lower in laboratory experiments with undisturbed cores than those calculated from mass balances. The difference was explained by the tight coupling between nitrification and denitrification in the lake stimulated by frequent oxidation of the sediment surface by wind-induced resuspension (Andersen 1977). Spatial heterogeneities in sediment composition due to variations in settling fluxes and lateral translocations of settled material were also important.

Sedimentation studies

The seasonal variation in sedimentation and the sediment-water exchange of nutrients were studied at depths in Lake Esrom (Kamp-Nielsen 1975a,b). Sedimentation peaked 1-2 weeks after maxima in primary production in September 1973 and April

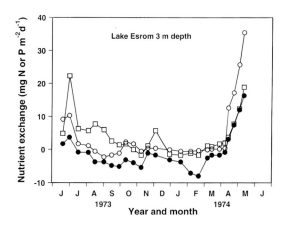

Figure 8.4a. Seasonal variation in release of phosphate-P(○) ammonia-N (●), and nitrate-N (□) at 3-m depth in Lake Esrom. Redrawn from Kamp-Nielsen (1975b).

1974 (Fig. 8.3). A third peak in sedimentation occurred during the autumn overturn (Lastein 1976) and was explained by resuspension events.

The exchange of nutrients was related to sedimentation, temperature and redox conditions (Fig. 8.4a,b). At 3-m depth a net sorption of phosphate prevailed throughout the year, except during the late spring when rapid mineralisation at the sediment surface of

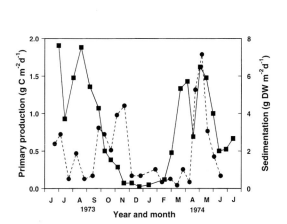

Figure 8.3. Seasonal variation in gross primary production (■) and gross sedimentation of dry matter (●) in Lake Esrom. Redrawn from Kamp-Nielsen (1975b).

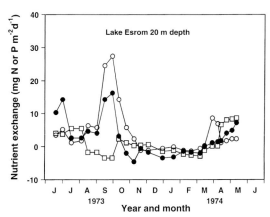

Figure 8.4b. Seasonal variation in release of phosphate-P (○), ammonia-N (●), and nitrate-N (□) at 20 m depth in Lake Esrom. Redrawn from Kamp-Nielsen (1975b).

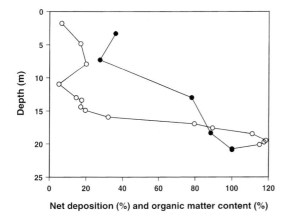

Figure 8.5. Net deposition of organic matter (gross sedimentation – mineralisation) (●) and content of organic matter (○) in the surface 10 cm of the sediment at various depths in Lake Esrom. Both are as percentages of the deposition and content at 20 m depth. Redrawn from Kamp-Nielsen (1977a).

the sedimentated spring algal bloom exceeded the sorption. At 20-m depth a secondary net release occurred during the anaerobic period in the late summer. Ammonia release followed the spring sedimentation peak both at 3-m and at 20-m depth, and the subsequent nitrification was manifested as increased

nitrate release. When oxygen was depleted at 20-m depth, nitrification stopped and denitrification dominated.

Besides the temporal variations, the spatial variation of sedimentation and sediment composition were also studied to quantify the importance of sediment transport and sediment focusing (Fig. 8.5). I compared the content of organic matter at different depths relative to the content at the deepest station in Lake Esrom with the relative net deposition, which I calculated by subtracting the mineralisation from the sedimentation (Kamp-Nielsen 1977a). The accumulation of organic matter was highest at the base of the slope of the lake basin at 18 – 20 m depth. Sediment focusing in the deepest part with highest accumulation in depressions in the profundal zone was demonstrated (Kamp-Nielsen and Hargrave 1978), and the accumulation at the base of steep bottom gradients was verified and compared with that of Bedford Basin, Nova Scotia (Fig. 8.6a,b, Hargrave and Kamp-Nielsen 1977).

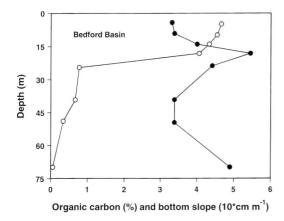

Figure 8.6a. Organic carbon in the surface 0.5 cm of the sediment (●) and the bottom slope (○) at various depths in Bedford Basin, Nova Scotia. Redrawn from Hargrave and Kamp-Nielsen (1977).

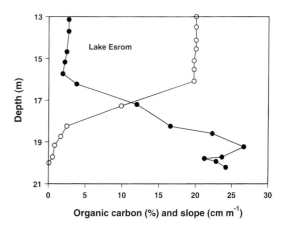

Figure 8.6b. Organic carbon in the surface 0.5 cm of the sediment (●) and the bottom slope (○) at various depths in Lake Esrom. Redrawn from Hargrave and Kamp-Nielsen (1977).

Mass balances and nutrient kinetics in coastal waters

Coastal waters differ from lakes in several aspects. Coastal waters have a boundary to the adjacent sea, their estuarine character creates gradients, resuspension of sediments is important for both horizontal redistribution of nutrients and for the redox conditions in the surface sediments. Submerged macrophytes are often important primary producers, the benthic community is dominated by large efficient filtrators, and nitrogen is often more important than phosphorus as the major regulator of primary production.

In 1985-1986 we established a mass balance for several sections of Roskilde Fjord. The advective/dispersive transport between sections and the exchange across the open boundary were calculated from a traditional box-model, using the salinity as a conservative element and showed a significant transport of sediments from the inner parts to the outer parts. This transport was presumably due to the reduced depths and reduced vegetation cover in the inner, hypertrophic parts. We demonstrated a tight coupling between phytoplankton blooms, sedimentation and bacterial activity (Flindt and Nielsen 1992), while sediment mineralisation, sediment oxygen uptake and nitrification/denitrification depended on the redox conditions (Flindt and Kamp-Nielsen 1992, Kamp-Nielsen 1992). Gross sedimentation exceeded net deposition measured by ^{210}Pb-dating by a factor 50-100, indicating substantial resuspension events, and supported the hypothesis of transport of sediment from the inner to the outer parts.

To evaluate the nutrient dynamics in the sediments, especially the influence of oxygen depletion, rooted macrophytes and resuspension events, we focused on the study of porewater profiles. In marine sediments, dynamic changes in the zonation of oxic, sub-oxic, and anoxic processes require a high temporal and spatial resolution of the porewater profiles. We developed a dialyzer with a vertical resolution of separated chambers of 4 mm and an equilibration time of about 8 hours which allowed daily sampling (Kamp-Nielsen and Flindt 1994). In sediments with *Zostera marina,* an aerobic environment is created in the rhizosphere by oxygen leaking

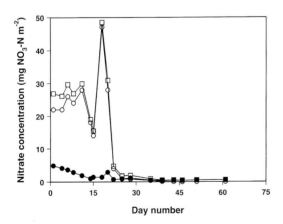

Figure 8.7a. Temporal variations in nitrate-N concentrations in an unvegetated sediment from Roskilde Fjord. The pool in the interstitial water (●), the pool in the overlying water (○), and the total pool in the mesocosm (□). Redrawn from Flindt (1994).

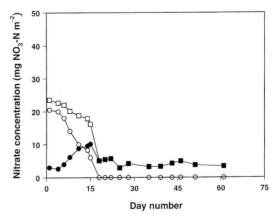

Figure 8.7b. Temporal variations in nitrate-N concentrations in a *Zostera*-vegetated sediment from Roskilde Fjord. The pool in the interstitial water (●), the pool in the overlying water (○), and the total pool in the mesocosm (□). Redrawn from Flindt (1994).

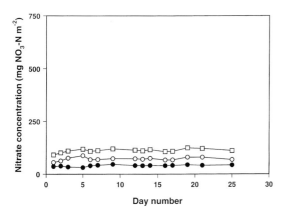

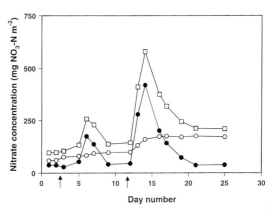

Figure 8.8a. Temporal variations in nitrate-N concentrations in a sediment core from Frederiksværk Broad, Roskilde Fjord. The pool in the interstitial water (●), the pool in the overlying water (○), and the total pool in the core (□). Redrawn from Flindt and Kamp-Nielsen (1997).

Figure 8.8b. Temporal variations in nitrate-N concentrations in a sediment core from Frederiksværk Broad, Roskilde Fjord with resuspension at day No 3 and No 12. The pool in the interstitial water (●), the pool in the overlying water (○), and the total pool in the core (□). Redrawn from Flindt and Kamp-Nielsen (1997).

from the roots (Flindt 1994). These conditions promote a nitrification and a subsequent denitrification in the suboxic environment outside the nitrification zone. By integrating over depth and accounting for net plant uptake, we estimated a denitrification loss in the *Zostera* bed of 1.5 mmol N m^{-2} d^{-1} compared to a loss of 0.9 mmol N m^{-2} d^{-1} from the bare sediment (Fig. 8.7a,b). The net accumulation of phosphorus was 0.3 mmol P m^{-2} d^{-1} compared to a net release of 0.08 mmol P m^{-2} d^{-1} from the non-vegetated sediment.

The resuspension study included both field and laboratory measurements. Two platforms were raised in Roskilde Fjord, one at 2.5-m depth in a *Zostera* bed, and one at 4.5-m depth on a non-vegetated site. At wind speeds less than about 15 m s^{-1}, the resuspension fluxes were lowest from the vegetated bottom due to damping of the turbulence in the macrophyte beds. At very high wind speeds, however, we found the highest resuspension fluxes from the vegetated sediments due to the lower depth and the higher concentrations of organic material accumulated on the surface. The highest recorded resus-

pension event by a 25 m s^{-1} northerly wind yielded a resuspension rate of 26 kg DW m^{-2} d^{-1} and 1.8 kg DW m^{-2} d^{-1} at the non-vegetated site. This showed that *Zostera* beds accumulate easily resuspendable material during moderate turbulence, but at high turbulence the accumulated material is redistributed.

Simulation of a 25 m s^{-1} storm in a laboratory experiments by a piston-driven wave simulator made it possible to study the effects of resuspension on porewater profiles and sediment processes with a high resolution in time and space. Phosphate sorption increased by 0.2 – 0.3 mg P m^{-2} over two days after a resuspension, but after the downwardly mixed oxygen had been consumed, the phosphate was desorbed. Nitrification increased to about 0.2 g N m^{-2} d^{-1} after resuspension and was followed by denitrification at approximately the same rate (Fig. 8.8a,b, Flindt and Kamp-Nielsen 1997). Using bathymetric maps and wind statistics we calculated the frequency and potential area influenced by resuspension from the Bretschneider equation (Bretschneider 1952) and estimated that about 25% of the total annual denitrification loss in Roskilde Fjord could derive from increased nitrification and denitrification following resuspension.

Nutrient transport and dynamics across the land – sea boundaries

During the transport from the terrestrial environment through streams, lakes and into coastal waters, nutrients are temporarily stored or permanently lost in a variety of sinks. The loss processes are different for nitrogen and phosphorus, and the physical, chemical and biological factors controlling these processes vary between aquatic systems and are reflected in changing nitrogen:phosphorus ratios.

The changing N:P ratio can be demonstrated for the River Langvad drainage basin and the adjacent Roskilde Fjord (Table 8.3). In the 1st order streams, the N:P ratio varies from 10 to 40 with the lowest values in streams with point sources of sewage. During the passage of streams, temporary deposition of nutrients occurs during the low flow period in the summer, but since the Danish lowland streams are intensely regulated, stream erosion and deposition of nutrients are insignificant on an annual scale. However, net losses of nitrogen occur by denitrification in the sediments which reduces the N:P ratio moderately along the streams.

Table 8.3. Nitrogen: phosphorus ratios at ecosystems boundaries in the Roskilde Fjord basin.

Boundary	N : P ratio
Land – 1st order stream	10 – 40
1st order stream – head water	20
Head water – lake	14
Lakes – Roskilde Fjord	9
Roskilde Fjord – the sea	5

At the entrance to the lakes in the River Langvad basin, the N:P ratio in the headwater is about 14, which is close to the average of the inlet ratio of 13 for the 37 survey lakes in the 'Water Action Programme' in 1989 (Kristensen et al. 1990a). Flowing through four closely linked lakes the N:P ratio is further reduced to about 9 at the inlet to Roskilde Fjord. Phosphorus and nitrogen are deposited in the lakes, and nitrogen is permanently lost by denitrification. Both in the River Langvad lakes and the national survey lakes the average annual loss of nitrogen was about 30% during passage of a lake. The phosphorus retention was 56% in the Langvad lakes but slightly negative (- 1.4%) in the survey lakes, many of which demonstrated a net internal loading in the recovery period following a reduction in loading. The much higher values found for Lake Fure and Lake Esrom (Table 8.1) demonstrate the importance of water residence time for the retention.

Along the length of Roskilde Fjord, the N:P ratio was further reduced to about 5 after a retention of 88% of the nitrogen load and 74% of the phosphorus load. However, this N:P ratio will increase in the coming years, since the phosphorus load from point sources is being significantly reduced, but the response is somewhat delayed by internal loading from accumulated sediment P-pools in the fjords (Kaas et al. 1996).

Relatively more nitrogen than phosphorus is usually lost during the flow of water through aquatic systems. This response explains why nitrogen is overtaking the role as the major limiting factor as we approach the open sea. However, if we carry out comparative studies across ecosystem types (wetlands, rivers, lakes and coastal waters) or within ecosystem types, it is extremely difficult to predict the net losses and the N:P ratios for single systems, due to the temporal and spatial variation of the many factors determining the cycling and deposition of nutrients. The residence time is short in regulated rivers and the retention of nutrients becomes insignificant. In lakes with long residence time, we have higher retention, but anaerobic sediment surfaces in hypertrophic and stratified lakes may reduce denitrification losses due to the uncoupling of the nitrification/denitrification mechanism. In coastal waters, the denitrification losses increase due to an improved oxygen regime in the sediment surface promoted by resuspension and macrophyte coverage. The phosphorus retention is, however, lower than in lakes since iron is inactivated by iron sulphides resulting from high sulphate reduction rates.

The overwhelming complexity of mechanisms and scales involved in the nutrient dynamics, especially during changed loading, advocates the use of mathematical models to improve predictions among and within ecosystems.

Recipient water quality planning and modelling

In 1973 the Environmental Protection Law passed the Danish parliament. In this law, the ambitious principle of 'recipient quality planning' was implemented for the protection and management of surface waters. In most other countries, water quality is regulated by effluent standards, or, in the case of phosphorus, in some countries by maximum concentrations in detergents (Sas 1989). According to the principle of recipient quality planning, an objective for each water body is set politically by the county councils based on usage criteria. The water quality criteria matching the objective are subsequently defined by the county administration supplemented with time schedules and measures for meeting these objectives.

The establishment of the relationships between loading scenarios and the resulting water quality required the use of predictive models. This demand was a challenge and an opportunity for limnologists to integrate

experimental results into whole-lake dynamics at various temporal and spatial scales.

Mathematical lake models can roughly be divided in two types: empirical steady state models based on statistical treatment of data from a large number of lakes and theoretical dynamic models derived from detailed, heuristic sub-models of physical transport, nutrient kinetics and population dynamics (Ahlgren et al. 1988). The first type is recognisable by its simplicity, analytical tractability and its strategic properties, whereas the second type is recognisable by its realistic description, predictive precision and tactical properties (Visser and Kamp-Nielsen 1996).

Empirical steady state models

In 1968, Vollenweider described the relationship between nutrient loading and lake trophic state (Vollenweider 1968). By plotting areal loads versus lake depths on double logarithmic scales he could draw straight lines representing the critical loads for the transitions between oligotrophic, mesotrophic and eutrophic lakes. Such plots have frequently been used for a graphical representation of lake trophic state as a function of different loading scenarios (e.g. Imboden 1974). Parallel to his empirical relations, Vollenweider (1964) developed a series of simple, theoretical models starting with:

$$V(dP/dt) = (1 - f)M - rPV \qquad (8.1)$$

where V is the lake volume, P is the lake phosphorus concentration, f is the fraction of influent phosphorus lost through the outflow, M is input of phosphorus and r is the sedimentary loss coefficient.

The steady state solution of (8.1) is:

$$P = (1 - f)M/Arz = L'/rz \qquad (8.2)$$

where A is the lake surface area, z is the lake mean depth and L'(=(1-f)M/A) is the net areal loading.

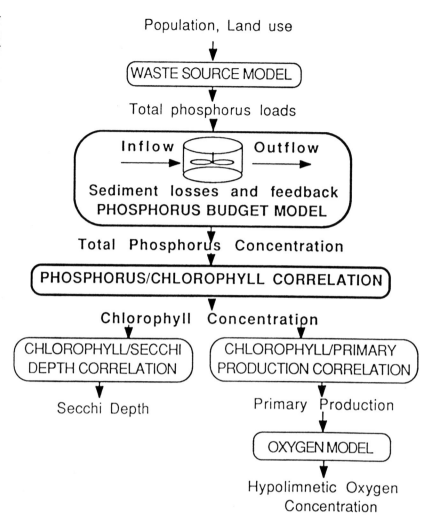

Figure 8.9. Sequence of simple models used for the prediction of lake eutrophication. Redrawn from Ahlgren et al. (1988).

The model was revised by Vollenweider (1969):

$$V(dP/dt) = M - rPV - QP \qquad (8.3)$$

with the steady state solution:

$$P = (L/z)/(I/r_w + r) \qquad (8.4)$$

where L is the gross areal loading and r_w (=V/Q) is the hydraulic residence time.

The sediment loss coefficient r cannot be expected to be the same for all lakes. Vollenweider (1975) found from mass balances of Swiss lakes an inverse relationship between r and z:

$$r = ln\ 5.5 - 0.85\ ln\ z \qquad (8.5)$$

with the approximation:

$$r = 10/z \qquad (8.6)$$

Later Vollenweider (1976) proposed a relationship between r and r_w in spite of its dimensional inconsistency:

$$r = 1/\sqrt{r_w} \qquad (8.7)$$

By substituting (8.7) into (8.4) we obtain:

$$P = (L/q_s)/(1 + \sqrt{r_w}) \qquad (8.8)$$

Table 8.4. Phosphorus-chlorophyll models (from Ahlgren et al. 1988).

Model	Reference
Linear models:	
chl a = 1.19 TP - 7.3	Schindler et al. (1978)
chl a = 0.58 TP + 4.2	Megard (1978)
chl a = 0.55 TP - 4.8	Edmonson and Leman (1981)
chl a = 0.42 TP - 0.93	Berge et al. (1980)
Logarithmic models:	
chl a = 0.735 $TP^{1.58}$	Sakamoto (1966)
chl a = 0.072 $TP^{1.45}$	Dillon and Rigler (1974)
chl a = 0.028 $TP^{0.98}$	Vollenweider and Kerekes (1982)
chl a = 0.111 $TP^{1.41}$	Brøgger and Heintzelmann (1976)
Models with maximum level:	
chl a = 150 $(1-e^{(-0.000867\,TP-0.000011\,TP\,exp\,2)})$	Ahl and Wiederholm (1977) (shallow lakes)
chl a = 50$(1-e^{(-0.0026\,TP-0.0001\,TP\,exp\,2)})$	Ahl and Wiederholm (1977) (deep lakes)
chl a = 40.1/$(1+130\,e^{(-0.114\,TP)})$	Straskraba (1980)
Model with N and P concentrations:	
log(chl a)=0.6531 log TP+0.548 log TN-1.517	Smith (1982)

where $q_s(=Q/A)$ is the water discharge height.

Equation (8.8) was applied to a large number of lakes in the OECD-study of Vollenweider and Kerekes (1982), and regression analysis suggested the modification:

$$P = 1.55[(L/q_s)/(1 + \sqrt{r_w})]^{0.82} \qquad (8.9)$$

Chapra (1980) and Ahlgren et al. (1988) proposed a linkage of simple sub-models for the prediction of lake trophic states from their phosphorus loading (Fig. 8.8.). In this procedure, the next step is to model the phosphorus-chlorophyll relationship. The underlying assumption behind this model is that all potentially available phosphorus is stochiometrically incorporated in phytoplankton biomass. The models proposed for this relationship are linear, logarithmic, saturation and N- and P-dependent models, but most of the models predict an increasing chlorophyll:P ratio with increasing P-concentrations (Table 8.4).

During the 1970's, mass balances and corresponding nutrient and chlorophyll concentrations became available for a number of Danish lakes, and some of the simple models were tested on these data (Brøgger and Heintzelmann 1979). The Kirchner and Dillon (1975) model was best at predicting phosphorus retention under steady state conditions for lakes with a water discharge height below 20 m y^{-1}, and the model by Larsen and Mercier (1976) was the best for lakes with a water residence time below 2.5 y and above 40 y. In the following years many more similar models were developed, and a later test on a much larger Danish dataset representing 131 lake years showed that the Vollenweider-model (8.8) had the lowest median of per centage standard deviation (Kristensen et al. 1990b).

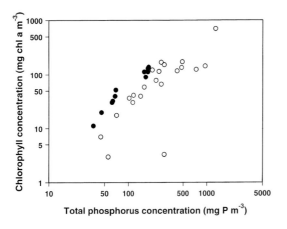

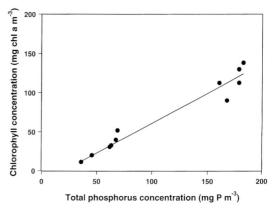

Figure 8.10. Chlorophyll a concentrations in July – August vs. the annual average concentrations of total phosphorus in Danish lakes. Lakes with N:P ratios > 10 and epilimnetic concentrations of phosphate-P < 10 mg m^{-3} (●) and all other lakes (○). Redrawn from Brøgger and Heintzelmann (1976).

Figure 8.11. The same data as in Fig. 8.10, but with linear scales and only the P-limited lakes. The line is from a linear regression. Redrawn from Brøgger and Heintzelmann (1976).

The subsequent description of the phosphorus-chlorophyll relationship by Brøgger and Heintzelmann (1979) used data from 43 Danish lakes (Fig. 8.10) and showed large variations, but after screening of the data for nitrogen limitation (N:P ratio <10) and for other limiting factors (PO_4-P > 10 µg l^{-1}), the level of chlorophyll in the photic layer during July-August could be predicted reasonably well from the annual average total P concentration (Fig. 8.11):

$$[chl\ a] = 0.111\ [P]^{1.406}\ mg\ m^{-3} \qquad (8.10)$$

After exclusion of humic lakes, a highly significant relation between Secchi depth transparency (SD) and chlorophyll (mg m^{-3}) was shown:

$$SD^{-1} = 0.093[chl\ a]^{0.622}\ m^{-1} \qquad (8.11)$$

The much larger dataset in the 1990-survey (Kristensen et al. 1990b) did not produce more precise models, since many shallow lakes with resuspension and some with submerged macrophytes had been added.

However, these models only predict the steady state conditions. In most cases the time-dependent responses to changed nutrient loading are also important. A simple time-dependent solution of the one-box model provide the dilution response to a reduced loading, but it still assumes that the per centage net retention of phosphorus will be the same.

We studied the response of lakes to reduced phosphorus load by comparing mass balances and phosphorus-chlorophyll relations for nine shallow and nine deep European lakes, several years before and after a significant reduction in phosphorus load had taken place (Sas 1989). In all the shallow lakes, a net P-release occurred in the years following the load reduction, but the net release decreased over time and returned to net-retention after a few years. The main factor describing the net release was the sediment concentration of total phosphorus; more than 1 mg P g^{-1} dry sediment resulted in a net release after the first year. In the deep lakes, a net release was never observed, and this was explained by a relatively higher input of mineral material resulting in much lower sediment concentrations. Apparently, the retention was more likely governed by sorption equilibria with the P-concentrations in the overlying water both before and after load reduction.

In comparing the steady state lake phosphorus concentrations before and after load reduction, we applied Eq. 8.9 to each case and found a generally weaker response to reduced loading than that predicted by Vollenweider and Kerekes (1982). On average, we calculated an exponent of $0.64^{\pm}0.13$ (mean $^{\pm}$SE), which was significantly lower than the exponent of $0.82^{\pm}0.07$ found in the OECD-study. Out of curiosity we also performed the regression on all data points and found a value of $0.81^{\pm}0.11$, which demonstrated how dangerous it is to use results from a general descriptive relation across spatial variations in predicting the temporal behaviour of a single lake.

We also compared the chlorophyll responses to reduced phosphorus loading with the OECD-study. The exponent of the logarithmic response of chlorophyll to phosphorus concentrations was $0.96^{\pm}0.12$ in the OECD-study and $0.77^{\pm}0.23$ in our study with a 7% probability of being identical. After screening for true P-limitation, the average exponent for the response of single lakes increased to $1.02^{\pm}0.17$. The exponents were significantly different in the shallow $(1.26^{\pm}0.26)$ and the deep lakes $(0.62^{\pm}0.45)$. In a few cases we could also estimate the limit concentration of phosphorus below which there was a linear response in chlorophyll. Thus, provided that there is true phosphorus limitation, lake chlorophyll concentration does respond to reduced phosphorus concentrations according to the OECD-model, but shallow and deep lakes respond differently.

Empirical models have their strength in revealing large-scale properties of lakes (Peters 1986), and they have demonstrated the importance of the nutrient exchange between sediment and water. However, when we come to the seasonal behaviour, transient loading situations, control by other factors than phosphorus and sophisticated interactions between organisms, their simplicity certainly set limitations (Lehman 1986). Nevertheless, their value as a predictive tool and their precision may be significantly improved if the data sets are sufficiently large to allow screening for hidden variables.

Dynamic models

The limitations of the empirical models and the promising results by Lorentzen et al. (1976) from modelling of the loading history of Lake Washington stimulated the wider application of dynamic models. Such models range from simple two-compartment models like the Lorentzen-model with lake phosphorus and sediment phosphorus as the only state variables, to complicated models with several biotic and abiotic variables. The more complicated models have been viewed with scepticism due to their intractability, lack of realism and calibration of unknown parameters (Scheffer and Beets 1994). Moreover, only very few models have been carefully validated on independent datasets. In our application of dynamic models at the Freshwater Biological Laboratory we have tried to meet some of this criticism.

Models should not be any more complicated than is necessary for their purpose. With an increasing number of parameters and state variables, the uncertainty increases. The optimum number of state variables is apparently about twenty for lake models (Constanza and Sklar 1985). The models should be constructed by linkage of heuristic sub-models which have been verified and validated. The parameter values should be experimentally determined and, if necessary, calibrated within pre-set realistic ranges, preferably on a dataset with the highest possible resolution in time and space. The most convincing validation is provided when the predictions have been published previously.

The simple, dynamic models of Lake

Glumsø considered only the flushing and the sediment water exchange of phosphorus (Jørgensen et al. 1973, Kamp-Nielsen et al. 1985), but since nitrogen was also limiting in Lake Glumsø, the model was extended to a complete bio-geochemical eutrophication model with compartments for phosphorus, nitrogen and carbon on different biotic and abiotic levels (Jørgensen 1975). In comparison with the existing models (Chen and Orlob 1975, Di Toro et al. 1971, Thomann et al. 1975), which used Monod-kinetics for algal growth and first order kineticts for the sediment-water exchange of nutrients, the Glumsø-model introduced the two-step uptake-growth kinetics based on sub-models developed by Nyholm (1977a,b). The sediment-water exchange of phosphorus was described by a partition of the settled material into exchangeable and non-exchangeable phosphorus, assuming that the concentration of phosphorus in the deeper part of the sediment represented the non-exchangeable concentration. The exchangeable part was mineralised by a rapid temperature-dependent process at the sediment surface and a slower process in the sediment. The mineralised phosphate equilibrated with the sediment matrix by a redox-dependent sorption process and diffused vertically according to the concentration gradient. The sub-model was developed by experimentally derived kinetics (Kamp-Nielsen 1975, Jørgensen et al. 1975) and was validated on seasonal variations in porewater concentrations (Fig. 8.12) using measured temperature, oxygen concentration and sedimentation/resuspension data as forcing functions (Kamp-Nielsen 1977b).

In the search for the balance between optimum complexity and applicability, various sub-models were developed and tested. The importance of including several pools of sedimentary phosphorus was demonstrated by simulating a 90% reduction of the phosphorus load to Lake Glumsø with a range of sediment-water exchange models (Kamp-Nielsen

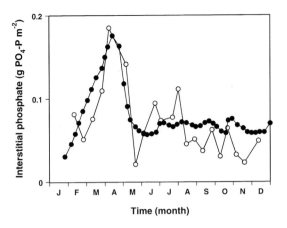

Fig. 8.12. Simulated (●) and observed (○) variations in the pool of porewater phosphate-P in Lake Esrom. Redrawn from Kamp-Nielsen (1977b).

1980a) and by simulation of the long-term behaviour of Lake Lyngby (Jørgensen et al. 1978). A version supplemented by an experimentally derived sub-model for nitrogen metabolism was implemented in a water quality model for lakes in the Upper Nile Basin (Kamp-Nielsen et al. 1981).

The sediment sub-models developed at that stage had spatially unlocalized pools, constant process rates and steady state diffusion assumptions. An increased resolution in time and space could only be achieved by a multilayer model. Such a model was developed with essentially the same equations in fifty 1-2 millimetre layers in the top 10 centimetres of the sediment as in the previous models, but supplemented with an age-dependent mineralisation constant (Kamp-Nielsen 1978). The simulations with such a complex model required the solution of numerous differential equations for each time-step, and the uncertainty increased to a level exceeding the gain in articulation, so further use of such complex models was abandoned (Jørgensen et al. 1982).

Further improvements of the Glumsø-model included automatic optimisation of

Table 8.5 Predicted and observed transparency, primary production and chlorophyll a concentrations in Lake Glumsø during recovery after a 90% reduction of phosphorus load. (From Jørgensen et al. 1986).

	Predicted	Observed
Minimum transparency (cm)		
1st year	20	20
2nd year	30	25
3rd year	45	50
Maximum primary production ($g\ C\ m^{-2}\ d^{-1}$)		
1st year	9.5	5.5
2nd year	6.0	11.0
3rd year	5.0	6.2
Maximum chlorophyll a concentration ($mg\ m^{-3}$)		
1st year	750	800
2nd year	520	550
3rd year	320	380

calibrations based on intensive measurements (Kamp-Nielsen 1980b, Jørgensen et al. 1981). For the first time a previously published prediction (Jørgensen et al. 1978) of the consequences of a reduced nutrient loading was validated in the literature (Jørgensen et al. 1986a, Kamp-Nielsen 1986). As a result of sewage diversion, high phosphorus loadings to Lake Glumsø were reduced by 73%. The measured results are shown in Table 8.5 together with the predicted values for a 92% reduction (from Jørgensen 1976). The overall standard deviation between measured and predicted values was 31% with values of 2% for dissolved, inorganic nitrogen, 10% for phytoplankton biomass, but 26% for zooplankton and 27% for total phosphorus. These results were much more precise than those predicted by simple steady state models (Table 8.6). However, the predicted phytoplankton biomass was overestimated in the first year and underestimated in the second year. Before the sewage diversion, *Scenedesmus sp.* dominated the phytoplankton with more than 80% of the biomass. Model verifications showed that a substantial wintering innoculum of *Scenedesmus sp.* outcompeted a spring bloom of diatoms although silicate concentrations were high (Christensen et al. 1986). After the sewage diversion, the spring dominance of *Scenedesmus sp.* was replaced by diatoms (*Stephanodiscus sp* and *Nitschia sp.*), and a revised version of the Glumsø-model introduced diatoms and silicate as new state variables together with a more sophisticated description of the denitrification process (Salomonsen and Jensen 1990). These changes reduced the average standard deviation to 17% with most significant improvements on the simulations of phytoplankton biomass and total phosphorus.

The application of the Glumsø-model to a

wide range of other lakes and coastal waters showed that the model had to be modified in accordance with specific characteristics of the ecosystem (Jørgensen et al. 1986b). In stratified lakes, the physical circulation was introduced; for shallow, wind-exposed lakes, a resuspension term was introduced, and in hypertrophic lakes, nitrogen-fixing cyano-bacteria were added as state variable. In a number of cases, a validation was possible either by hindcasts or forecasts, and overall standard deviations varying from 7% to 36% were achieved.

These examinations of the generality of the Glumsø-model showed that the basic structure of the model works over a wide range of aquatic ecosystems, provided that the biological structure is identified and the model partially reconstructed. But we also observed that significant changes in nutrient loading changed the biological structure. To account for the structural changes we have to use models which can change their biological structure according to the fitness of species (Jørgensen 1986, 1992, 1996). This can be done by having all possible species 'sleeping' in the model, but the resultant complexity and derived uncertainty limit the application. Or we can discretise the model and introduce or exclude species according to their fitness to the ambient conditions, eventually using allometric criteria (Reynolds 1989, 1996). Finally, we can use goal functions which optimise the holistic properties of the entire ecosystem by selecting species from a 'library'. Such goal functions can be taken from network theory like the indirect effect-theory (Patten 1990) and the ascendancy-theory (Ulanowicz 1986) or from non-equilibrium thermodynamics and information theories like minimum entropy-production (Prigogine 1984), maximum power (Odum 1982) or maximum exergy (Mejer and Jørgensen 1979, Jørgensen 1989, 1992, 1996).

Nutrient dynamics and modelling

Many of the physical, chemical and biological mechanisms involved in the generalisation and transformation of nutrients are non-linear. The classical hypothetico-deductive approach by falsification of a null-hypothesis through experiments, as formulated in the 'strong interference' method (Platt 1964), has been criticised due to the possibility of multiple causality (Scheffer and Beets 1994). Observed phenomena can be explained from many mechanisms, and different mechanisms can be responsible for a similar phenomenon, while several mechanisms may act simultaneously to produce the same phenomenon as that produced by a single mechanism (Quinn and Dunham 1983).

Mathematical models as a tool for overcoming the problems of precise description and multiple causality can be approached in both a strategic and a tactic way (Nisbeth and Gurney 1982), but the conflict between analytical tractability and realism always underlies the choice. Strategic models are based on empirical relationships across a large number of ecosystems, but as shown for the lake phosphorus models, their predictive power is often poor. However, by introducing a sediment phosphorus pool, the predictive power may be significantly increased, though the resolution in time remains low.

On the other hand, complex tactic simulation models have been questioned (Rigler 1982) due to unrealistic parameter values and the conservative nature of the biological structure. New generations of tactic models, however, with strategic use of simplifications and goal functions for a dynamic structure may increase their value as tools for testing holistic properties across ecosystems (Ross et al. 1993, 1994, Visser and Kamp-Nielsen 1996).

At the Freshwater Biological Laboratory, the use of both simple steady-state models and complex, dynamic models has been based on experimentally derived submodels and has been a fruitful way of identifying the relative importance of ecologically important mechanisms. We believe that this will contribute to realistic models with improved predictive power and draw more attention to the value of models as scientific tools in ecosystem research.

Literature cited

Ahl, T., and T. Wiederholm. 1977. Svenska Vattenkvalitetskriterier. Eutrofierande ämnen. Statens Naturvårdsverk. PM-series 918. Stockholm, Sweden.

Ahlgren, I, T. Frisk, and L. Kamp-Nielsen. 1988. Empirical and theoretical model of phosphorus loading, retention and concentration vs. lake trophic state. Hydrobiologia 170: 285-303.

Andersen, J.M. 1974. Nitrogen and phosphorus budgets and the role of sediments in six shallow Danish lakes. Archiv für Hydrobiologie 74: 528-550.

Andersen, J.M. 1977. Rates of denitrification of undisturbed sediment from six lakes as a function of nitrate concentration, oxygen and temperature. Archiv für Hydrobiologie 80: 147-159.

Baldinger, F. 1957. Das Hallwilersee-Projekt als Beispiel einer grosszügigen Seesanierung. Schweizerisches Zeitschrift für Hydrologie 19: 18-36.

Berg, K., K. Andersen, T. Christensen, F. Ebert, E. Fjerdingstad, C. Holmquist, K. Korsgaard, G. Lange, J.M. Lyshede, H. Mathiesen, G. Nygaard, S. Olsen, C.V. Overstrøm, U. Røen, A. Skadhauge, and E. Steemann-Nielsen. 1958. Furesøundersøgelser 1950 – 54. Folia Limnologica Scandinavica 10: 1-189.

Berge, D., S. Rognerud, and M. Johannesen. 1980. Videreutvikling af fosforbelastningsmodeller for store skiktede insjöer. Pages 39-42 *in* Norsk Institut for Vannforskning. Årbok 1979. Oslo, Norway

Bretschneider, C.L. 1952. The generation and decay of wind waves in deep water. Transactions of the American Geophysical Union 33: 381-389.

Brøgger Jensen, J. and F. Heintzelmann. 1979. Sørestaurering. Simple stofbalancemodellers anvendelse i recipientplanlægning. Miljøprojekt 16. Miljøstyrelsen, Copenhagen, Denmark.

Chapra, S.C. 1980. Application of the phosphorus loading concept to the Great Lakes. Pages 135-152 *in* C. Loehr, C.S. Martin and W. Rast, editors. Phosphorus management-strategies for lakes. Ann Arbor Science Publishers, Ann Arbor, USA.

Chen, C.W. and G.I. Orlob. 1975. Ecological simulation for aquatic environments. Pages 475-588 *in* B.C. Patten, editor. System analysis and simulation in ecology. Vol 3. Academic Press, New York, USA.

Christensen, T. and J. Windolf-Nielsen. 1986. Glumsømodellen. Dokumentation. Miljøstyrelsen, Copenhagen, Denmark.

Constanza, R. and F.H. Sklar. 1985. Articulation, accuracy and effectiveness of mathematical models: A review of freshwater wetland applications. Journal of Ecological Modelling 27: 45-69.

Dillon, P.J., and F.H. Rigler. 1974. The phosphorus-chlorophyll relationship in lakes. Limnology and Oceanography 19: 767-773.

DiToro, D.M., R.V. Thomann, and D.J. O`Connor. 1971. A dynamic model of phytoplankton populations in the Sacramento-San Joaquin Delta. Page 131 *in* R.F. Gould, editor. Advances in Chemistry. Series 106: Non-equilibrium systems in natural water chemistry. American Chemical Society. Washington DC. USA.

Domogalla, B.P., C. Juday, and W.H. Peterson. 1925. The forms of nitrogen found in certain lake waters. Journal of Biological Chemistry 63: 269-285.

Edmonson, W.T., and J.T. Lehman. 1981. The effect of changes in nutrient income on the condition of Lake Washington. Limnology and Oceanography 26: 1-29.

Einsele, W. 1936. Über die Beziehungen des Eisenkreislaufs zum Phosphatkreislauf im eutrophen See. Archiv für Hydrobiologie 29: 664-686.

Einsele, W. 1941. Die Umsetzung von zugeführtem, anorganischem Phosphat im eutrophen Seen und ihre Rückwirkung auf seinen Gesamthaushalt. Zeitung für Fischerei 39: 407-488.

Flindt, M.R. 1994. Measurements of nutrient fluxes and mass balances by on-line *in situ* dialysis in a *Zostera marina* bed culture. Verhandlungen Internationale Vereinigung für Theoretische und Angewandte Limnologie 25: 2259-2264.

Flindt, M.R. and J.B. Nielsen. 1992. Heterotrophic bacterial activity in Roskilde Fjord sediment during an autumn sedimentation peak. Hydrobiologia 235/236: 283-293.

Flindt, M.R. and L. Kamp-Nielsen. 1997. The influence of sediment resuspension on nutrient metabolism in the eutrophic Roskilde Fjord. Verhandlungen Internationale Vereinigung für Theoretische und Angewandte Limnologie 26. In press.

Gessner, F. 1934. Phosphat und Nitrat als Produktionsfaktoren der Gewässer. Verhandlungen Internationale Vereinigung für Theoretische und Angewandte Limnologie 7: 525-538.

Hargrave, B.T., and L. Kamp-Nielsen. 1977. Accumulation of sedimentary organic matter at the base of steep bottom gradients. Pages 168-174 *in* H.L. Golterman, editor. Proceedings of an international symposium on the interactions between sediments and freshwater. Dr. W. Junk b.v. Publishers, Haag, The Netherlands.

Hasler, A.D. 1947. Eutrophication of lakes by domestic drainage. Journal of Ecology 28: 383-395.

Hayes, F.R., J.A. McCarter, M.L. Cameron, and D.A. Livingstone. 1952. On the kinetics of phosphorus exchange in lakes. Journal of Ecology 40: 202-216.

Hutchinson, G.E., and V.T. Bowen. 1947. A direct demonstration of the phosphorus cycle in a small lake. Proceedings of the National Academy of Science USA 33: 148-153.

Hutchinson, G.E., and V.T. Bowen. 1950. Limnological studies in Connecticut: IX. A quantitative radiochemical study of the phosphorus cycle in Linsley Pond. Ecology 31: 194-203.

Imboden, D. 1974. Phosphorus model of lake eutrophication. Limnology and Oceanography 19: 297-304.

Jonasson, P.M. 1972. Ecology and production of the profundal benthos in relation to phytoplankton in Lake Esrom. Oikos Supplementum 14: 1-148.

Jonasson, P.M., E. Lastein, and A. Rebsdorf. 1974. Production, insolation, and nutrient budget of eutrophic Lake Esrom. Oikos 25: 255-277.

Juday, C., E.A. Birge, G.I. Kemmerer, and R.J. Robinson. 1928. Phosphorus content of lake waters of Northeastern Wisconsin. Transactions of the Wisconsin Academy of Science 23: 233-248.

Jørgensen, S.E. 1976. A eutrophication model for a lake. Journal of Ecological Modelling 2: 147-165.

Jørgensen, S.E. 1986. Structural dynamic model. Journal of Ecological Modelling 31: 16-28.

Jørgensen, S.E. 1992. Developments of models able to account for changes in species composition. Journal of Ecological Modelling 62: 195-208.

Jørgensen, S.E. 1996. The application of ecosystem theory in limnology. Verhandlungen Internationale Vereinigung für Theoretische und Angewandte Limnologie 26.181-193.

Jørgensen, S.E., O.S. Jacobsen, and I. Høi. 1973. A prognosis for a lake. Vatten 29: 382-404.

Jørgensen, S.E., L. Kamp-Nielsen, and O.S. Jacobsen. 1975. A sub-model for the anaerobic mud-water exchange of phosphate. Journal of Ecological Modelling 1: 133-146.

Jørgensen, S.E., H.F. Mejer, and M. Friis. 1978. Examination of a lake model. Journal of Ecological Modelling 4: 257-278.

Jørgensen, S.E., L.A. Jørgensen, L. Kamp-Nielsen, and H.F. Mejer. 1981. Parameter estimation in eutrophication modelling. Journal of Ecological Modelling 13: 111-129.

Jørgensen, S.E., L. Kamp-Nielsen, and H.F. Mejer. 1982. Comparison of a simple and a complex sediment phosphorus model. Journal of Ecological Modelling 16: 99-124.

Jørgensen, S.E., L. Kamp-Nielsen, T. Christensen, J. Windolf-Nielsen, and B. Vestergaard. 1986a. Validation of a prognosis based on a eutrophication model. Journal of Ecological Modelling 32: 165-182.

Jørgensen, S.E., L. Kamp-Nielsen, and L.A. Jørgensen. 1986b. Examination of the generality of eutrophication models. Journal of Ecological Modelling 32: 251-266.

Kaas, H., F. Møhlenberg, A. Josefsson, B. Rasmussen, D. Krause-Jensen, H.S. Jensen, L.M. Svendsen, J. Windolf, A.L. Middelboe, K. Sand-Jensen, and M.F. Pedersen. 1996. Marine områder. Danske fjorde – status over miljøtilstand, årsagssammenhænge og udvikling. Vandmiljøplanens Overvågningsprogram 1995. Danmarks Miljøundersøgelser. Faglig rapport nr 179. Copenhagen, Denmark.

Kamp-Nielsen, L. 1974. Mud-water exchange of phosphate and other ions in undisturbed sediment cores and factors affecting the exchange rates. Archiv für Hydrobiologie 73: 218-237.

Kamp-Nielsen, L. 1975a. A kinetic approach to the aerobic, sediment-water exchange of phosphorus in Lake Esrom. Journal of Ecological Modelling 1: 153-160.

Kamp-Nielsen, L. 1975b. Seasonal variation in sediment-water exchange of nutrient ions in Lake Esrom. Verhandlungen Internationale Vereinigung für Theoretische und Angewandte Limnologie 19: 1057-1065.

Kamp-Nielsen, L. 1977a. Horizontal and temporal variation in sedimentation in Lake Esrom. Pages 1-8 *in* J.M. Andersen, O.S. Jacobsen, and L. Kamp-Nielsen, editors. Proceedings of the 5th Nordic Sediment Symposium. Freshwater Biological Laboratory, Hillerød, Denmark.

Kamp-Nielsen, L. 1977b. Modelling the temporal variation in sedimentary phosphorus fractions. Pages 277-285 *in* H.L. Golterman, editor. Proceedings of an International Symposium on the Interactions between Sediments and Freshwater. Dr. W. Junk b.v. Publishers, Haag, The Netherlands.

Kamp-Nielsen, L. 1978. Modelling the vertical gradients in sedimentary phosphorus fractions. Verhandlungen International Vereinigung für Theoretische und Angewandte Limnologie 20: 720-727.

Kamp-Nielsen, L. 1980a. The influence of sediments

on changed phosphorus loading to hypertrophic Lake Glumsø. Pages 29-36 *in* J. Barica, and L.R. Mur, editors. Developments in Hydrobiology 2. Dr. W. Junk b.v. Publishers, Haag, The Netherlands.

Kamp-Nielsen, L. 1980b. Intensive measurements of sedimentation in Lake Esrom and Lake Glumsø. Pages 44-56 *in* L. Kamp-Nielsen, editor. Proceedings of the 8th Nordic sediment symposium. Freshwater Biological Laboratory, Hillerød, Denmark

Kamp-Nielsen, L. 1985. Modelling of eutrophication processes. Pages 5-16 *in* R. Bernardi, E.F. Frangipane, R. Marchetti, G. Margaritora, A. Misiti, R. Passino, and R. Vismara, editors. International Congress on Lake Pollution and Recovery. Andis, Rome, Italy.

Kamp-Nielsen, L. 1986. Modelling the recovery of hypertrophic Lake Glumsø, Denmark. Hydrobiological Bulletin 20: 245-255.

Kamp-Nielsen, L. 1992. Benthic-pelagic coupling of nutrient metabolism along an estuarine eutrophication gradient. Hydrobiologia 235/236: 457-470.

Kamp-Nielsen, L., L.A. Jørgensen, and S.E. Jørgensen. 1981. A sediment-water exchange model for lakes in the Upper Nile Basin. Pages 557-582 *in* D.M. Bubois, editor. Progress in ecological engineering and management by mathematical modelling. Editions cebedoc, Liege, Belgium.

Kamp-Nielsen, L., and B.T. Hargrave. 1978. Influence of bathymetry on sediment focusing in Lake Esrom. Verhandlungen Internationale Vereinigung für Theoretische und Angewandte Limnologie 20: 714-719.

Kamp-Nielsen, L., and M.R. Flindt. 1994. On-line recording of porewater profiles from *in situ* dialysis. Verhandlungen Internationale Vereinigung für Theoretische und Angewandte Limnologie 25: 151-156.

Kirchner, W.B., and P.J. Dillon. 1975. An empirical method of estimating the retention of phosphorus in lakes. Journal of Water Resources Research 11: 182-183.

Kristensen, P.B., B. Kronvang, E. Jeppesen, P. Græsbøll, M. Erlandsen, A. Rebsdorf, A. Bruhn, and M. Søndergaard. 1990a. Ferske vandområder – vandløb, kilder og søer. Vandmiljøplanens overvågniongsprogram. Danmarks Miljøundersøgelser. Faglig rapport nr 5. Miljøministeriet. Danmarks Miljøundersøgelser, Copenhagen, Denmark.

Kristensen, P., J.P. Jensen, and E. Jeppesen. 1990b. Eutrofieringsmodeller for søer. NPO-forskning fra Miljøstyrelsen. Nr C9. Miljøministeriet. Miljøstyrelsen, Copenhagen, Denmark.

Larsen, D.P., and H.T. Mercier. 1976. Phosphorus retention capacity of lakes. Journal of Fisheries Research Board of Canada 33: 1742-1750.

Lastein, E. 1976. Recent sedimentation and resuspension of organic matter in eutrophic Lake Esrom, Denmark. Oikos 27: 44-49.

Lehman, J.T. 1986. The goal of understanding in limnology. Limnology and Oceanography 31: 1160-1166.

Lorenzen, M.W., D.J. Smith, and L.V. Kimmel. 1976. A long-term phosphorus model for lakes: application to Lake Washington. Pages 75-91 *in* R.P. Canale, editor. Modelling biochemical processes in aquatic ecosystems. Ann Arbor Science Publishers Inc, Ann Arbor, USA.

Megard, R.P. 1972. Phytoplankton, photosynthesis, and phosphorus in Lake Minnetonka, Minesota. Limnology and Oceanography 17: 68-87.

Mejer, H.M., and S.E. Jørgensen. 1979. Exergy and ecological buffer capacity. Pages 829-847 *in* S.E. Jørgensen, editor. State-of-the-art in ecological modelling. Fair-Print a/s, Roskilde, Denmark

Miljøstyrelsen. 1995. Vandmiljø-95. Redegørelse fra Miljøstyrelsen nr 3. Miljø- og energiministeriet. Miljøstyrelsen, Copenhagen, Denmark.

Mortimer, C.H. 1941. The exchange of dissolved substances between mud and water in lakes I. Journal of Ecology 29: 280-329.

Mortimer, C.H. 1942. The exchange of dissolved substances between mud and water in lakes II. Journal of Ecology 30: 147-201.

Naumann, E. 1919. Några synpunkter angående limnoplanktons ekologi med särskild hänsyn till fytoplankton. Svensk Botanisk Tidskrift 13: 129-163.

Naumann, E. 1930. Die Haupttypen der Gewässer in produktionsbiologischer Hinsicht. Verhandlungen Internationale Vereinigung für Theoretische und Angewandte Limnologie 5: 72-74.

Nisbeth, R.M., and W.S.C. Gurney. 1982. Modelling fluctuating populations. Wiley. Chichester. England.

Nygaard, G. 1938. Hydrobiologische Studien über dänische Teiche und Seen. Archiv für Hydrobiologie 32: 523-692.

Nyholm, N. 1977a. Kinetics of phosphate-limited algal growth. Journal of Biotechnology and Bioengineering 19: 467-492.

Nyholm, N. 1977b. Kinetics of nitrogen-limited algal growth. Progress in Water Technology 8: 247-358.

Odum, H.T. 1982. Pulsing power and hierarchy. Pages 33-60 *in* W. Mitsch, R. Ragade, R. Bosserman, and J. Dillon, editors. Energetics and systems. Ann Arbor Science Publishers, Ann Arbor, USA.

Olsen, S. 1958. Phosphate adsorption and isotopic exchange in lake muds. Experiments with ^{32}P. Verhandlungen Internationale Vereinigung für Theoretische und Angewandte Limnologie 13: 915-922.

Olsen, S. 1964. Phosphate equilibrium between reduced sediments and water. Verhandlungen Internationale Vereinigung für Theoretische und Angewandte Limnologie 15: 333-341.

Olsen, S., and K. Larsen. 1948. Er Furesøen ved at ændre karakter? Et bidrag til diskussionen om søens eutrofiering. Lystfisker-Tidene 60: 467.

Patten, B.C. 1990. Network ecology – indirect determination of life – environment relationship in ecosystems. Chapter 12 *in* M. Higashi, and T.P. Burns, editors. The network perspective. Cambridge University Press, London, England.

Pearsall, W.H. 1932. Phytoplankton in the English lakes. II. The composition of the phytoplankton in relation to dissolved substances. Journal of Ecology 20: 241-262.

Platt, J.R. 1964. Strong interference. Science 146: 347-353.

Peters, R.H. 1986. The role of prediction in limnology. Limnology and Oceanography 31: 1143-1159.

Priogine, I. 1955. Thermodynamics of irreversible processes. Interscience Publishers, New York, USA.

Reynolds, C.S. 1987. The response of phytoplankton communities to changing lake environments. Schweizerisches Zeitschrift für Hydrologie 49: 220-236.

Reynolds, C.S. 1996. The plant life of the pelagic. Verhandlungen Internationale Vereinigung für Theoretische und Angewandte Limnologie 26: 97-113.

Ross, A.H., W.S.C. Gurney, M.R. Heath, S.J. Hay, and E.W. Henderson. 1993. A strategic simulation model of a fjord ecosystem. Limnology and Oceanography 38: 128-153.

Ross, A.H., W.S.C. Gurney, and M.R. Heath. 1994. A comparative study of the ecosystem dynamics of four fjords. Limnology and Oceanography 39: 318-343.

Quinn, J.F., and A.E. Dunham. 1983. On hypothesis testing in ecology and evolution. American Naturalist 122: 602-617.

Rigler, F.H. 1956. A tracer study of the phosphorus cycle in lake water. Ecology 37: 550-562.

Rigler, F.H. 1982. Recognition of the possible: An advantage of empiricism in ecology. Canadian Journal of Fisheries and Aquatic Science 39: 1323-1331.

Rigler, F.H. 1985. Nutrient kinetics and the new typology. Verhandlungen Internationale Vereinigung für Theoretische und Angewandte Limnologie 19: 197-210.

Sakamoto, M. 1966. Primary production by phytoplankton community in some Japanese lakes and its dependence on lake depth. Archiv für Hydobiologie 62: 1-28.

Salomonsen, J., and J.J. Jensen. 1990. Modellering af strukturdynamik i Glumsø under reduceret næringssaltbelastning. MS thesis. Freshwater Bio-

logical Laboratory, Hillerød, Denmark.

Sand-Jensen, K. 1995. Furesøen gennem 100 år. Naturens Verden 5: 176-187.

Sand-Jensen, K. 1997. The eutrophication of Lake Fure. Chapter 3 *in* K. Sand-Jensen and O. Pedersen, editors. Freshwater Biology – Priorities and Development in Danish Research. Gad, Copenhagen, Denmark.

Sas, H. 1989. Lake restoration by reduction of nutrient loading: expectations, experiences, and extrapolations. Academia Verlag Richarz Gmbh, Sankt Augustin, Germany.

Scheffer, M. and J. Beets. 1994. Ecological models and the pitfall of causality. Hydrobiologia 275/276: 115-124.

Schindler, D.W., E.J. Fee, and R. Ruszcynski. 1978. Phosphorus inputs and its consequences for phytoplankton standing crop and production in experimental lakes area and in similar lakes. Journal of Fisheries Research Board of Canada 35: 190-196.

Smith, V.H. 1982. The nitrogen and phosphorus dependence of algal biomass in lakes: An empirical and theoretical analysis. Limnology and Oceanography 27: 1101-1112.

Steemann-Nielsen. 1973. Hydrobiologi. Polyteknisk Forlag, Lyngby, Denmark.

Straskraba, M. 1980. The effects of physical variables on freshwater production: analyses based on models. Pages 13-84 *in* C.D. Le Cren and R.H. Lowe-McConnel, editors. The functioning of freshwater ecosystems. International Biological Programme 22. Cambridge University Press, Cambridge, England.

Thienemann, A. 1921. Seetypen. Die Naturwissenschaften 18: 1-3.

Thienemann, A. 1932. Tropische Seen und Seetypenlehre. Archiv für Hydrobiologie Supplementum 9: 205-231.

Thomann, R.V., D. DiToro, R.P. Winfield, and D.J. O`Connor. 1975. Mathematical modelling of phytoplankton in Lake Ontario. Part 1. U.S. Environmental Protection Agency, Corvallis, USA.

Ulanowicz, R.E. 1986. Growth and development. Ecosystem phenomenology. Springer Verlag. New York. USA.

Visser, A.W. and L. Kamp-Nielsen. 1996. The use of models in eutrophication studies. Pages 221-243 *in* B.B. Jørgensen and K. Richardson, editors. Eutrophication in coastal marine ecosystems. Coastal and estuarine studies 52. American Geophysical Union, Washington DC, USA.

Vollenweider, R.A. 1964. The correlation between area inflow and lake budget. Proceedings of the German Limnological Conference, Lunz. Germany.

Vollenweider, R.A. 1968. The scientific basis of lake

and stream eutrophication, with special reference to phosphorus and nitrogen as eutrophication factors. Technical report DAS/DSI68.27. The Organization for Economic Co-operation and Development, Paris, France.

Vollenweider, R.A. 1969. Possibilities and limits of elementary models concerning the budget of substances in lakes. Archiv für Hydrobiologie 66: 1-36.

Vollenweider, R.A. 1975. Input-output models. With special reference to the phosphorus loading concept in limnology. Schweizerische Zeitschrift für Hydrologie 37: 53-84.

Vollenweider, R.A. 1976. Advances in defining critical loading levels for phosphorus in lake eutrophication. Memorie dell`Istituto Italiano di Idrobiologia 33: 53-83.

Vollenweider, R.A. and J. Kerekes. 1982. Eutrophication of waters, monitoring, assessment and control. The Organisation for Economic Co-operation and Development, Paris, France.

9 Bacteria and dissolved organic carbon in lakes

By Morten Søndergaard

New measurements of bacterial secondary production have identified bacterial utilization of dissolved organic carbon (DOC) as a main metabolic pathway in aquatic systems. The carbon demand of pelagic bacteria in lakes can be as high as or higher than phytoplankton primary production, and input of organic substrates from terrestrial and littoral sources must be included to fully comprehend water column metabolism. Dissolved organic carbon is the largest organic carbon pool in most lakes, but temporal variations in concentrations are small compared with the high removal rates by bacteria. Only a small part of the large concentration of organic carbon is available for bacteria, despite the dominance of small molecules. Substrate for bacteria must accordingly be produced and removed at similarly high rates. Direct release of organic carbon by primary producers is an important source of substrate for bacteria, but recycling of organic carbon due to grazers and cell lysis can be equally important. The biological structure of the plankton is shown to be of decisive importance for the major routes of substrate production.

Decomposition of dead organic matter by microorganisms was described as a major metabolic pathway in lacustrine ecosystems in Lindeman's classical paper on trophodynamic interactions (1942). However, the dominant conceptual and predictive models of trophic relationships in the plankton remained characterized as a linear food chain among phytoplankton, zooplankton, and fish (Steele 1974, Cushing and Walsh 1976). This perception did not change until measurements of high production by the bacterioplankton were widely accepted.

A scientific revolution took place in the late 1970's when new methods made it possible to quantify bacterial abundance and production in aquatic environments. Measurements showed that bacterial biomasses were high and bacterial secondary production in the water column was 30 to 50% of phytoplankton primary production, and even higher in some lakes (reviewed by Børsheim 1992). These findings resulted in a rapid expansion of the research in aquatic microbial ecology and detailed studies on the role of heterotrophic bacteria and their substrate sources.

The demonstration of the importance of bacterioplankton as a particulate carbon link to higher trophic levels and a major respiratory sink also created a renewed interest in the production of bacterial substrates. Two major questions were, how dissolved bacterial substrates were produced and how bacterial biomass and production were controlled. Thus, research on the production and utilization of dissolved organic carbon (DOC) had a major renaissance due to the fact that only molecules in solution (monomers and oligomers) can be utilized by bacteria (Payne 1976). The initial interest in the role of DOC centred around dissolved compounds as a major nutrition for larger aquatic animals (Pütter 1911, Krogh 1930) and as a nutrient resource in lakes (Birge and Juday 1934).

In this chapter I will review current knowledge on bacterioplankton ecology, link bacterial activity with DOC dynamics, and present results on the most important processes producing bacterial substrates.

The microbial loop

August Krogh (1934a, b) predicted the presence of a 'microbial loop' in the oceans. However, it took another 40 years before Pomeroy (1974) published what became the classical conceptual food web model with bacteria at the base of a 'microbial loop' scavenging DOC with high uptake affinity and thereby return DOC to new bacterial biomass. The bacteria were eaten by grazers, notably flagellates, which were in turn eaten by larger ciliates in a complex food web dominated by small organisms (Sieburth 1976). Azam et al. (1983) in their model on

trophic relationships and interactions in the pelagic included bacterial production as a quantitatively important process fuelled by the release of DOC by phytoplankton, and to a much smaller extent the release of DOC by animals. Microheterotrophs were considered the dominant mortality factor for bacterioplankton in the early days of 'the microbial loop', but recent results have also shown vira and cladocerans to be major mortality factors (Bergh et al. 1989). *Daphnia* and other cladocerans can efficiently graze bacteria and occupy a dominant bacteriovorous position in freshwaters. A few *Daphnia* per litre can virtually remove all flagellates and ciliates and create a direct *route* of bacterial biomass to higher trophic levels (Riemann & Christoffersen 1993, Christoffersen 1997). Predatory bacteria are suggested as a mortality factor in very recent carbon flow scenarios (Cole and Caraco 1993).

The paper by Azam et al. (1983) was preceded by a series of results, which made it conceivable that bacterial production and metabolism were quantitatively important. High production of bacterioplankton was predicted by high uptake rates of amino acids

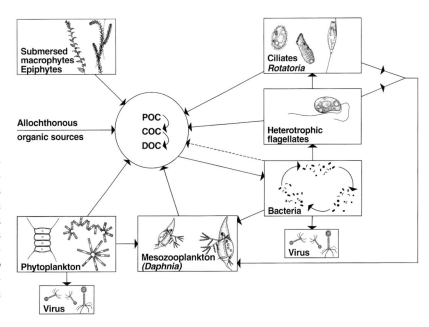

Figure 9.1. A conceptual model of potential substrate sources for bacterioplankton in lakes. The bacteria compartment includes 3 levels. Bacteria consume bacteria at two levels, either by predation or utilization of lysed cells.

 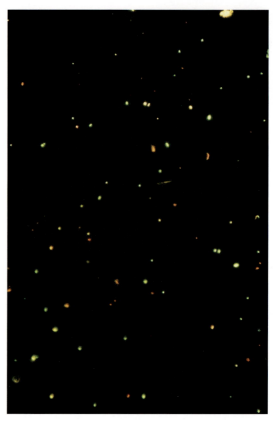

Figure 9.2. Micrographs of bacteria cells stained with DNA staining dyes and viewed in an epifluorescence microscope. A: DAPI stained sample from Lake Stigsholm with a morphologically diverse bacterial community.

B: Live/dead stained bacteria cells (Molecular Probes) in a culture from Lake Esrum. The green/yellow fluorescent cells were living cells with an intact membrane. The red 'cells' were dead, but contained DNA. Søndergaard (unpublished data)

and sugars measured with radiotracer technique (Wright and Hobbie 1966, Crawford et al. 1974). The use of single substrate uptake to measure bacterial activity could not solely be used to estimate total bacterial production, as the natural concentrations of the substrates in question were not known and the number of potential substrates were extremely high. The results, however, inclined Pomeroy (1974) to include DOC and microheterotrophic processes in his ocean food web model. Independent and reliable data showing dominance of respiration by very small organisms (< 5µm) in the sea (Williams 1981a) also suggested that the linear planktonic food web model had to be changed in

accordance with empirical results (Sieburth 1977, Williams 1981b).

A 'contemporary' planktonic food web (Fig. 9.1) includes loss of organic carbon from all trophic levels, and also from bacteria. Dissolved organic carbon lost at all trophic levels is recycled by heterotrophic bacteria. The organic carbon can initially originate from the watershed or be produced within the lake ecosystem. Bacterial mortality is caused by grazing and virus infection.

Bacterioplankton

The most important stepping stones for reaching the new insight concerning the role of bacterioplankton in organic carbon cycling were the development of relatively easy and accurate methods for measuring bacterial abundance and growth. Hobbie et al. (1977) developed a method for counting bacteria directly with an epifluorescent microscope after DNA staining with a fluorescent dye (Acridine Orange). Compared with the traditional microbiological plate-count technique, the direct count showed the natural abundance and biomass of bacterioplankton to be about 100-fold higher than previously found. A few years later Hagstrøm et al. (1979) and Fuhrman and Azam (1980) presented new methods for estimating bacterial growth and opened a new avenue for the investigation of bacterial production *in situ*. Calculated *in situ* growth rates were, however, much lower than expected from experiments with cultures. Recent results have shown the reason to be that most (50 to 95%) bacteria in a natural sample stained with the most commonly used DNA-stain DAPI (Fig. 9.2) are inactive or dead (Choi et al. 1996). A new staining method with an ability to distinguish between live and dead bacteria can most probably provide a direct count of active bacteria (Fig. 9.2). With this technique it is possible to measure the active biomass of the bacterioplankton and to calculate more accurate *in situ* growth rates.

A summary of bacterioplankton ecology
Research over the past 20 years has demonstrated the overall importance of bacterioplankton for pelagic carbon dynamic and metabolism. This knowledge can be summarized in several sections:

1. Bacterioplankton biomass and production is positively related to phytoplankton biomass and production, both across systems and within different experimental studies (Riemann and Søndergaard 1986a, Billen et al. 1990, Simon et al. 1992). However, the phytoplankton biomass varies about 50-fold more than bacterial biomass, both globally and locally. The control of bacterioplankton abundance and biomass is apparently much stronger than the control of phytoplankton. The empirical relationship between bacterial and phytoplankton production ratios across a trophic gradient is curvilinear with higher values in oligotrophic systems than in eutrophic systems, and decreasing along the trophic gradient (Cole and Caraco 1993). The reason for the curvilinear relationship is unknown, but as a consequence the planktonic recycling of organic carbon and nutrients is more efficient in oligotrophic systems. At chlorophyll levels below 0.5 and 1 mg m^{-3} in marine and limnetic systems respectively, the bacterial biomass generally exceeds the phytoplankton biomass (Simon et al. 1992).

2. Lakes support more bacterial biomass relative to phytoplankton than marine systems. The most obvious reason for the higher biomass in lakes is the input of organic carbon from other sources than pelagic primary production. Organic matter from the catchment (allochthonous) and from decomposing emergent and submersed vegetation in the littoral zone are major sources which support a high bacterial biomass in lakes (Wetzel and Søndergaard 1997).

3. Bacterioplankton production averages 30% of phytoplankton primary production in both lakes and seas (Cole et al. 1988), but sometimes exceeds phytoplankton primary production (Søndergaard 1993, Coveney and Wetzel 1995). Bacterial gross production (carbon demand), calculated from measurements of net production and an average growth efficiency of 40% (Søndergaard and Theil-Nielsen 1997), identify bacteria as one of the main pathways for organic carbon metabolism in aquatic systems.

4. Bacterioplankton biomass and production have seasonal patterns related to the development of phytoplankton biomass and production, but with temperature as the global modulator and predictor of activity (Ducklow & Carlsson 1992, Søndergaard

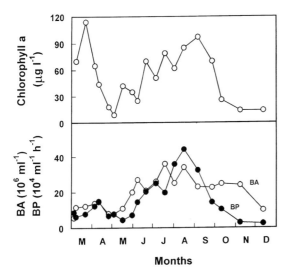

Figure 9.3. Seasonal variations in bacterial abundance and production (lower panel) and chlorophyll (upper panel) in Frederiksborg Slotssø, 1990. Bacterial production was measured by ^{3}H-thymidine incorporation into DNA. Data compiled from Søndergaard (1993) and unpublished.

1993). A typical seasonal pattern of bacterial abundance and production and phytoplankton biomass in a eutrophic Danish lake (Frederiksborg Slotssø) is shown in Fig. 9.3. Unlike the diatom spring bloom, the summer bloom of cyanobacteria is closely traced in time by the biomass and production of the bacterioplankton. The weak coupling of phytoplankton and bacteria in spring is probably due to low water temperatures and the greater loss of diatoms by sedimentation, which removes organic carbon from the water column and displaces decomposition to the sediment.

5. Bacterial mortality is primarily caused by grazers and viruses (Bergh et al. 1989). Flagellates (Fenchel 1982), ciliates (Christoffersen et al. 1990) and cladocerans (Sanders et al. 1989) are the most prominent grazers. Predatory bacteria are known, but their importance for bacterioplankton mortality is unknown (Cole and Caraco 1993).

6. The factors that control bacterioplank-

ton growth and biomass over time scales of days and weeks in different aquatic ecosystems are difficult to evaluate. Top-down control by grazers (Jeppesen et al. 1997) and resource control (bottom-up) by either inorganic nutrients (Morris and Lewis 1992) or organic carbon (Kirchman 1990, Kristiansen et al. 1992) have been found. Billen et al. (1990) argued strongly for a general resource control due to a positive, global relationship between bacterial biomass and production, but the hypothesis needs further confirmation. However, a bottom-up control near the base of food webs does agree with the general theory on trophic interactions (McQueen et al. 1986) and with the close correlation between abundance and production in the data from Frederiksborg Slotssø (Fig. 9.3).

7. Heterotrophic bacteria can only transport small molecules directly across the membrane (Payne 1976), so the action of enzymes working outside the cell (ectoenzymes) is a prerequisite for the decomposition and utilization of polymeric compounds (proteins, polysaccharides, and hemicellulose). Hydrolysis is considered the rate limiting step for polymeric organic matter decomposition and bacterial growth rates (Chrôst 1990, Billen 1991). The diversity and potential capacity of the ectoenzymes in a given bacterioplankton community are high (Martinez et al. 1996), but not necessarily expressed at all times. Several studies have verified that bacterial demand for energy and nutrition can be satisfied by the action of ectoenzymes processing polymeric DOC (Chrôst 1990, Middelboe et al. 1995) and directly from particle hydrolysis (Smith et al. 1992, Middelboe and Søndergaard 1995). Ectoenzymes are present in the periplasmic space or surface bound and cleave polymeric material to monomers or oligomers in close proximity to the cell. The hydrolysis products are readily available for transport across the cell membrane, but can also diffuse into the bulk water phase. The ectoenzymatic proteases of att-

Figure 9.4. Biomass (upper panel), production (mid panel) and potential aminopeptidase activity (lower panel) in Frederiksborg Slotssø, 1995. B_F = free bacteria in the filter size fraction < 20 μm, B_M = bacteria attached to *Microcystis*. Bacterial production was measured by ^{3}H-Leucine incorporation into proteins. Data supplied by Jakob Worm.

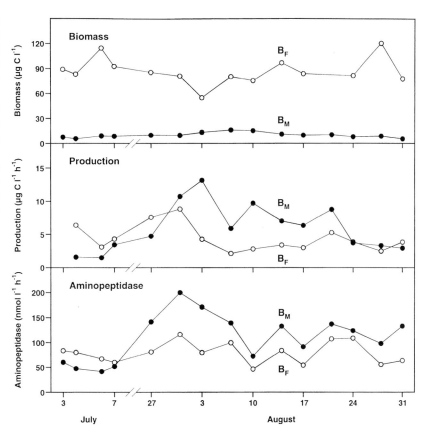

ached bacteria have been shown to release amino acids, which can support free-living bacteria taking up amino acids with a higher affinity but lower capacity than attached bacteria (Smith et al. 1992).

8. The small size (0.2 to 2 μm, but some are filamentous) and very large surface area per biovolume make bacteria very competitive at low nutrient concentrations because molecular diffusion totally dominates transport to such small particles (Kiørboe 1996). Colonization of phytoplankton cells and particularly buoyancy regulated cyanobacteria can be more advantageous than living as single, free cells. Phytoplankton cells produce organic carbon, and an attached bacteria thus lives in a microzone with a high and continuous production of substrates. The low sedimentation rate of cyanobacteria keeps the attached bacterial community suspended for

a long time, as opposed to attachment on other phytoplankton cell with high sedimentation rates. Bacteria associated with *Microcystis* in Frederiksborg Slotssø accounted for 10% of the bacterioplankton biomass, but for 55% of the production and 56% of the aminopeptidase activity (Fig. 9.4, Worm and Søndergaard 1997). The attached community is very active and adapted to the high-nutrient environment of a cyanobacterial colony. The constant number of bacteria cells per colony, despite a very fast growth (1-2 cell divisions per day), indicates that *Microcystis* is an incubator for the delivery of free-living bacteria.

High bacterioplankton production

An illustrative example to demonstrate the quantitative flow of carbon to the bacterioplankton in lakes was the seasonal study of

bacterial production in eutrophic Frederiksborg Slotssø and mesotrophic Buresø, Denmark. The annual bacterial carbon demands
(BP_{gross}) were about 250 g C m^{-2} in Frederiksborg Slotssø and 180 g C m^{-2} in Buresø,
compared with a net phytoplankton primary
production (PP) of about 400 and 130 g
C m^{-2} (Søndergaard 1993). Bacterial carbon
demand – even accounting for organic recycling – was high and stress the view of the
bacterioplankton as a pivotal biological component of organic carbon metabolism.

Evidently, comparisons between phytoplankton and bacterioplankton production are
only relevant if bacterial substrate supplied
by other primary producers than phytoplankton or by allochthonous sources are negligible. The ratio between phytoplankton primary production and planktonic respiration is
often less than 1 in oligotrophic and mesotrophic lakes (del Giorgio and Peters 1994),
and many humic lakes have very low ratios
because of high bacterial production due to
utilization of the allochthonous organic input
(Børsheim and Andersen 1987, Hessen 1992,
Tranvik 1992). Coveney and Wetzel (1996)
estimated that bacterioplankton production
exceeded that of phytoplankton between 1.3-
and 3.4-fold in oligotrophic Lawrence Lake
and suggested that the high bacterioplankton
production depended on dissolved organic
material originating from a very productive
benthos (macrophytes, epiphytes and micro-
phytobenthos), from emergent vegetation
fringing the lake, and from terrestrial sources.
The bacterioplankton carbon demand also
exceeded phytoplankton production in Buresø (see above), emphasizing the possible
influence of non-phytoplankton organic
sources.

Bacterial production measurements are
subject to greater uncertainty than phytoplankton primary production measurements.
Measurements of BP_{gross} can be biased by
many assumptions and conversion factors
before the 'final' values of carbon demand

are reached. The conversion of bacterial
abundance to biovolume and biomass, ^{3}H-
thymidine incorporation to cell production
(the most used method for measuring bacterial production), and estimates of bacterial
growth efficiency are among the most important potential errors. The methodological pit-
falls are reviewed by Riemann and Bell
(1990) and Søndergaard and Theil-Nielsen
(1997) and will not be treated here. However,
in most cases the ratios of BP_{gross}:PP are
neither very low (1-5%) nor very high (>>
100%). This result lends some confidence to
the methods used.

Dissolved organic carbon in lakes
Dissolved organic carbon is an operational
name and only relates to the method used for
measuring organic carbon in a filtrate free of
particles. The filter pore sizes used are normally between 0.2 and 0.5 µm. In reality
there is a continuum from truly dissolved
compounds to colloids and to particles
(Kirchman 1993), so the word dissolved is
somewhat misleading. Sub-micron organic
particles (0.2 – 0.7 µm) are accordingly
included in most DOC measurements and
can amount to 20% of the total DOC in very
eutrophic lakes. In most lakes and coastal
areas, sub-micron particles are about 5-8% of
total DOC (Middelboe and Søndergaard
1995). For the sake of simplicity, DOC is
here defined as the concentration of organic
carbon in a 0.7 µm filtrate (a Whatman GF/F
filter), and particulate organic carbon (POC)
is the material retained by this filter.

*Distribution and variability of organic
carbon pools*
The distribution of organic carbon in two
lakes of contrasting productivity is shown in
Fig. 9.5. In the eutrophic Frederiksborg Slotssø, the concentration of total organic carbon
during summer was about 18 mg C l^{-1}. The
concentration of DOC was 13 mg C l^{-1} of
which 30% were colloids (COC) and sub-

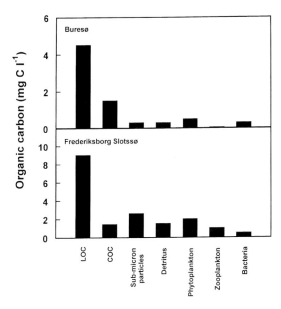

Figure 9.5. Distribution of organic carbon in Frederiksborg Slotssø and Buresø during summer. LOC = low molecular weight compounds (< 10,000 daltons), COC = colloidal organic carbon (> 10,000 daltons < 0.2 μm), sub-micron particles are > 0.2 μm < 0.7 μm, Detritus = dead particles > 0.7 μm. Data compiled from Søndergaard et al. (1995), Middelboe and Søndergaard (1995) and Søndergaard (unpublished data).

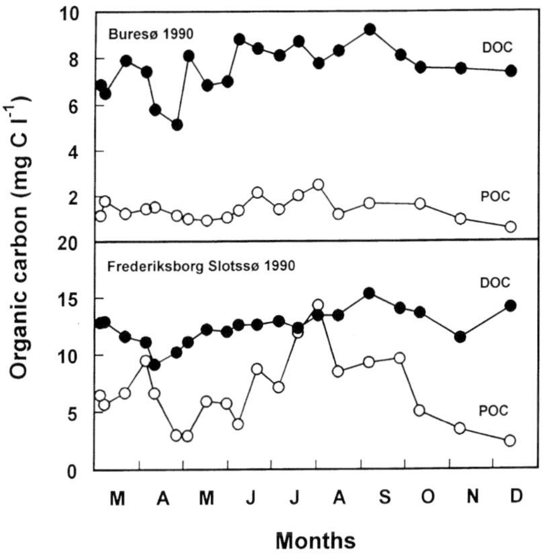

Figure 9.6. The seasonal variability of dissolved organic carbon (DOC) and particulate organic carbon (POC) in Buresø (upper panel) and Frederiksborg Slotssø (lower panel) in 1990. Søndergaard (unpublished data).

micron particles. Colloids are defined as molecules > 10,000 daltons (about 1 nm) and < 0.2 μm (Kirchman 1993). Particulate organic carbon (POC) is distributed among organisms and dead particles (particulate detritus). Phytoplankton is often the major component of POC in eutrophic lakes, but the distribution varies seasonally (Søndergaard et al. 1995). In the example from Frederiksborg Slotssø the DOC:POC ratio was about 3, but seasonally it ranged between 12 and < 1 (Fig. 9.6). In the mesotrophic Buresø, the concentrations of both DOC and especially POC were much lower than in Frederiksborg Slotssø, and the DOC:POC ratio varied seasonally between 5 and 12 (Fig. 9.6). The contribution of colloids and sub-micron particles to the total DOC was about 25% in Buresø, however, the distribu-

tion differing from the distribution found in Frederiksborg Slotssø. The concentration of COC was similar in the two lakes, but the concentration of sub-micron particles was 10-fold lower in Buresø. Molecules < 10,000 daltons dominated the DOC in both lakes. The dominance of small molecules in aquatic systems is universal with 70- 75% of DOC in molecules smaller than 10,000 daltons (Allen 1976, Benner et al. 1992, Søndergaard 1993).

The seasonal variability in the concentration of DOC is much smaller than for POC. In Frederiksborg Slotssø, POC varied 15-fold, while DOC only varied 1.5-fold (Fig. 9.6). The variability of POC was largely controlled by the dynamic behaviour of phytoplankton, zooplankton and bacteria. It is apparent that DOC was the largest organic pool with some seasonal variations and the highest values in autumn. When POC decreased by about 12 mg C l⁻¹ from August to November, DOC only increased a few mg C l⁻¹. The organic carbon was 'lost' either by sedimentation or by respiration in the water column. The loss of POC from the water column during this period was 60 g C m⁻², DOC gained 10 g C m⁻², and the bacterial carbon demand calculated with a growth efficiency of 40% (Middelboe and Søndergaard 1993) was 80 g C m⁻² (Fig. 9.3, data from Søndergaard 1993). In this simple and rather superficial analysis, there is no room for sedimentation and respiratory loss by grazing. As POC was totally dominated by cyanobacteria (*Microcystis spp.*), loss by grazing may be considered small (Christoffersen et al. 1990). We are still some 25 g C deficient in accounting for bacterial carbon demand. The deficiency cannot be explained by input of bacterial substrate from terrestrial sources and the littoral zone, because these sources are negligible in Frederiksborg Slotssø (Andersen and Jacobsen 1979). The apparent paradox can be solved, however, if organic recycling within the bacterioplankton community is included in the scenario. The use of three levels of bacterial interactions, as suggested in the model presented by Cole and Caraco (1993, see Fig. 9.1), decreased net bacterial carbon demand to about 45 g C m⁻². Organic recycling in this context involves bacteria consuming bacteria. Either bacteria utilize organic compounds released during cell lysis by viral attack and bacteriovorous zooplankton, or by bacterial predation on bacteria. We are now as close to an explanation of the fate of POC in the cyanobacterial bloom as we can get with these measurements, and there is room for sedimentation and non-bacterial respiratory losses. The recycling model does not change bacterial production and respiration, but the organic requirements from phytoplankton production decline.

Concentration of DOC along productivity gradients

DOC is the largest pool of organic carbon in almost any aquatic system, and it is positively related with phytoplankton productivity (Fig. 9.7). Bacterial production and biomass are also positively related with (lake) productivity (Billen et al. 1990, Simon et al. 1992), unless a substantial part of bacterial growth relies on allochthonous organic sources, like in many humic lakes (Tranvik 1992). Thus, a positive relationship between bacterial production and DOC should be expected. The slopes of the linear regression lines of DOC versus phytoplankton production for lakes and marine areas are not significantly different, but the intercept for lakes

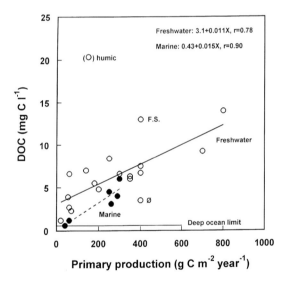

Figure 9.7. Relationship between phytoplankton primary production and summer concentrations of DOC in Danish lakes and coastal areas. Two data points from an oligotrophic blue-water region of the Mediterranean Sea were included. Linear regression equations were calculated separately for lakes and coastal waters (the humic lake was excluded). The calculated intercepts are theoretical values as the lowest possible concentration is the deep ocean of about 0.6 mg C l⁻¹. The annotated markers are: F. S. = Frederiksborg Slotssø and Ø = Ørn Sø. Data from Søndergaard (1990) and unpublished.

is higher (Fig. 9.7). It should be noted that the concentration of DOC in deep oceanic waters and in some groundwater streams sets the lower limit of about 0.4 to 0.6 mg C l^{-1}. Furthermore, there is a tendency for the oligotrophic, seepage lakes to approach oceanic DOC concentrations and the high-productivity marine locations to approach the DOC level in lakes. The DOC rich marine locations include the Danish coastal waters and the Baltic Sea which, like the majority of the Danish lakes, have a large terrestrial input providing a higher background DOC-level than found in aquatic systems totally dominated by autochtonous organic sources. The general conclusion is that the concentration of DOC is controlled by the level of primary production and modulated by external sources. Dissolved organic carbon entering lakes from terrestrial catchments usually contains aromatic compounds that have largely resisted microbial decomposition in soils and rivers (Kaplan and Newbold 1993). These compounds increase the level of DOC in lakes and coastal areas, but are not directly related to system productivity.

Decomposition of phytoplankton and submersed macrophytes is characterized by the fact that a small, but substantial, amount of the organic carbon remains as un-decomposed DOC for long periods and can be classified as recalcitrant (Chen and Wangersky 1996), even after UV-B exposure (Thomas and Lara 1995), which may otherwise aid decomposition of organic matter (Wetzel et al. 1996). If the recalcitrant organic products are a constant percentage of the biota, the relationship between productivity and DOC should be positive and linear unless special conditions (e.g. variable DOC sources, hydraulic residence time) generate excess DOC or remove more DOC than expected. The marked data points in Fig. 9.7 with unexpected high and low DOC are from lakes with long and short hydraulic retention times respectively. The high DOC concentra-

tion in Frederiksborg Slotssø, totally dominated by phytoplankton production and with a long hydraulic residence time (> 4 years), is in accordance with recent results on slow microbial degradation of part of the DOC-pool in cultures of natural assemblages of phytoplankton (Chen and Wangersky 1996). Thus, about 15% of the DOC had a decay rate of only 0.003 day^{-1}. A high steady state concentration could accordingly be present in a lake with a low dilution rate and a high DOC input, like Frederiksborg Slotssø. The low DOC concentration in Ørn Sø accords with a very short hydraulic residence time (19 days), but the lake also has an extensive precipitation of particles associated with a high input of iron. DOC might co-precipitate with the iron.

DOC concentrations in humic lakes are high and dominated by the terrestrial input of coloured and aromatic compounds. The highest concentration measured in a Danish lake is about 20 mg C l^{-1}, but values as high as 50 mg C l^{-1} are known from Sweden (Tranvik 1988). The DOC range observed for non-humic Danish lakes is between 2 and 15 mg C l^{-1}. The oceanic range is from 0.6 mg C l^{-1} in the deep aphotic zone to 1.5 mg C l^{-1} in the photic zone, while coastal waters have higher concentrations depending on terrestrial inputs and autochthonous production.

Temporal DOC variations

The concentration of DOC is so large that, theoretically, it should be able to sustain bacterial production for a long period. In Frederiksborg Slotssø and Buresø the average theoretical turnover time for the entire DOC pool would be 90 and 140 days respectively, and 15-20 days during peak bacterial activity. However, the relatively static behaviour of the DOC pool (Fig. 9.6), despite the dominance of small molecules (Fig. 9.5) and the high bacterial carbon demand, implies that a high flux of labile DOC is kept at low concentrations, and that the large DOC pool is dominated by recalcitrant and refractory compounds with balanced input and removal. Microbial utilization is thought to be the main process in removing DOC, but other processes are implied. Coagulation of DOC to POC and subsequent sedimentation (Kiørboe 1996) and photochemical oxidation (Granéli et al. 1996) are probably the two most important abiotic removal processes.

It is pertinent here to remind the reader that production and removal of DOC is not always in perfect balance. Measurements in the productive layer of open oceans have shown accumulation of carbon-rich dissolved organic material (C/N ratio of twice the Redfield value of 6.6) in connection with phytoplankton blooms and over time scales of 50-100 days (Carlson et al. 1994, Williams 1995). The implication of this finding is that a large amount of the carbon exported from the euphotic zone is in a dissolved form and not in particles as currently believed. The apparent imbalance between production and removal has been hypothesized to be caused by either nitrogen limitation of the bacteria or by a general low availability of DOC for bacterial decomposition.

Seasonal accumulations of DOC have not, to my knowledge, been convincingly described for lakes. One reason is that allochthonous DOC generally increases the background concentration, and small changes in concentration against a high background value are very difficult to detect. Another reason is that temporal variability in terrestrial runoff is high and can be very difficult to account for in an in-lake DOC budget (Kaplan and Newbold 1993). Finally, compared with the oligotrophic oceans, lakes are generally nutrient-rich, and although phosphorus limitation of lake water bacteria has been described (Morris and Lewis 1992), nutrients might not limit DOC decomposition in most lakes.

The above arguments can only lead to the conclusion that the bulk amount of DOC is recalcitrant or refractory, and only a small fraction measured at any given time is readily available to bacterial utilization. Is this true, and if so, how is a high bacterial production obtained?

Sources of bacterial substrate

The sources of dissolved bacterial substrate are either allochthonous or autochtonous (Fig. 9.1). The autochtonous source derives from the primary producers (phytoplankton, macrophytes and epiphytes), and three major routes have been suggested to feed bacteria DOC at the base of the microbial loop (Riemann and Søndergaard 1986b). 1) Loss of extracellular organic carbon (EOC) by healthy phytoplankton was for a long time considered the quantitatively most important route (Søndergaard and Jensen 1986), but the position of EOC has been replaced by 2) the supply of substrates produced during grazing at all trophic levels (Jumars et al. 1989). 3) The third pathway was autolysis of dead organisms, including bacteria. 4) A pathway not mentioned by Riemann and Søndergaard (1986b) is the direct solubilization of particulate detritus by attached bacteria. Although recycling of organic carbon within the bacterial community, either by predation or utilization of dead bacteria, cannot be viewed as a primary source, it has to be considered as a substrate supply for bacterial production (Cole and Caraco 1993). Before we examine the abovementioned biological processes in some detail, the hypothesis that the DOC pool is mainly refractory should be tested.

The bioavailability of DOC

When a water sample is filtered and inoculated with an assemblage of natural bacteria, the standing stock of DOC can in most cases support growth. The addition of inorganic nutrients is sometimes needed to overcome nutrient limitation (Søndergaard and Middelboe 1995). The only known exceptions are the apparent lack of growth in water from the

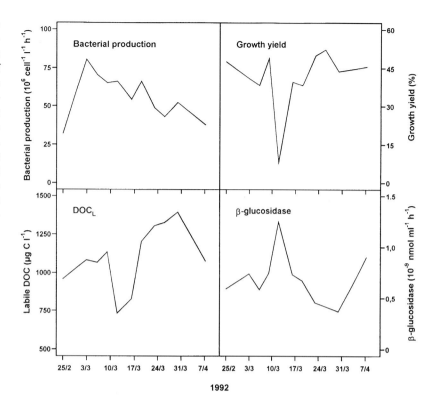

Figure 9.8. Measurements of bacterial production, bacterial growth yield, concentration of labile DOC, and potential β-glucosidase activity during the diatom spring bloom in Frederiksborg Slotssø in 1992. Data from Middelboe and Søndergaard (1993) reproduced with permission from the American Society for Microbiology.

deep oceans (Barber 1968) and in patches of DOC depleted ocean surface waters (Carlson and Ducklow 1996). The yield of bacterial biomass plus the carbon lost by respiration is a measure of labile organic carbon (DOC_L). Most bioassays from lakes, rivers, and marine areas have unequivocally demonstrated low values of DOC_L relative to DOC. The median DOC_L was 12% in lakes, 25% in rivers, and 14% in marine areas (reviewed by Søndergaard and Middelboe 1995). The high value in rivers was explained by anthropogenic influences (e.g. organic carbon from wastewater).

Frederiksborg Slotssø can provide an example for the temporal variability of DOC_L ranging from 0.6 to 1.5 mg C l^{-1} and averaging 9% of DOC (Søndergaard and Borch 1992). Temporal changes in concentrations could account for less than 15% of the bacterioplankton carbon demand, and even a complete exhaustion of DOC_L could only sustain

bacterial production for 2-3 days (Søndergaard et al. 1995). The only known case where the decrease of the standing pool of DOC_L over time could explain all bacterioplankton carbon demand was found in a study in Frederiksborg Slotssø with more frequent sampling (Middelboe and Søndergaard 1993). Within three days, DOC_L decreased 0.4 mg C l^{-1} concomitant with an increase in the cell specific ectoenzymatic activity and a lowering of the growth efficiency from 40 to 10% (Fig. 9.8). About 40% of the labile DOC pool was mobilized for growth by ectoenzymes, but at the expense of a very low growth efficiency. Bacterial production steadily declined over a period encompassing the dynamic changes in growth efficiency and DOC_L concentrations (Fig. 9.8).

The chemical composition of the DOC_L used is not known, but the measured response in specific β-glucosidase activity suggests polysaccharides released from a declining diatom community. Diatoms are known to release large quantities of polysaccharides during senescence and to enhance bacterial β-glucosidase activity and affinity (Fogg 1983, Chrôst 1989). Concomitant measurements of mono- and polysaccharides over the same period showed large fluctuations in dissolved polysaccharides (700-1200 µg C l^{-1}) and a constant decline in monosaccharides from about 500 to 200 µg C l^{-1} (Middelboe et al. 1995). Such data are supportive, but not conclusive, to the identification of the specific compounds utilized. The dynamic situation was short-lasting; within a week DOC_L, the specific ectoenzymatic activity, and the bacterial growth yield had all returned to previous levels (Fig. 9.8).

Although the presented results show the DOC_L to have a relatively dynamic behaviour, the overall conclusion is that bacterial utilization controls the variations. The labile fraction of DOC turns over within few days, and a high bacterial activity must rely on a continued flux of organic carbon, which in turn is kept at low concentrations by efficient hydrolysis and uptake systems.

Release of extracellular organic carbon (EOC) by primary producers

Loss of organic carbon by apparently healthy algae and submersed macrophytes has been known for some time (Harder 1917, Krogh et al. 1930), but unfortunately these early results were biased by methodological problems. With the introduction of the ^{14}C-method (Steemann Nielsen 1952) it became possible – with some confidence – to investigate the release of extracellular organic carbon under natural conditions. In his original articles Steemann Nielsen mentioned that the production of radiolabelled dissolved organic matter during incubation could be a potential problem and would, if not accounted for, lead to an underestimation of primary production. The results he obtained in algal cultures made him conclude that such organic losses

were a minor problem. Later it was realized that most microalgae in culture seem to lose much less EOC (1-2% of primary production) than natural phytoplankton assemblages, although severe nutrient and light stress in cultures can enhance loss rates (reviewed by Fogg 1983, Søndergaard and Jensen 1986, Williams 1990).

Measurements of very high planktonic EOC loss rates with the ^{14}C-method in the late 1960's and early 1970's – some as high as 50-70% of the particulate primary production – aroused suspicions of severe methodological problems and errors. Sharp (1977) examined previous results and experimental protocols and especially criticized the lack of proper corrections for large blank values and weak counting statistics in samples with very low counts originating from radiolabelled EOC. The international scientific community responded to the critique, and new experimental protocols were developed. Although most published values of EOC were lower after 1977, the process remained as a real phenomenon. In the most recent cross-system review, Baines and Pace (1991) estimated from 225 observations that the average per cent extracellular release (PER) was 13% of total carbon fixation. Previous suggestions of a general decline in PER with increasing productivity were not supported by the data. Interestingly though, PER tended to be very low in highly eutrophic lakes. This result might explain the curvilinear relationship between the ratio of bacterial and phytoplankton production versus system productivity. The data analyzed by Baines and Pace (1991) also suggested that EOC accounted for less than half of the organic carbon required for bacterial growth in most pelagic systems.

The release of EOC has, since the first reliable measurements of bacterial production, been forwarded as a prime source of substrate for the bacterioplankton (Azam et al. 1983). Many of the released products are

Table 9.1. Carbon flow scenario in experimental enclosures in eutrophic Lake Hylke (20 May – 2 June, 1983). Data compiled from Riemann and Søndergaard (1986a). Enclosure treatments are explained in the text. PP = particulate phytoplankton production, BP = bacterial production, EOC = bacterial uptake of and EOC left in solution, Zooplankton = calculated DOC released by zooplankton, Lysis = estimated bacterial production from cell lysis. Units: g C m^{-2} period^{-1}.

Enclosure treatment	PP	BP	EOC	Zooplankton	Lysis
Control	3.0	1.8	0.3	1.5	0
Fish	7.8	2.9	1.6	0	1.3
Fish + nutrients	9.7	2.9	2.2	0	0.7

small molecules < 1,000 daltons (Søndergaard and Schierup 1982, Sundh 1989). Amino acids and different types of saccharides seem to predominate, but other compounds have also been identified (Søndergaard and Jensen 1986). Extracellular organic carbon is generally characterized by its availability for fast and efficient bacterial uptake (Cole et al. 1982, Chrôst and Faust 1983, Søndergaard et al. 1985) and some of the released products are species specific. Prolonged exposure of a natural bacterioplankton community to a given pool of algal EOC selects for bacteria physiologically adapted for growth on compounds available in that pool (Bell 1983). The demonstration of physiological adaptation to a given algal EOC pool suggests transport-limited uptake and that bacterial growth becomes limited by the rate of uptake of available substrates (Jensen 1985).

The quantitative importance of EOC for growth of the bacterioplankton is experimentally difficult to determine *in situ*, although many have tried. Perhaps the most important reason is that in a tracer experiment with $^{14}CO_2$ and several interacting biological compartments (phytoplankton, zooplankton, bacteria), the pathways of a labelled organic molecule cannot be traced. Labelled bacteria recovered after incubation for several hours might have utilized both EOC and radiolabelled organics released from grazed algal

cells. A very practical problem is the separation of the bacterioplankton from other particles, including picoalgae, which overlap in size.

Published results on the importance of EOC for bacterial production are extremely variable and range from almost zero to > 100% (Søndergaard and Jensen 1986, Sundh 1991). To circumvent some of the abovementioned problems, Riemann and Søndergaard (1986a) tried experimentally to show the importance of the pelagic biological structure and trophic interactions for the production of bacterial substrates. In a mesocosm study they created dominance of three different pathways of organic carbon to bacteria: EOC, zooplankton grazing, and cell lysis. The different biological structures were achieved by removing mesozooplankton by means of planktivorous fish (Enclosure: Fish, where cell lysis induced by nutrient limitation was considered the dominant loss process), amendments with both fish and nutrients (Enclosure: Fish + nutrients, where release of EOC dominated), and a Control enclosure with high mesozooplankton grazing (dominance of zooplankton mediated DOC release).

Measurements of phytoplankton and bacterial production, and bacterial uptake of EOC integrated over time and depth, are presented in Table 9.1 together with calculations of bacterial substrate originating from meso-zooplankton grazing and cell lysis. Assuming insignificant utilization of the ambient DOC pool, the bacterioplankton carbon-budgets were balanced by cell lysis. Release of organic carbon by the mesozooplankton was estimated as 30% of the measured ingestion (Lampert 1978).

The main conclusions from this experiment were: 1) Release of EOC was the dominant process providing bacterial substrate during a period with a growing phytoplankton community and low grazing losses (enclosure amended with fish and nutrients), 2) Organic recycling by mesozooplankton grazing was the dominant bacterial substrate source in periods with top-down control of the phytoplankton (Control enclosure), and 3) In the experiment with low grazing and a decreasing phytoplankton biomass due to nutrient limitation, cell lysis seemed to be the major route of organic carbon to the bacterioplankton. The interpretation of the scenario is open to criticism, as carbon for bacterial respiration was not accounted for. However, the conclusions were supported by a detailed carbon budget obtained during a spring diatom bloom and the subsequent clearwater phase in Frederiksborg Slotssø (Markager et al. 1994).

Release of EOC by phytoplankton is real, but highly variable, and contradictory results concerning the actual quantity of EOC make generalizations difficult. One reason for the uncertainty is that we do not know the physiological mechanisms or environmental factors in control of the release (Williams 1990). It is unclear whether EOC release is related to the photosynthetic activity of phytoplankton or is simply controlled by diffusion through a semipermeable membrane and thus related to the surface:volume ratio of an algal cell. The latter model was suggested but not proven by Bjørnsen (1986). Accordingly, the smallest algae (picophytoplankton < 2 μm in diameter) with the largest surface area to vol-

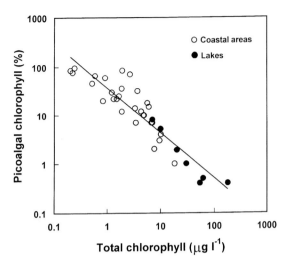

Figure 9.9. Relationship between total chlorophyll and the relative contribution of picophytoplankton. Surface water samples from Danish lakes and coastal waters. Data combined and redrawn from Søndergaard (1991) and Søndergaard et al. (1991).

ume ratio should have the highest loss per unit of biomass (Bjørnsen 1986).

Empirical and experimental support of Bjørnsen's hypothesis is scarce. Results showing substantially higher relative release rates of EOC by picoalgae (29% of primary production) than by larger size classes (4-5%) were recently presented by Malinsky-Rushansky and Legrand (1996). The results support the measurements of relatively higher EOC values in oligotrophic as opposed to eutrophic systems (Thomas 1971, Søndergaard and Jensen 1986). Picophytoplankton are extremely important in oligotrophic areas, and their importance decreases along a productivity gradient (Fig. 9.9), yet a high release of EOC in these environments due to the dominance of picophytoplankton is difficult to separate from the influence of environmental variables. It should also be emphasized that Baines and Pace (1991) did not find a higher relative release in oligotrophic areas, nor evidence to support the

Figure 9.10. The development of chlorophyll and per cent extracellular release (PER) by phytoplankton in a mesocosm experiment in Lake Almind. Two 7 m³ mesocosms were added nutrients (N + P) and two served as controls. Redrawn after data from Søndergaard et al. (1988).

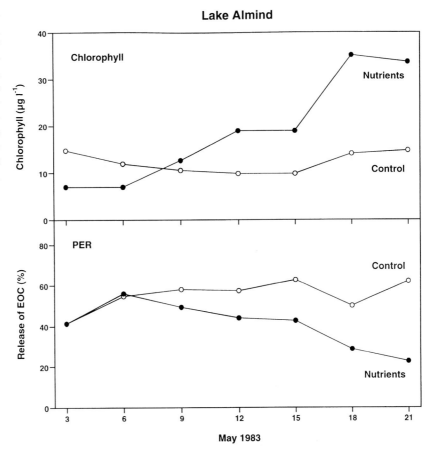

notion that the release is controlled by algal size.

The availability of nutrients has also been hypothesized to influence the release of EOC (Fogg 1983). Nutrient limitation should result in a relatively high release. An experiment in oligo-mesotrophic Lake Almind, Denmark, showed that the addition of inorganic nutrients (N and P) resulted in an increase in the phytoplankton biomass and a decrease in percent release (Fig. 9.10, Søndergaard et al. (1988)). As the species composition did not change during the experiment, it was suggested that a nutrient-related change in the physiological behavior of the phytoplankton with respect to release of EOC was induced. A similar influence on EOC release by phosphorus-limited phyto-

plankton was found in a recent study in the Northern Adriatic Sea (Obernosterer and Herndl 1995). However, the severe nutrient limitation apparently prevented bacterial utilization of the released EOC, which coagulated to particle-like mucilage.

Baines and Pace (1991) concluded that phytoplankton EOC products, although of quantitative importance, on an average can support less than half the carbon required to account for the bacterioplankton production. This result based on seasonal averages does not contradict the many published results on a high quantitative importance of EOC during shorter time periods (Cole et al. 1982, Søndergaard et al. 1988, Lignell 1990, Sundh 1991).

Despite the fact that release of EOC by submersed macrophytes has not been investigated with the intensity and effort devoted to studying the phytoplankton it is safe to conclude that all published measurements have indisputably shown the percentage release to be low. Submersed angiosperms release 1 and maximally 5% of their primary production as EOC (Søndergaard 1981). In contrast, high release (15-20%) by marine macroalgae is not uncommon (Søndergaard 1990).

Zooplankton grazing as a DOC source

The apparent inability of EOC release to support the major part of measured bacterioplankton production left a gap in the microbial loop with respect to bacterial substrate. Jumars et al. (1989) tried to close the microbial loop by suggesting that incomplete ingestion, digestion and absorption by animals, notably zooplankton, were the key processes in bacterial substrate production. The organic material left the gut, or the vacuole in protists partly digested and was readily available for bacterial utilization. Their idea was not new (Lambert 1978, Roman et al. 1988), but the conceptual principles to explain high bacterial production at times of high zooplankton activity provided clarity. As Jumars et al. (1989) rejected any measured EOC release rates above 5-10% of the primary production, they created room for alternative pathways for DOC production. Perhaps their strongest argument which cast doubt on the importance of EOC was the impossible task of interpreting radiotracer studies with several interacting compartments.

Previous experiments have demonstrated a significant loss of dissolved material in association with mesozooplankton feeding on phytoplankton (Lampert 1978, Güde 1988, Arndt et al. 1992). The loss has been ascribed to direct excretion, leakage from faecal pellets, and from broken algae during 'sloppy feeding'. The experimental designs have often included grazing on radiolabelled phytoplankton followed by a subsequent detection of labelled dissolved organic prod-

ucts and bacteria. By this method it is possible to quantify the amount of carbon released in relation to the amount ingested. The results range widely, but 10-20% of ingested algal carbon is an average observation for species of *Daphnia* (Lampert 1978, Olsen et al. 1986).

A more direct method for studying the importance of zooplankton-mediated DOC release is to use bacterial growth as a detector. Different bacterial yields in experiments with zooplankton, zooplankton plus algae, and algae alone can provide an estimate of how much substrate is made available by grazing zooplankton. Combinations of trophic complexity in serially connected chemostats were used by Arndt et al. (1992) to demonstrate enhanced bacterial growth in chemostats with herbivorous zooplankton. One weakness of this approach is that many herbivores also graze the bacteria, so a strict quantitative interpretation is not possible, but

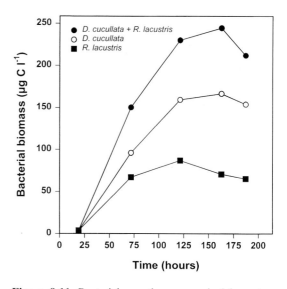

Figure 9.11. Bacterial growth response in lake-water after 3 hours of *Daphnia cucullata* grazing on *Rhodomonas lacustris* (ZP), and in water with either *Rhodomonas* or *Daphnia* added. The water was sterile filtered before an inocculum of natural bacteria was added. Reconstructed from Hygum et al. (1997).

tentative budgets of carbon flow can be constructed. Other studies have used specific organic compounds to trace the effects of grazers. Increased concentrations of dissolved amino acids have been observed as a function of zooplankton activity (Riemann et al. 1986), but results from *in situ* measurements have been unclear. Most probably due to the simultaneous and very efficient bacterial uptake of amino acids.

A 'classical' type of experiment using bacterial yield to detect zooplankton mediated release of DOC is presented in Fig. 9. 11. *Daphnia cucullata* isolated from Frederiksborg Slotssø grazed on *Rhodomonas lacustris* for 3 hours, and the subsequent increase in bacterial biomass was followed over time. Parallel samples with algae and zooplankton as sole organisms in filtered and aged water were used as controls. There was a significantly higher bacterial biomass in the zooplankton plus algae bottle than in the algae bottle. The presence of *Daphnia* clearly increased the concentration of substrate available for bacterial growth, and the release averaged 57% (range 26-78%) of the ingested algal carbon. A similar experimental protocol used *in situ* during a clear-water phase in Frederiksborg Slotssø demonstrated that the mesozooplankton community released a high-quality substrate for bacteria. Combining these results with the measurements of algal biomass, bacterial production and estimated community grazing rates in May (Markager et al. 1994), it was possible to account for 35-60% of the bacterial carbon demand by the zooplankton-mediated release of substrate and thereby verify the potential importance of this carbon pathway.

Cell lysis

Loss of control of the cell membrane is immediately followed by a diffusive efflux of water-soluble organic material. The disintegration of cells can be aided by autolysis caused by the enzymes in the cell. The rapid

loss of organic material following cell death has been demonstrated for all types of aquatic organisms. This is probably why DOC release by lysis has been an attractive explanation for bacterial substrate sources, although knowledge of the process under natural conditions is limited. Support for fast DOC release and subsequent bacterial utilization has been obtained in experiments in which the organisms were killed by heat treatment or freezing or exposed to toxic fumes and then allowed to lyse in sterile water or in water with bacteria. Such experimental results are difficult to extrapolate to natural conditions. However, empty phytoplankton cells infested with bacteria and fungi are routinely observed under the microscope, so death and lysis are real phenomena, probably caused by environmental stress,

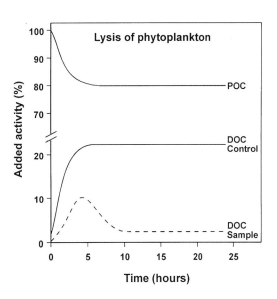

Figure 9.12. Release of DOC by an immobilized, dead phytoplankton community from Lake Mossø in August 1984. The algae were labelled to isotopic equilibrium before being immobilized on a filter and killed by acid fumes. Release of radiolabelled DOC was followed over time in bottles with sterile lake-water (Control) and water with a natural bacterial community. All values are unified to the particulate radioactivity at time zero. Reconstructed from Hansen et al. (1986) and Riemann and Søndergaard (1986b).

nutrient deficiency or attack by pathogenic microorganisms.

The initial release of dissolved organic compounds during cell lysis is fast. This process is not temperature-dependent and has been explained as simple water extraction and diffusion (Cole et al. 1984). The released products are effectively utilized by bacteria. The speed of the following decomposition process is largely controlled by the invasion of bacteria and fungi, their ectoenzymatic activity, the temperature and the chemical nature of the organic material. The C:N ratio and the amount of structural compounds are of decisive importance for the decay rates (Enriquez et al. 1993).

The fast release of DOC by dead algal cells and the DOC uptake by bacteria is shown in Fig. 9.12. Initially, the bacterial uptake capacity was lower than the loss rate, but as soon as the loss rate decreased, the DOC concentration was reduced to a constant level of about 20% of the steady state concentration in the sterile control bottle. This value represented the recalcitrant pool of DOC from cell lysis and was identical to the level of recalcitrant DOC measured by Chen and Wangersky (1996) in algal cultures. The subsequent decomposition proceeded much more slowly and was due to bacterial colonization of the dead algal cells. Upon cell death, from 20 to 40% of the phytoplankton carbon was lost to the water within a few hours, and most of the DOC was immediately utilized by bacteria. In cyanobacterial communities the initial loss amounted to about 43% of the total cell carbon, while about 20% was lost for communities dominated by diatoms and green algae.

In recent experiments, the ratio of dissolved and particulate esterase has been used to estimate autolysis during a *Phaeocystis* bloom in the North Sea (Brussaard et al. 1995). During the bloom, the entire bacterial production and carbon demand could be explained by autolysis of *Phaeocystis* caused by virus. Contrary to this, the loss of cells during the diatom bloom was due to sedimentation and not to lysis, and the bacterial carbon demand was explained by release of EOC (Brussaard et al. 1995). These results confirm that cell lysis can be a major source of dissolved organic carbon to the bacterioplankton and support the results from Lake Hylke (see Table 9.1) and the interpretation of the origin of bacterial carbon sources during a diatom bloom and a clear-water phase in Frederiksborg Slotssø (Markager et al. 1994).

Concluding remarks

Bacterioplankton and DOC have a central position in current conceptual models of pelagic carbon metabolism. All methodological problems concerning measurements of bacterial growth, biomass and production have not yet been solved but seem to be constrained within an ecologically reasonable error.

The ecological importance of bacteria is based on their utilization of dissolved organic material and is dependent on an astonishing diversity of ectoenzymes able to decompose almost any complex molecule in nature – given sufficient time. The diversity of ectoenzymes is a unique feature defining the functional importance of bacteria. However, it is not known if bacteria in aquatic systems are functionally diverse or if most species have the capability to express a broad spectrum of ectoenzymes. The question is whether planktonic bacteria are specialists or generalists? Do they have a predictable seasonal species succession like phytoplankton and zooplankton, and are different species (functional groups) to be found in different environments? The current working hypotheses suggest that the answer to each of these questions is yes, but experimental results are needed. Future research on bacterioplankton, therefore, has to include work at the level of single cells and enable the identification of the function(s) of active bacteria.

Most of the DOC in aquatic systems is recalcitrant, but with a dynamic component, as organic carbon metabolized by bacteria has to be part of the DOC pool for shorter or longer periods. Aside from the most obvious function as a substrate for bacterial uptake, DOC has other important functions in the aquatic ecosystem. The concentration and colour of DOC control the light climate in lakes and oceans, except for situations with a high concentration of particles. DOC chelates metals and facilitates their movement through the environment in solution, and DOC is the principal UV-B absorbing component. Changes in climate and exploitation of land can alter the DOC concentrations in lakes and consequently these important functions (Schindler et al. 1996). For the aquatic microbial ecologist, however, the focus remains on understanding the interactions between DOC and bacteria.

Literature cited

Allen, H.L. 1976. Dissolved organic matter in lakewater: Characteristics of molecular weight size-fractionations and ecological implications. Oikos 27: 64-70.

Andersen, J.M., and O. Jacobsen. 1979. Production and decomposition of organic matter in eutrophic Frederiksborg Slotssø, Denmark. Archiv für Hydrobiologie 85: 511-542.

Arndt, H., H. Güde, M. Macek, and K. O. Rothhaupt. 1992. Chemostats used to model the microbial food web. Archiv für Hydrobiologie, Ergebnisse der Limnologie 37: 187-194.

Azam, F., T. Fenchel, J. G. Field, J. S. Gray, L.A. Meyer-Reil, and F. Thingstad. 1983. The ecological role of water-column microbes in the sea. Marine Ecology Progress Series 10: 257-263.

Baines, S. B., and M. L. Pace. 1991. The production of dissolved organic matter by phytoplankton and its importance to bacteria: patterns across marine and freshwater systems. Limnology and Oceanography 36: 1078-1090.

Barber, R. T. 1968. Dissolved organic carbon from deep water resists microbial oxidation. Nature 220: 274-275.

Bell, W. H. 1983. Bacterial utilization of algal extracellular products. 3. The specificity of algal-bacterial interaction. Limnology and Oceanography 25: 1007-1020.

Bergh, Ø., K. Y. Børsheim, G. Bratbak, and M. Heldal. 1989. High abundances of virus found in aquatic environments. Nature 340: 467-468.

Benner, R., J. D. Pakulski, M. McCarthy, J. I. Hedges, and P. G. Hatcher. 1992. Bulk chemical characteristics of dissolved organic matter in the ocean. Science 255: 1561-1564.

Billen, G. 1991. Protein degradation in aquatic environments. Pages 123-143 *in* R. J. Chróst, editor. Microbial enzymes in aquatic environments. Springer-Verlag. New York.

Billen, G., P. Servais, and S. Becquevort. 1990. Dynamics of bacterioplankton in oligotrophic and eutrophic aquatic environments; bottom-up or top-down control. Hydrobiologia 207: 37-42.

Birge, E.A., and C. Juday. 1934. Particulate and dissolved organic matter in inland lakes. Ecological Monographs 4: 440-474.

Bjørnsen, P.K. 1986. Phytoplankton exudation of organic matter: why do healthy cells do it. Limnology and Oceanography 33: 151-154.

Børsheim, K.Y. 1992. Growth rate and mortality of bacteria in aquatic environments. Doctoral thesis, The University of Trondheim, Norway.

Børsheim, K. Y., S. Andersen. 1987. Grazing and food selection by crustacean zooplankton compared to production of bacteria and phytoplankton in a shallow Norwegian mountain lake. Journal of Plankton Research 9: 367-379.

Brussaard, C.P.D., R. Riegman, A. A. M. Noordeloos, G. C. Cadée, H. Witte, A. J. Kop, G. Nieuwland, F. C. van Duyl, and R. P. M. Bak. 1996. Effects of grazing, sedimentation and phytoplankton cell lysis on the structure of a coastal pelagic food web. Marine Ecology Progress Series 123: 259-271.

Carlson, C.A., and H. W. Ducklow. 1996. Growth of bacterioplankton and consumption of dissolved organic carbon in the Sargasso Sea. Aquatic Microbial Ecology 10: 69-85.

Carlson, C.A., H. W. Ducklow, and A. F. Michaels. 1994. Annual flux of dissolved organic carbon from the euphotic zone in the northwestern Sargasso Sea. Nature 371: 405-408.

Chen, W., and P. J. Wangersky. 1996. Rates of microbial degradation of dissolved organic carbon from phytoplankton cultures. Journal of Plankton Research 18: 1521-1533.

Choi, J.W., E. B. Sherr, and B. F. Sherr. 1996. Relation between presence/absence of a visible nucloid and metabolic activity in bacterioplankton cells. Limnology and Oceanography 41: 1161-1168.

Christoffersen, K. 1997. Abundance, size and growth of heterotrophic nanoflagellates in eutrophic lakes with contrasting *Daphnia* and macrophyte densities. In press *in* E. Jeppesen, Ma. Søndergaard, Mo. Søndergaard, and K. Christoffersen, editors. The structuring role of submerged macrophytes in lakes. Springer Verlag, New York, USA.

Christoffersen, K., B. Riemann, L. R. Hansen, A. Klysner, and H. B. Sørensen. 1990. Qualitative importance of the microbial loop and plankton community structure in a eutrophic lake during a bloom of cyanobacteria. Microbial Ecology 20: 253-272.

Christoffersen, K., B. Riemann, A. Klysner, and M. Søndergaard. 1993. Potential role of fish predation and natural populations of zooplankton in structuring a plankton community in eutrophic water. Limnology and Oceanography 38: 561-573.

Chrôst, R.J. 1989. Characterization and significance of β-glucosidase activity in lake water. Limnology and Oceanography 34: 660-672.

Chrôst, R.J. 1990. Microbial ectoenzymes in aquatic environments. Pages 47-78 *in* J. Overbeck, and R. J. Chrôst, editors. Aquatic Microbial Ecology, Biochemical and Molecular Approaches. Springer Verlag. New York.

Chrôst, R.J., and M. A. Faust. 1983. Organic carbon release by phytoplankton: its composition and utilization by bacterioplankton. Journal of Plankton Research 5: 477-493.

Cole, J.J., and N. F. Caraco. 1993. The pelagic microbial food web of oligotrophic lakes. Pages 101-111 *in* T. E. Ford, editor. Aquatic microbiology. An ecological approach. Blackwell Scientific Publishers, Oxford, England.

Cole, J.J., S. Findlay, and M. C. Pace. 1988. Bacterial production in freshwater and saltwater ecosystems: a cross-system overview. Marine Ecology Progress Series 43: 1-10.

Cole, J.J., G. E. Likens, and J. E. Hobbie. 1984. Decomposition of planktonic algae in an oligotrophic lake. Oikos 42: 257-266.

Cole, J.J., G. E. Likens, and D. L. Strayer. 1982. Photosynthetically produced dissolved organic carbon: An important carbon source for planktonic bacteria. Limnology and Oceanography 27: 1080-1090.

Coveney, M.F., and R. G. Wetzel. 1995. Biomass, production, and specific growth rate of bacterioplankton and coupling to phytoplankton in an oligotrophic lake. Limnology and Oceanography 40: 1187-1200.

Crawford, C.C., J. E. Hobbie, and K. L. Webb. 1974. The utilization of dissolved free amino acids by estuarine microorganisms. Ecology 55: 551-563.

Cushing, D.H., and J. J. Walsh. 1976. The ecology of the seas. Blackwell Scientific Publishers, Oxford, England

Ducklow, H.W., and C. A. Carlson. 1992. Oceanic bacterial production. Advances in Microbial Ecology 12: 113-181.

del Giordio, P.A., and R. H. Peters. 1994. Patterns in planktonic P:R ratios in lakes: Influence of trophy and dissolved organic carbon. Limnology and Oceanography 39: 772-787.

Enriquez, S., C. M. Duarte, and K. Sand-Jensen. 1993. Pattern in decomposition rates among photosynthetic organisms: the importance of detritus C:N:P content. Oecologia 94: 457-471.

Fenchel, T. 1982. Ecology of heterotrophic microflagellates. IV. Quantitative occurrence and importance as bacterial consumers. Marine Ecology Progress Series 9: 35-41.

Fogg, G.E. 1983. The ecological significance of extracellular products of phytoplankton photosynthesis. Botanica Marina 26: 3-14.

Furhman, F.A., and F. Azam. 1980. Bacterioplankton secondary production estimates for coastal waters of British Columbia, Antarctica, and California. Applied and Environmental Microbiology 39: 1085-1095.

Granéli, W., M. J. Lindell, and L. Tranvik, L. 1996. Photo-oxidative production of dissolved inorganic carbon in lakes of different humic content. Limnology and Oceanography 41: 698-706.

Güde, H. 1988. Direct and indirect influence of crustacean zooplankton on bacterioplankton of Lake Constance. Hydrobiologia 159: 63-73.

Hagstrøm, C., U. Larsson, P. Hørstedt, and S. Normark. 1979. Frequency of dividing cells, a new approach to the determination of bacterial growth rates in aquatic environments. Applied and Environmental Microbiology 37: 805-812.

Hansen, L., G. F. Krog, and M. Søndergaard. 1986. Decomposition of lake phytoplankton. 1. Dynamics of short-term decomposition. Oikos 46: 37-44.

Harder, R. 1917. Ernährungsphysiologische Untersuchungen an Cyanophyceen hauptsächlich dem endophytischen *Nostoc punctiforme*. Zeitschrift für Botanik 9: 145.

Hessen, D. O. 1992. Dissolved organic carbon in a humic lake: Effects on bacterial production and respiration. Hydrobiologia 229: 115-123.

Hobbie, J.E., R. J. Daley, and S. Jasper. 1977. Use of Nuclepore filters for counting bacteria by fluorescence microscopy. Applied and Environmental

Microbiology 33: 1225-1228.

Hygum, B.H., J. W. Petersen, and M. Søndergaard. 1997. Dissolved organic carbon released by zooplankton grazing activity – a high-quality substrate pool for bacteria. Journal of Plankton Research 19: 97-111.

Jensen, L.M. 1985. Characterization of native bacteria and their utilization of algal extracellular products by a mixed-substrate kinetic model. Oikos 45: 311-322.

Jeppesen, E., Ma. Søndergaard, Mo. Søndergaard, K. Christoffersen et al. Cascading trophic interactions in the littoral zone of a shallow lake: the role of fish and submersed macrophytes. Submitted.

Jumars, P.A., D. L. Penry, J. A. Baross, M. J. Perry, and B. W. Frost. 1989. Closing the microbial loop: dissolved carbon pathway to heterotrophic bacteria from incomplete ingestion, digestion and absorption in animals. Deep Sea Research 36: 483-495.

Kaplan, L.A., and J. D. Newbold. 1993. Biogeochemistry of dissolved organic carbon entering streams. Pages 139-165 *in* T. E. Ford, editor. Aquatic microbiology. An ecological approach. Blackwell Scientific Publishers, Oxford, England.

Kirchman, D. L. 1990. Limitation of bacterial growth by dissolved organic matter in the subarctic Pacific. Marine Ecology Progress Series 62: 47-54.

Kirchman, D.L. 1993. Particulate detritus and bacteria in marine environments. Pages 321-341 *in* T. E. Ford, editor. Aquatic microbiology. An ecological approach. Blackwell Scientific Publishers, Oxford, England.

Kiørboe, T. 1996. Material flux in the water column. *In* B. B. Jørgensen, and K. Richardson, editors. Eutrophication in Coastal Marine Ecosystems. American Geophysical Union, Washington, DC. Coastal and Estuarine Studies 52: 67-94.

Kristiansen, K., H. Nielsen, B. Riemann, and J. A. Fuhrman. 1992. Growth efficiency of freshwater bacterioplankton. Microbial Ecology 24: 145-160.

Krogh, A. 1930. Über die Bedeutung von gelösten organischen Substanzen bei der Ernährung von Wassertieren. Zeitschrift für vergleichenden Physiologie 12: 668-681.

Krogh, A. 1934a. Conditions of life in the ocean. Ecological Monographs 4: 421-429.

Krogh, A. 1934b. Conditions of life at great depth in the ocean. Ecological Monographs 4: 430-439.

Krogh, A., E. Lange, and W. Smith. 1930. On the organic matter given off by algae. Biochemical Journal 24: 1666-1671.

Lampert, W. 1978. Release of dissolved organic carbon by grazing zooplankton. Limnology and Oceanography 23: 831-834.

Lignell, R. 1990. Excretion of organic carbon by phytoplankton: its relation to algal biomass, primary productivity and bacterial secondary production in the Baltic Sea. Marine Ecology Progress Series 68: 85-99.

Lindeman, R.L. 1942. The trophic-dynamic aspect of ecology. Ecology 23: 399-418.

Malinsky-Rushansky, N.Z., and C. Legrand. 1996. Excretion of dissolved organic carbon by phytoplankton of different sizes and subsequent bacterial uptake. Marine Ecology Progress Series 132: 249-255.

Markager, S., B. Hansen, and M. Søndergaard. 1994. Pelagic carbon metabolism in a eutrophic lake during a clear-water-phase. Journal of Plankton Research 16: 1247-1267.

Martinez, J., D. C. Smith, G. F. Steward, and F. Azam. 1996. Variability in ectohydrolytic enzyme activities of pelagic marine bacteria and its significance for substrate processing in the sea. Aquatic Microbial Ecology 10: 223-230.

McQueen, D.J., J. R. Post, and E. L. Mills. 1986. Trophic relationships in freshwater pelagic ecosystems. Canadian Journal of Fisheries and Aquatic Sciences 43: 11-17.

Middelboe, M., and M. Søndergaard. 1993. Bacterioplankton growth yield: Seasonal variations and coupling to substrate lability and β-glucosidase activity. Applied and Environmental Microbiology 59: 3916-3921.

Middelboe, M., and M. Søndergaard. 1995. Concentration and bacterial utilization of sub-micron particles and dissolved organic carbon in lakes and a coastal area. Archiv für Hydrobiologie 133: 129-147.

Middelboe, M., M. Søndergaard, Y. Letarte, and N. H. Borch. 1995. Attached and free-living bacteria: Production and polymer hydrolysis during a diatom bloom. Microbial Ecology 29: 231-248.

Morris, D.P., and W. M. Lewis. 1992. Nutrient limitation of bacterioplankton growth in Lake Dillon, Colorado. Limnology and Oceanography 37: 1179-1192.

Obernosterer, I., and G. J. Herndl. 1995. Phytoplankton extracellular release and bacterial growth: dependence on the inorganic N:P ratio. Marine Ecology Progress Series 116: 247-257.

Olsen, Y., K. M. Vårum, and A. Jensen. 1986. Some characteristics of the carbon compounds released by *Daphnia*. Journal of Plankton Research 8: 505-517.

Pütter, A. 1911. Die Ernährung der Wassertiere durch gelöste organische Verbindungen. Pflüger's Archiv für der gesamten Physiologie 137: 595-621.

Payne, J.W. 1976. Peptides and microorganisms. Advances in Microbial Physiology 13: 55-113.

Pomeroy, L.R. 1974. The ocean's food web, a changing paradigm. Bioscience 24: 499-504.

Riemann, B., and R. T. Bell. 1990. Advances in estimating bacterial biomass and growth in aquatic systems. Archiv für Hydrobiologie 118: 385-402.

Riemann, B., and K. Christoffersen. 1993. Microbial trophodynamics in temperate lakes. Marine Microbial Food Webs 7: 69-100.

Riemann, B., N. O. G. Jørgensen, W. Lampert, and J. A. Furhman. 1986. Zooplankton induced changes in dissolved free amino acids and in production rates of freshwater bacteria. Microbial Ecology 12: 247-258.

Riemann, B., and M. Søndergaard. 1986a. Regulation of bacterial secondary production in two eutrophic lakes and experimental enclosusres. Journal of Plankton Research 8: 519-536.

Riemann, B., and M. Søndergaard, editors. 1986b. Carbon dynamics in eutrophic, temperate lakes. Elsevier Science Publishers, Amsterdam, The Netherlands.

Roman, M.R., H.W. Ducklow, J.A. Fuhrman, C. Garside, P.M. Gilbert, T.C. Marlone, and G.B. McManus. 1988. Production, consumption and nutrient cycling in a laboratory mesocosm. Marine Ecology Progress Series 42: 39-52.

Sanders, R. W., K. G. Porter, S. L. Bennet, and A. E. DeBiase. 1989. Seasonal patterns of bacterivory by flagellates, ciliates, rotifers and cladocerans in a freshwater planktonic community. Limnology and Oceanography 34: 673-687.

Schindler, D. W., P. J. Curtis, B. R. Parker, and M. P. Stainton. 1996. Consequences of climate warming and lake acidification for UV-B penetration in North American boreal lakes. Nature 379: 705-708.

Sharp, J.H. 1977. Excretion of organic matter by marine phytoplankton. Do healthy cells do it? Limnology and Oceanography 22: 381-399.

Sieburth, J.M. 1976. Bacterial substrates and productivity in marine ecosystems. Annual Review of Ecology and Systematics 7: 259-286.

Sieburth, J. M. 1977. International Helgoland Symposium: Convenor's report on the informal session on biomass and productivity of micro-organisms in planktonic ecosystems. Helgoländer Wissenschaftliges Meeresuntersuchungen 30: 697-704.

Simon, M., B. C. Cho, and F. Azam. 1992. Significance of bacterial biomass in lakes and the ocean: comparison to phytoplankton biomass and biogeochemical implications. Marine Ecology Progress Series 86: 103-110.

Smith, D., M. Simon, A. L. Alldredge, and F. Azam. 1992. Intense hydrolytic enzyme activity on marine aggregates and implications for rapid particle disso-lution. Nature 359: 139-142.

Steele, J.H. 1974. The structure of marine ecosystems. Blackwell Scientific Publishers, Oxford, England.

Steemann Nielsen, E. 1952. The use of radioactive carbon (^{14}C) for measuring organic production in the sea. Journal de Conseil Permanent International pour l'exploration de la Mer 18: 117-140.

Sundh, I. 1989. Characterization of phytoplankton extracellular products (PDOC) and their subsequent uptake by heterotrophic organisms in a mesotrophic forest lake. Journal of Plankton Research 11: 463-486.

Sundh, I. 1991. The dissolved organic carbon released from lake phytoplankton – biochemical composition and bacterial utilization. P.h. D.-thesis, Uppsala University, Sweden.

Søndergaard, M. 1981. Kinetics of extracellular release of ^{14}C-labelled organic carbon by submerged macrophytes. Oikos 36: 331-347.

Søndergaard, M. 1990. Extracellular organic carbon (EOC) in the genera *Carpophyllum* (Phaeophyceae): Diel release patterns and EOC lability. Marine Biology 104: 143-151.

Søndergaard, M. 1991.Phototrophic picoplankton in temperate lakes: Seasonal abundance and importance along a trophic gradient. Internationale Revue der Gesamten Hydrobiologie 76: 505-522.

Søndergaard, M. 1993. Organic carbon pools in two Danish lakes. Flow of carbon to bacterioplankton. Verhandlungen Internationale Vereinigung Theoretische und Angewandte Limnologie 25: 593-598.

Søndergaard, M., and N. H. Borch. 1991. Decomposition of dissolved organic carbon (DOC) in lakes. Archiv für Hydrobiologie. Ergebnisse der Limnologie 37: 9-19.

Søndergaard, M., B. Hansen, and S. Markager. 1995. Dynamics of dissolved organic carbon lability in a eutrophic lake. Limnology and Oceanography 40: 46-54.

Søndergaard, M., and L. M. Jensen. 1986. Phytoplankton. Pages 27-126 *in* B. Riemann, and M. Søndergaard, editors. Carbon dynamics in eutrophic, temperate lakes. Elsevier Science Publ. Amsterdam.

Søndergaard, M., L. M. Jensen, and G. Ærtebjerg. 1991. Picoalgae in Danish coastal waters during summer stratification. Marine Ecology Progress Series 79: 139-149.

Søndergaard, M., and M. Middelboe. 1995. A cross-system analysis of labile dissolved organic carbon. Marine Ecology Progress Series 118: 283-294.

Søndergaard, M., B. Riemann, and N. O. G. Jørgensen. 1985. Extracellular organic carbon (EOC) released by phytoplankton and bacterial production. Oikos 45: 323-332.

Søndergaard, M., B. Riemann, L. M. Jensen, N. O. G. Jørgensen, P. K. Bjørnsen, M. Olesen, J. B. Larsen, O. Geertz-Hansen, J. Hansen, K. Christoffersen, A.-M. Jespersen, F. Andersen, and S. Bosselmann. 1988. Pelagic food web processes in an oligotrophic lake. Hydrobiologia 164: 271-286.

Søndergaard, M., and H.-H. Schierup. 1982. Release of extracellular organic carbon during a diatom bloom in Lake Mossø: molecular weight fractionation. Freshwater Biology 12: 313-320.

Søndergaard, M., J. Theil-Nielsen, K. Christoffersen, L. Schlüter, E. Jeppesen, Ma. Søndergaard. 1997. Bacterioplankton and carbon turnover in a dense macrophyte canopy. In press *in* E. Jeppesen, Ma. Søndergaard, Mo. Søndergaard, and K. Christoffersen editors. The structuring role of submerged macrophytes in lakes. Springer Verlag, New York, USA.

Søndergaard, M., and J. Theil-Nielsen. 1997. Bacterial growth efficiency in lakewater cultures. Aquatic Microbial Ecology. In press.

Thomas, D.N., and R. J. Lara. 1995. Photodegradation of algal derived dissolved organic carbon. Marine Ecology Progress Series 116: 309-310.

Thomas, J.P. 1971. Release of dissolved organic matter from natural populations of marine phytoplankton. Marine Biology 11: 311-323.

Tranvik, L. 1988. Availability of dissolved organic carbon for planktonic bacteria in oligotrophic lakes of differing humic content. Microbial Ecology 16: 311-322.

Tranvik, L. 1992. Allochthonous dissolved organic matter as an energy source for pelagic bacteria and the concept of the microbial loop. Hydrobiologia 29: 107-114.

Wetzel, R.G., P. G. Hatcher, and T. S. Bianchi. 1996. Natural photolysis by ultraviolet irradiance of recalcitrant dissolved organic matter to simple substrates for rapid bacterial metabolism. Limnology and Oceanography 40: 1369-1380.

Wetzel, R.G., and M. Søndergaard. 1997. The role of submersed macrophytes for the microbial community and dynamics of dissolved organic carbon in aquatic ecosystems. In press *in* E. Jeppesen, Ma. Søndergaard, Mo. Søndergaard, and K. Christoffersen editors. The structuring role of submerged macrophytes in lakes. Springer Verlag, New York, USA.

Williams, P. J. le B. 1981a. Microbial contribution to overall marine plankton metabolism: direct measurements of respiration. Oceanologica Acta 4: 359-364.

Williams, P.J.le B. 1981b. Incorporation of microheterotrophic processes into the classical paradigm of the planktonic food web. Kieler Meeresforschung Sonderheft 5: 1-28.

Williams, P.J.le B. 1990. The importance of losses during microbial growth: commentary on the physiology, measurement and ecology of the release of dissolved organic material. Marine Microbial Food Webs 4: 175-206.

Williams, P.J.le B. 1995. Evidence for the seasonal accumulation of carbon-rich dissolved organic material, its scale in comparison with changes in particulate material and the consequential effect on net C/N assimilation ratios. Marine Chemistry 51: 17-29.

Worm, J., and M. Søndergaard. 1997. Dynamics of heterotrophic bacteria attached to the cyanobacterium *Microcystis*. Aquatic Microbial Ecology. In press.

Wright, R. T., and J. E. Hobbie. 1966. Use of glucose and acetate by bacteria and algae in aquatic ecosystems. Ecology 47: 447-468.

10 Zooplankton: growth, grazing and interactions with fish

By Kirsten Christoffersen and Suzanne Bosselmann

Studies of lake zooplankton at the Freshwater Biological Laboratory have developed from descriptions and classification of species to analysis of their distribution, reproduction and behaviour. More recent studies have focused on feeding biology and production of zooplankton. Current research is centred on structural and functional aspects of whole lake ecosystems by integrating analysis of phytoplankton, picoplankton, zooplankton and fish. We review these studies in the light of Danish investigations and emphasize regulations of pelagic food webs through the grazing impact of the macro-zooplankton.

Studies of limnetic zooplankton organisms in Denmark date back more than 200 years. The pioneering work by Johan Christian Lange and Otto Fredrik Müller resulted in excellent descriptions and drawings of many crustacean species from microscopical observations. Lange (1756) published a guide to the aquatic research of that time ("Science of natural waters"). A more comprehensive study of the taxonomy and morphology of Danish Cladocera by Müller (1867) includes most of the species known today. An example of the very detailed, hand-coloured drawings, published by O. F. Müller in 1785, is given (Fig. 10.1).

Ecological studies of the zooplankton started when the Freshwater Biological Laboratory was opened in 1897. Wesenberg-Lund (1904, 1908) made numerous studies of the seasonal distribution of Cladocera and Copepoda in lakes and ponds and described temporal cyclomorphosis, reproduction potential and feeding biology. Subsequently, Berg and Nygaard (1929) conducted quantitative studies of zooplankton succession and related the appearance and disappearance of

zooplankton species to changes in the succession and chemical composition of phytoplankton (Krogh and Berg 1931).

Experimental studies were initiated by Berg (1931, 1936) who realized that *Daphnia* species were easy to collect and keep in cultures. He found that the specific body weight of parthenogenetic females of *D. cucullata* was higher than that of sexual females. The results supported the "depression hypothesis", namely that the transition from parthenogenesis to gamogenesis was caused by environmental conditions. Berg (1936) also found that different food conditions and temperatures did not induce any change in the formation of helmets of *D. cucullata*, though *Daphnia* with helmets were frequently found in their natural habitat during summer. His experiments could not simulate the stress factor due to fish predation that has later been found to induce the formation of high helmets in some species of *Daphnia* (Jacobs 1987).

Studies of zooplankton population dynamics and secondary production are important disciplines in studies of energy formation and

Figure 10.1 Zooplankton illustrated by O. F. Müller (1985). This example shows *Daphnia rectirostris* [*Moina beachiata*] (1-3), *Daphnia pennata* [*Daphnia magna*] (4-7), *Daphnia longispina* (9-10) and *Daphnia sima* [*Simocephalus vetulus*] (11-12). The present names are in brackets.

transfer in aquatic ecosystems (Rigler and Downing 1984). Regulation of zooplankton production by environmental factors (e.g. temperature, food availability and oxygen) has been one of the main research fields of zooplankton ecologists (Downing 1984).

Measurements of zooplankton secondary production were, for example, part of the studies of Esrom Sø (Bosselmann 1975, Jónasson 1977). Lately, energy flow between different compartments in the pelagic-benthic system became a tool for analyzing the ecosystem function (Jónasson et al. 1974, Andersen and Jacobsen 1979, Riemann and Søndergaard 1986). Regulation of secondary produktion and the interactions between various components of the food web in lakes may be controlled by productivity of the phytoplankton and by predator control of zooplankton population structure and composition (Carpenter 1988, Riemann and Christoffersen 1993).

The purpose of this chapter is to illustrate the important ecological role of zooplankton using recent Danish studies of lake zooplankton ecology. We discuss and give examples of seasonal and spatial distribution, allometric relationships between zooplankton biomass and grazing rates, measurements of grazing activities, and interactions with fish.

Temporal and spatial variation
Seasonal variation
The seasonal succession of zooplankton is controlled mainly by temperature, food availability and predation. The seasonal pattern of zooplankton succession is predictable in temperate lakes (Sommer et al. 1986, Gliwicz and Pijanowska 1989), though high variability may be seen.

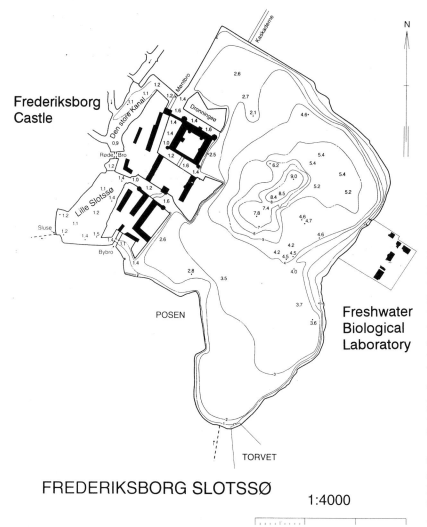

Frederiksborg Castle

Freshwater Biological Laboratory

POSEN

TORVET

FREDERIKSBORG SLOTSSØ

1:4000

Figure 10.2 Bathymetric map of Frederiksborg Slotssø from echo soundings in 1996 (reproduced with permission from T. Høy). The lake is small 0.23 km² with a mean depth of 3.2 m and a maximum depth of 9 m. There is a stable thermocline at 3-5 m depth and an underlying anoxic hypolimnion from mid May to September. Frederiksborg Slotssø is highly productive (500 g C m⁻² y⁻¹) with nutrient inputs from rich soils in the catchment area and from anthropogenic activities, especially during the last 200 years. All external nutrient inputs to the lake have stopped, but a large internal pool of nitrogen and phosphorus in the sediments keeps algal turbidity high (Andersen and Jacobsen 1979). Pollen analysis (O. S. Jacobsen pers. comm.) indicates that the lake's history dates back 6,000 years (i.e the Stone Age).

For stratified Danish eutrophic lakes a very general picture can be drawn. An accelerated growth of herbivorous zooplankton (rotifers, copepods and small cladocerans) is observed in early spring concomitant with an increase in temperature. Calanoid and cyclopoid copepods can maintain growth and reproduction at temperatures below 10°C whereas many cladocerans have growth optimum above 15°C with maximum growth rates at 20-25°C (Muck and Lampert 1984). The differences imply that the copepod populations increase earlier in the year in temperate lakes than those of cladocerans. The simultaneous increase in irradiance stimulates the growth of small green algae and diatoms. The phytoplankton is grazed intensively, which can eventually lead to a reduction of phytoplankton biomass and promote a clear-water phase in late spring, even under eutrophic conditions (Lampert et al. 1986, Jespersen et al. 1988). Subsequently, zooplankton production may be limited until a high phytoplankton biomass is reestablished; the latter comprises a diverse summer community which includes chlorophytes, cryptophytes and dinoflagellates. The higher summer temperatures will stimulate the growth of zooplankton populations provided edible phytoplankton species are available. If colonial cyanobacteria and large dinoflagellates dominate, the diet of the zooplankton must be supplemented by other food sources. An examination of the zooplankton community during a bloom of *Aphanizomenon* and *Microcystis* in eutrophic Frederiksborg Slotssø revealed that the zooplankton population structure remained constant (Fig. 10.2). Zooplankton grazing was directed instead against picoplankton as the net phytoplankton consisted of inedible filamentous cyanobacteria (Christoffersen et al. 1990). Presumably, the cyanobacteria were non-toxic strains. There are, however, numerous examples of lethal effects of cyanobacteria on zooplankton populations (Christoffersen 1996).

Many cladocerans and calanoid copepods develop population maxima in the summer. Populations of cyclopoid copepods can also form maxima in summer, when they do not enter a diapause induced by food limitation (Nilssen 1978, Hansen and Santer 1995). Predation by fish and invertebrates (e.g. *Chaoborus*, *Hydracarina*, and *Leptodora*) can remove a significant part of the zooplankton production at this time and severely reduce the population density and mean body size (e.g. Christoffersen 1990, Jeppesen et al., 1990, 1997). Small cladocerans, copepods, ciliates and heterotrophic nanoflagellates, are frequently found during periods of intensive fish predation (Christoffersen et al. 1993).

Most zooplankton populations decrease during autumn as primary production and temperature decrease. Some species, however, can obtain a high growth rate based on renewed growth of diatoms. Cladocerans produce resting eggs that allow them to survive winter in the sediment until they hatch in the following spring.

Cladocerans are particularly successful in establishing large populations during spring and summer. This success is based on their ability to utilize a wide spectrum of food particles (bacteria, picoalgae protozoans and phytoplankton) and to achieve high growth rates at high food concentrations (Muck and Lampert 1984, Lampert 1987, Jürgens 1994, Wickham et al. 1993). Calanoid copepods feed selectively on a narrow food spectrum of small phytoplankton species (DeMott 1989, Sterner 1989). Calanoids, on the other hand, can maintain growth at lower temperatures and lower food concentrations than cladocerans due to lower maintenance costs (Peters and Downing 1984). Cyclopoid copepods are omnivorous and/or predacious and feed on a variety of food particles including phytoplankton, detritus, protozoans and other zooplankton, but they seem to require higher food concentrations than cladocerans and calanoid copepods to ensure maximum rates of reproduction (Hansen and Santer 1995). Periods of high and diverse food availability will, therefore, favour cladocerans and cyclopoid copepods, while copepods are more successful during periods of low food availability.

Seasonal pattern of zooplankton in Frederiksborg Slotssø

The seasonal abundance of the zooplankton community in eutrophic Frederiksborg Slotssø has been investigated on several occasions (Fig. 10.2). A comparison of the zooplankton abundance during 1926, 1976 and 1988 is shown in Fig. 10.3. The winter community is low in numbers and poor in species. It comprises populations of *Eudiaptomus graciloides* and cyclopoid copepods such as *Cyclops strenuus* and *Mesocyclops leuckarti*. The abundance of these species increases at the onset of the spring phytoplankton bloom in February-March. *Bosmina* spp. start to grow at water temperatures of 8 to 10 °C and form the first peak in abundance within a few weeks. *Bosmina* originate from the few wintering specimens or hatch from resting eggs in the sediments. The populations of *B. coregoni* and *B. longirostis* can attain more than

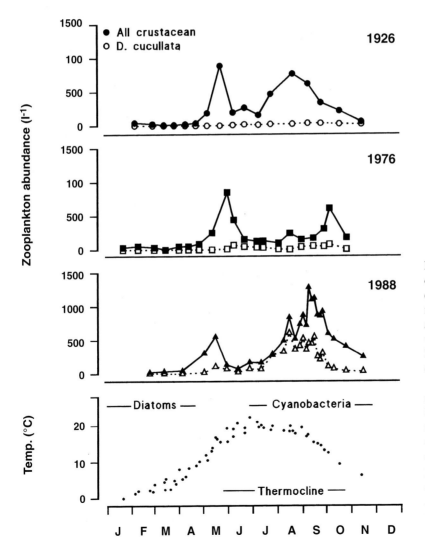

Figure 10.3 Abundance of zooplankton and the contribution of the dominant cladoceran (*Daphnia cucullata*) as well as water temperature in Frederiksborg Slotssø during 1926, 1976 and 1988. The periods of the spring phytoplankton bloom, the cyanobacterial bloom and the thermocline are tentatively indicated in the bottom panel. Data from Berg and Nygaard (1929), Andersen and Jacobsen (1979) and Christoffersen (unpublished data).

2,000 individuals l⁻¹. Their dominance lasts until approximately the end of May, and they are replaced by species of *Daphnia* and *Chydorus* as the temperature increases. The abundance of zooplankton is low in June-July, which is also the period when fish fry appears (Jeppesen et al. 1997). Massive blooms of toxic cyanobacteria may also pre-vent the zooplankton from attaining high summer biomasses (Chistoffersen et al. 1993, Christoffersen 1996). The second peak in zooplankton abundance appears in July-August, or even later in some years (e.g. 1976). *Daphnia cucullata* is the dominant zooplankton species in terms of abundance and biomass during the summer. Other cla-docerans (*Diaphanosoma brachyurum, Ceri-odaphnia quadrangula*) are present, but in lower densities. Calanoid and cyclopoid copepods slowly replace the cladocerans as water temperature decreases in the autumn.

The three year-long studies in Frederiks-borg Slotssø show considerable differences in the contribution of *D. cucullata* (the domi-nant daphnid) to the total zooplankton abun-dance. Very few individuals (< 30 l⁻¹) were recorded in 1926 and 1976 compared to 1988 (max. 865 l⁻¹). There are two possible expla-nations. First, the sampling strategies were not identical in these studies. Only the epi-limnion (0 to 3 m depth) was sampled in 1926, which may have lead to underestimates of cladocerans as they tend to migrate down-ward during the day. This migratory behavi-our cannot explain the low abundance in *D. cucullata* in 1976, however, since most of the water column was sampled (0 to 5 m depth) as it also was in 1988. The second, and most likely, explanation is that the observed differ-ences are due to a biomanipulation of the lake in 1986-87 which resulted in a 60% decrease in the abundance of planktivorous fish. A subsequent increase in zooplankton abundance and a decrease in summer phyto-plankton biomass was observed (Riemann et al. 1990).

Vertical distribution and grazing activity

Many zooplankton species display strong diel vertical migration. The term diel vertical migration refers to the movements performed by zooplankton species between upper and lower layers of the water column during night and day, respectively. An array of dif-ferent patterns has been described in the liter-ature including nocturnal migrations (upward movements at night), twilight migration (sev-eral up- and downward movements during the night), reverse migration (downward dur-ing night) and combinations of these types (Haney 1993). The causes and consequences have frequently been debated (Lampert 1993).

Several *in situ* studies have shown that zooplankton feeding is coupled to vertical migrations, although the patterns described are often conflicting. For example, Haney and Hall (1975) observed a close relationship between high nocturnal individual grazing rate and vertical migration of large *Daphnia* species, whereas no diurnal periodicity in the feeding rate was observed for the migrating *Diaptomus*. In contrast, Lampert and Taylor (1985) reported that both qualitative and quantitative aspects of the community grazing were strongly coupled with vertical migra-tion, but they did not observe any periodicity in individual grazing rates. Lampert and Tay-lor (1985) emphasized that different mecha-nisms may operate in different ecosystems.

Studies of periodicity in the feeding activ-ity of individual organisms are limited by the fact that it is difficult to measure grazing rates of individual organisms without disturb-ing the grazer and its environment. Quantifi-cation of the phytoplankton pigments in the guts of herbivorous zooplankton by fluores-cence (the gut fluorescence technique) is, however, a suitable technique because it is very sensitive and measures the ingestion that has taken place immediately before the cap-ture of the animal (Christoffersen and Jesper-sen 1986). The technique was applied in a

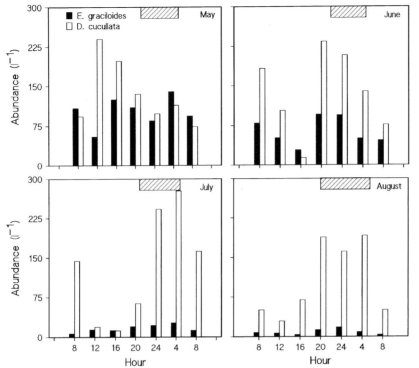

Figure 10.4 Variation in the abundance of *Eudiaptomus graciloides* and *Daphnia cucullata* at 1.5 m depth over 24 hours in May, June, July and August in Frederiksborg Slotssø. The period from sunset to sunrise is indicated by a hatched bar. Data from Jespersen and Christoffersen (1985).

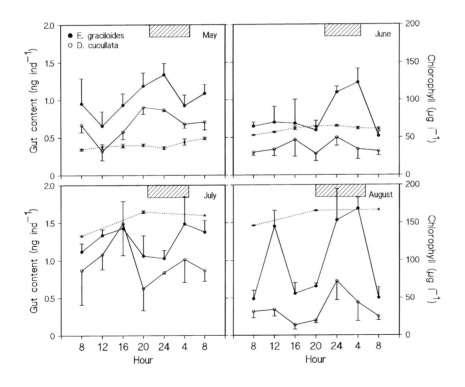

Figure 10.5 Variations in the gut contents of *Eudiaptomus graciloides* and *Daphnia cucullata* in Frederiksborg Slotssø in May, June, August and September 1984. The chlorophyll *a* concentration is indicated by asterisks and dotted lines. The period from sunset to sunrise is indicated by a hatched bar. Data from Jespersen and Christoffersen (1985) and Christoffersen and Jespersen (1986).

series of experiments in which the relationship between vertical migration and feeding activity was investigated (Jespersen and Christoffersen 1985). An upward migration was recorded for *D. cucullata* during the day in May and at night in June, July and August (see example from a depth of 1.5 m; Fig. 10.4). The animals were virtually absent from the surface during the day in July and August and a pronounced upward migration occurred after sunset. *E. graciloides* showed a weaker migration pattern. Measurements of the chlorophyll content of the animals' guts revealed a pronounced diel periodicity in the feeding activity of both species (Fig. 10.5). The gut contents were largest at night, except in July when daytime values were highest. In August, the gut contents had two distinct maxima, one at noon and one at midnight. The observed patterns of individual gut contents were consistent (i.e not significantly different) at depths from 0 to 4 m (Jespersen and Christoffersen 1985).

If the diel community grazing activity is averaged for the whole water column and separated into night-time and day-time values, the average grazing impact was almost identical during the two periods except in May (Fig. 10.6). However, since the individual ingestion rates showed almost no changes with depth, and the food concentration remained constant over 24 hours (Fig. 10.5), the hourly ingestion was about two-fold higher at night than during the day (Fig. 10.6). Thus, the impact on the phytoplankton biomass was much stronger at night than during the day except in May. These calculations assume, however, that the total number of zooplankton is constant in a given water column. This assumption is seldom fulfilled, and in this case an immigration of zooplankton from other regions of the lake was observed. Immigration of zooplankton elevated the total number by 25% and 11% respectively. However, periodicity in feeding activity is common, but the timing and ampli-

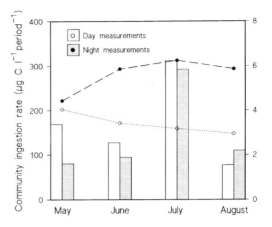

Figure 10.6 Diel vertical distribution of the community ingestion rates of zooplankton on phytoplankton in May, June, July and August in Frederiksborg Slotssø. Bars represent values averaged over night-time or daytime periods and lines represent the hourly rates during day or night. All values are integrated over depths from 0 to 5 m. Data as for Figs. 4 and 5.

tudes are not constant and show no clear correlation with the migration patterns.

Day-night changes in ingestion rates were maintained under different environmental conditions, but migration was recorded only on two of five occasions. These results show, that the two phenomena are not strongly coupled. This finding differs from those of Lampert and Taylor (1985) who found no differences in the feeding activity per unit biomass between day and night. They concluded that the grazing activity took place in the lower strata during day, and in the upper strata during night, but at similar rates.

The diel studies from Frederiksborg Slotssø indicate that erroneous community ingestion estimates may be obtained if diurnal periodicity in feeding activity is neglected. For example, if ingestion rates at noon in May and June (Fig. 10.5) are extrapolated to 24 hours, they result in underestimates of the daily ingestion by 30-40%. Extrapolation of midnight measurements would likewise lead to substantial overestimates.

Grazing measurements

Zooplankton grazing on phytoplankton has been studied intensively over the past decades. The development of *in situ* techniques has allowed direct quantification of the energy from phytoplankton that passes through zooplankton to higher trophic levels in lake ecosystems (Gliwicz 1968, Gulati et al. 1982, Bosselmann and Riemann 1986, Lampert 1988). Different techniques have been used in grazing experiments (see Peters 1984 for a review). Procedures involving the counting of particles before and after grazing have gradually been replaced by a variety of radiotracer techniques, i.e. labelling of natural or cultured food particles by isotopes (e.g. ^{14}C, ^{32}P) before administering them to natural zooplankton assemblages. The development of an *in situ* grazing chamber, where the tracer-food was injected into an enclosed water sample, minimized the manipulations of zooplankton populations during grazing experiments (Gliwicz 1968, Haney 1973). Later, the gut fluorescence technique, which was originally developed in marine investigations (Mackas and Bohrer 1976), was introduced as another suitable method (Christoffersen and Jespersen 1988, Christoffersen 1988). New techniques are currently being developed, and approaches using immunofluorescence to detect the uptake of specific food particles may prove useful in future grazing studies (Kandel et al. 1993, Ohman 1993, Gorant et al. 1994).

Comparison of two grazing methods

Zooplankton ingestion rates of phytoplankton were determined by two independent grazing methods in a mesocosm study in Frederiksborg Slotssø in June 1984. The experiment was designed to evaluate the role of grazers under the influence of nutrients and predators (Bosselmann and Riemann 1986).

Positive linear relationships between results obtained by the ^{14}C technique and the gut fluorescence technique were found (Fig.

10.7). A total number of 37 estimates of ingestion rates were involved in the regression analysis with ingestion rates ranging from 2 to 750 µg C l^{-1} d^{-1}. Despite differences in methodology and assumptions made in the calculations of ingestion rates, the results were closely related. The gut fluorescence

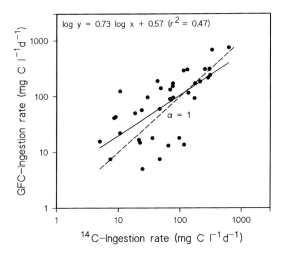

Figure 10.7 Regression analysis between measurements of zooplankton ingestion rates obtained by the gut fluorescence technique (GFC) and ^{14}C-labelled phytoplankton (^{14}C). Each point summarizes one grazing experiment (duplicate measurements with both methods). The tracer technique was based on labelling prefiltered 100 µm natural phytoplankton assemblages with inorganic $^{14}CO_2$ for 24 hours *in vitro*. Subsamples of labelled phytoplankton were mixed with lake water containing natural zooplankton densities (ratio of 1:5 by volume) and incubated for 20 or 30 min, a period set by preliminary time series experiments. Community grazing rates were calculated from the radioactivity of phytoplankton and zooplankton. The gut fluorescence technique was based on fluorescent measurements of gut pigments of the dominant grazers, *Eudiaptomus graciloides* and *Daphnia cucullata*. Ingestion rates were calculated from the amount of pigments ingested, multiplied with a species-specific and temperature-corrected gut evacuation rate. Community ingestion rates were estimated by extrapolation from the biomass of the dominant grazers (> 80% of the total biomass) and an assumption that the rest of the zooplankton community had similar ingestion rates. Data from Bosselmann and Riemann (1986), Christoffersen and Jespersen (1986) and Jespersen et al. (1988).

technique gave generally higher ingestion rates than the tracer technique, however. This discrepancy suggests the existence of a systematic error. First, both *E. graciloides* and *D. cucullata* exhibit diel changes in gut content with 30-40% lower values during the day than during the night in May and June (Fig. 10.6). If it is assumed that diel changes occurred in the experimental enclosures, most of the deviation between the results of the two methods may be due to the tracer experiments being carried out at noon, whereas samples for gut fluorescence analysis were obtained in the afternoon. Secondly, ingestion rates derived from labelled natural phytoplankton have been criticized because of the variable specific activity of the different algal species present (Gulati et al. 1982). If zooplankton ingest only a certain size fraction, this may lead to under- or overestimates of the feeding rates.

Bosselmann and Riemann (1986) demonstrated that ingestion rates of natural zooplankton populations obtained by the ^{14}C-procedure were similar to those calculated from zooplankton biomass if conventional relations between growth and ingestion were assumed. This correspondence may not always exist, however, because food limitation, inhibition by cyanobacteria, reproductive state and physical and chemical conditions can alter the feeding behaviour of zooplankton (see Peters 1984).

The gut fluorescence technique has previously been favourably compared to other *in situ* techniques (Ramcharan and Sprules 1988), although it tends to underestimate ingestion rates due to degradation of chlorophyll inside the digestive system. The technique requires a constant gut evacuation rate, which is present only at food concentrations above the incipient limiting level (Christoffersen 1988). In addition, the full application of the method on a community level involves the determination of evacuation rates for all species present. For logistic reasons it is seldom possible to examine all species, which means that community ingestion rates usually require extrapolation from few species. The measurements obtained in the experiments reported here included approximately 80% of the total herbivorous biomass.

Despite these possible limitations and uncertainties it was encouraging to find comparable estimates of community ingestion rates by two methods based on different principles. Direct comparisons of grazing methods have not been made very often. Peters (1984) evaluated several methods and found that long-term tracer methods (i.e. incubations that last up to 24 hours) tended to underestimate grazing rate and that short-term tracer techniques and cell counts before and after the incubation period yielded comparable results.

Community ingestion rates
Ingestion rates of zooplankton communities vary with the species composition and the body size of the grazers and are generally positively related to zooplankton biomass (Peters and Downing 1984, Lampert 1988). If this generality holds true also for lakes covering a nutrient gradient and at different seasons it should be possible to predict the impact of zooplankton grazing on phytoplankton. Empirical analyses of field and laboratory data have been used to establish such relationships (Peters and Downing 1984). Significant positive correlations between zooplankton biomass and ingestion rate have been found in most cases (Gulati et al. 1982, Lampert 1988, Cyr and Pace 1992).

Allometric relationships show that small organisms tend to have higher mass-specific metabolic rates than large organisms (Peters 1983). This result implies that zooplankton communities dominated by small organisms should graze more than communities dominated by large organisms. However, for several reasons this is too simple. Firstly, not all organisms fit this general model for instance,

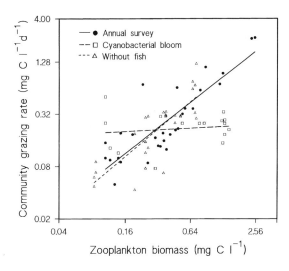

Figure 10.8 Relationship between zooplankton biomass and community ingestion rates under different conditions. Ingestion rates were measured as stated in Fig. 7. The regression equations were: $\log_{10}y = 0.979 \log_{10}x - 0.169$; $r^2 = 0.734$ (annual survey), $\log_{10}y = 0.0625 \log_{10} x - 0.642$; $r^2 = 0.020$ (cyanobacterial bloom) and $\log_{10}y = 1.047 \log_{10}x - 0.149$; $r^2 = 0.666$ (without fish). Data from Bosselmann and Riemann (1986), Jespersen and Christoffersen (1988) and Christoffersen et al. (1990).

copepods ingest less food than daphnids per unit biomass (Muck and Lampert 1984), which implies that copepod-dominated systems have lower grazing impacts on phytoplankton than daphnid-dominated systems. Secondly, Cyr and Pace (1993), who analyzed a large data set, found that community mass-specific grazing rate decreased with increasing mean body size, and also that community biomass increased with mean body size. Together, these findings suggest that communities of large-bodied zooplankton graze more intensively on phytoplankton than communities of small zooplankton.

An example of biomass-controlled grazing impact is given in Fig. 10.8. Community grazing rates were measured using ^{14}C-labelled natural algal assemblages and fluorescence analyses of the gut contents of the

dominant grazers (Bosselmann and Riemann 1986, Jespersen and Christoffersen 1986). Grazing rates were obtained at different zooplankton community structures (i.e. contrasting small and large organisms). Clearly the grazing rates were positively related to the zooplankton biomass irrespective of the community composition. Small species generally dominated at low zooplankton biomass and large species at high biomass. Regression coefficients (r^2) between zooplankton biomass and ingestion rates ranged from 0.68 to 0.87 except for the data obtained during a bloom of cyanobacteria. Cyanobacteria are generally not filtered or are inefficiently filtered and may contain toxins that prevent grazing (Fulton and Pearl 1987, DeMott et al. 1991, DeMott and Moxter 1991).

The measured community grazing rates cover almost the entire season and the grazing impact on the phytoplankton biomass varied from 0.5% to more than 300% of the biomass removed per day. In this eutrophic lake, the highest grazing impact takes place in early spring (36-380% d^{-1}) and in autumn (15-57% d^{-1}), while the remaining period is characterized by low grazing impact. However, intermediate to high grazing impacts (5% to 50%, occasionally up to 300%) were observed in summer when planktivorous fish were absent (as in the fish-free enclosures). These findings are in accordance with literature surveys of zooplankton grazing activity in which grazing impacts of 2 to 310% d^{-1} in lakes of different nutrient status and different fish community structure have been reported (e.g. Lampert 1988, Cyr and Pace 1992, Gulati et al. 1990, Jeppesen et al. 1996). Thus the herbivorous zooplankton community has a potential filtering capacity that allows phytoplankton biomass to be controlled in most seasons, but the grazing is modulated by many other factors (Sterner 1989).

Daphnia as key grazers

Daphnia are an important structuring element of the plankton community because of their strong effects on the abundance and composition of other plankton organisms (Jürgens 1994). This impact is caused by high individual grazing rates and potentially high population densities. With a typical summer abundance in eutrophic lakes of several hundreds per liter and a filtration capacity of 0.5 to 1 ml per *Daphnia* per hour, they can process the water column 1 to 5 times each day. Cells smaller than 20 μm in diameter (or length) are the preferred food items and are filtered at high rates, while cells smaller than 1 μm and larger than 50-100 μm are filtered at slower rates (Brendelberger 1991). Figure 10.9 illustrates the gut of *D. magna* packed with phytoplankton cells.

Spring and early summer blooms of diatoms, green algae and cryptophytes are often grazed at high efficiency, and a large fraction or the entire primary production can be removed on a daily basis. When the phytoplankton community is dominated by larger forms e.g. dinoflagellates and colonial cyanobacteria, it is less likely that *Daphnia* can control the phytoplankton biomass.

The ability of *Daphnia* species to control blooms of cyanobacteria has frequently been debated (De Bernardi and Guissani 1990). Co-existence of *Daphnia* and filamentous cyanobacteria has been described (e.g. Lynch and Shapiro 1981, Fig. 10.10) as well as the ability of *Daphnia* to graze colonial cyanobacteria (Schoenberg and Carlson 1984, Davidowicz et al. 1988). The general concept, however, is that large crustacean species tend to decrease in numbers during blooms of cyanobacteria and are replaced by smaller species, capable of feeding on pico- and nanoplankton (Burns et al. 1989, Christoffersen et al. 1990). A main reason for the decline of large crustaceans when cyanobacteria proliferate seems to be the difficulties associated with handling and avoiding the inedible cya-

nobacteria (Lampert 1987, Burns 1987). Single cells and small filaments can be ingested but provide an insufficient diet for the zooplankton because most cyanobacterial strains are difficult to digest and are less nutritious than for example chlorophytes (Ahlgren et al. 1990, Gliwicz 1990). The inhibition of feeding is illustrated in Figure 10.11. The gut fullness decreases with the increasing proportion of cyanobacteria in the food. The feeding activity is almost zero when 75% of the suspended particles are cyanobacteria. The inhibitory effect of cyanobacteria filaments on feeding is most pronounced for large zooplankton like many *Daphnia* species because smaller species are likely to select small food particles (Gliwicz 1990). This result implies that large zooplankton may starve during cyanobacterial blooms (Porter and McDonough 1984, Christoffersen et al. 1990). Other food sources such as bacteria, flagellates and small ciliates may then become important (Christoffersen et al. 1990, Pace et al. 1990, Weisse et al. 1990).

The influence of cladocerans on the microbial food web has been described in several studies (e.g. Arndt and Nixdorf 1991, Wickham et al. 1993, Jürgens et al. 1994). *Daphnia* are capable of consuming most of the picoplankton production, especially in mesotrophic and eutrophic lakes (Vaqué and Pace 1992, Sanders et al. 1994, Jürgens and Stolpe 1995). The consumption of bacteria by zooplankton, rotifers, ciliates and heterotrophic nanoflagellates, for example, was investigated during a cyanobacterial bloom in Frederiksborg Slotssø (Christoffersen et al. 1990). The zooplankton community, dominated by *D. cucullata*, consumed 21% to 87% of the bacterial production daily. Rotifers consumed on average 20% to 40%, while ciliates and heterotrophic nanoflagellates consumed 10% to 16% and 3% to 4% respectively. The adverse effects of daphnids on rotifers, ciliates, heterotrophic nanoflagellates, picoalgae and bacteria may prevail in most lakes as

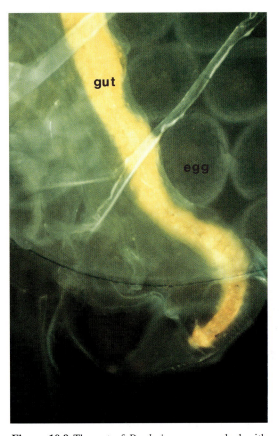

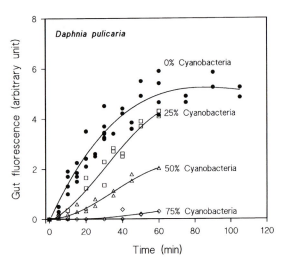

Figure 10.11 Accumulation over time in the gut contents (in arbitrary fluorescence units) of *Daphnia pulicaria* feed an algal suspension of 1 mg C l⁻¹ with different rations of an edible chlorophyte (*Scenedesmus*) and an inedible cyanobacteria (filamentous *Cylindrospermopsis*; non-toxic). Each point represents an average gut content from 5 to 10 animals. Christoffersen (unpubl. data).

Figure 10.9 The gut of *Daphnia magna* packed with fluorescently labelled phytoplankton cells (approx 10 x 30 μm). The gut contents appear as a yellowish mass. A few eggs are visible at the right side of the photo. Photographed with an Olympus camera attached to an epifluorescence microscope (400x magnification).

soon as the water temperature allows the cladocerans to grow. Inverse relationships between a high cladoceran biomass and low biomasses of pico- and microzooplankton (illustrated in Fig. 10.12) are typically found in lakes where the fish predation on large zooplankton is low or absent (Christoffersen 1993, Jürgens 1994, Christoffersen 1997).

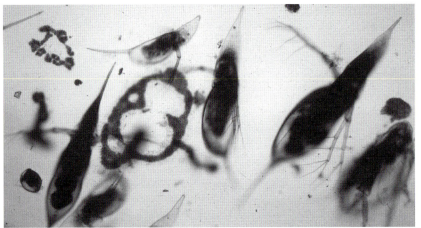

Figure 10.10 Mixed assemblages of cladocerans (*Daphnia cucullata*, *Diaphanosoma brachyurum* and *Chydorus coregoni*) and cyanobacteria (*Microcystis* spp.). Photographed with an Olympus camera attached to an inverted microscope (40x magnification).

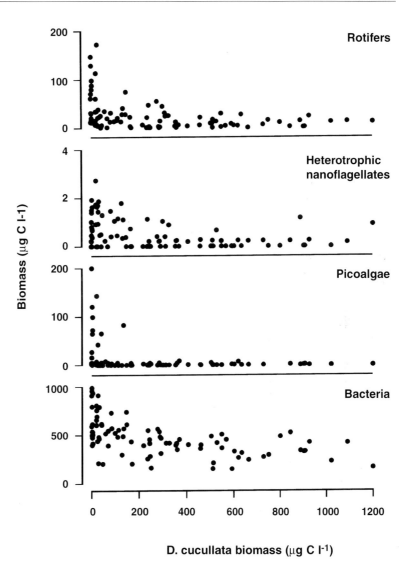

Figure 10.12 Biomasses of rotifers, heterotrophic nanoflagellates, picoalgae and bacteria as a function of the biomass of *Daphnia cucullata* in Frederiksborg Slotssø in 1989. Data from Christoffersen et al. (1993).

Thus cladocerans and *Daphnia* in particular can utilize a wide spectrum of food particles, have a high growth potential (provided the temperature is not limiting their growth) and can partly exclude other groups of zooplankton by predation or by mechanical interference through cell damage (Gilbert 1988). Consequently, *Daphnia* strongly influence the environment and will, when fish predation is low, channel more organic material through the food web than copepods (Cyr and Pace 1992, Riemann and Christoffersen 1993).

Interactions with fish

Ecological studies of zooplankton under natural conditions have emphasized that crustaceans act as links between primary producers and other trophic components in the pelagic food web (Wetzel 1975, Moss 1988). This key role arises from the ability of herbivorous zooplankton to graze phytoplankton and from their vulnerability to fish predation. Zooplankton secondary production is consumed by a variety of invertebrates (e.g. predatory zooplankton and *Chaoborus*) and fish. The spawning of planktivorous fish is synchronized with the development of zoo-plankton as the survival and growth of many fish larvae and yearlings depend on this food source. Most carnivorous fish feed on benthic and plant-associated animals or on fish older than one year, but they may switch back to a zooplankton diet if the food conditions are scarce (Persson et al. 1988).

Predation by planktivorous fish is for several reasons considered to be the most important threat to zooplankton (Gliwicz and Pijanowska 1989). Owing to their size, fish need large quantities of food. They can consume a large range of prey items, are flexible predators, and can change foraging strategies. Prey selection is mainly governed by 1) the size of prey which can be retained, 2) the seasonal availability of prey items, and 3) the feeding behaviour of the predator (Zaret 1980). Size preference is species- and age-specific and changes when the fish stock develops, e.g. when new cohorts of fish appear. Feeding behaviour is partly related to the availability of prey because a certain behavioural reaction must be stimulated by one or more of the above factors. Motility and size of the prey are additional important factors for visual predators such as roach. Brooks and Dodson

(1965) formulated the size-selective hypothesis which predicts that large zooplankton such as some species of *Daphnia* are more vulnerable to predation from visual predators than smaller zooplankton such as *Bosmina*.

Examination of the contents of fish stomachs or alimentary canals can reveal the composition and size of the food, provided that digestion has left remains which can be identified. The diet of the entire roach and bream populations (all size-classes) in Frederiksborg Slotssø has been studied (Michelsen et al. 1994). The ingested food of 1-2-year old roach and 3-4-year old bream is illustrated in Table 10.1. The roach had eaten mainly zooplankton and supplemted with detritus and benthos. The majority of the zooplankton remains were from *Daphnia* (90%), and the rest originated from *Leptodora* and *Chydorus*. The bream stomachs contained almost equal amounts of benthos and zooplankton plus a substantial amount of detritus. The zooplankton component consisted of *Alona* (50%), *Daphnia* (45%) and *Leptodora* (5%). The general picture is that bream within a transitional length of 15 – 20 cm shifted from a diet of benthic cladocerans (*Alona sp.*) to zooplankton and chironomids. This shift was coupled to a behavioural change from particulate to filter feeding. The biomass of chironomids was too low to sustain the consumption of larger bream (> 20 cm) which started feeding in the pelagic even during periods when the mean length and biomass of the preferred zooplankton, *D. cucullata*, were low. In contrast to bream, roach fed mainly on zooplankton. As they increased in size and the mean size of their branchial system increased, roach progressively shifted to larger zooplankton species. The importance of benthic animals in the diet of roach was minor, which may reflect low feeding efficiency on prey buried in the sediment. Detritus appeared in the diet of bream and roach during periods of low availability of animal food items. Feeding on detritus may provide an energetic advantage to

Table 10.1. The relative composition of prey items ingested by 12 cm roach and 25 cm bream (% of the total number of food items recorded). Each item was identified to the lowest possible taxonomic group from head capsules, carapaces, shells and other remains which were left undigested. Quantification of the relative contribution of each food group to the total content is based on AFDW after drying for 24 h at 60°C and burning for 2 h at 550°C. Data from Michelesen et al. (1994).

Food item	Roach	Bream
Benthos	6	28
Zooplankton	75	22
Detritus	6	45
Other sources	3	5

Figure 10.13 Distribution of zooplankton biomass in mesocosms with and without fish. Small cladocerans included *Bosmina* sp. and *Chydorus coregoni* and large cladocerans included *Daphnia cucullata* (the majority), *D. galeata* and *Diaphanosoma brachyurum*. Data from Christoffersen et al. (1993).

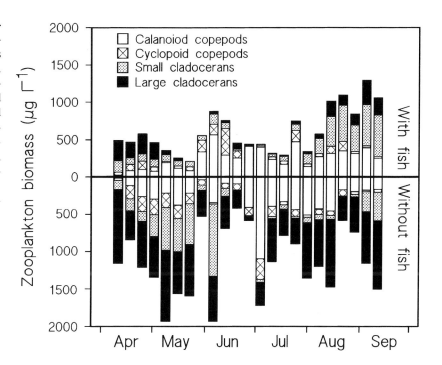

bream and roach and increase the carrying capacity for these species in lakes with much detritus (Persson 1983)

Abundance and biomass of planktivorous fish increase with increasing productivity of lake ecosystems which, consequently, leads to intensive predation on the zooplankton community (Persson et al. 1988, Jeppesen et al. 1997). Often, the pattern implies that the size distribution of the zooplankton changes towards dominance of less vulnerable size categories (Brooks and Dodson 1965, Gulati et al. 1990, Jeppesen et al. 1992). Predation by planktivorous fish has ecological implications which involve other trophic levels as well (Jeppesen et al. 1997), i.e. effects created by predators at one level may affect several levels of the food web – the "trophic cascade" (Carpenter et al. 1985, Riemann and Søndergaard 1986, Jeppesen et al. 1997). This thesis predicts that effects of size-selective predation in systems of high densities of planktivorous fish will force the zooplankton community towards small and inefficient

grazers unable to control phytoplankton growth (Benndorf et al. 1984). Instead, a microbial community of bacteria, flagellates, and ciliates may develop and serve as an alternative pathway for the degradation of organic material (Stockner and Porter 1988). A considerably reduced biomass of planktivorous fish in a eutrophic lake will result in a higher biomass of zooplankton, especially *Daphnia*, which in turn may graze more efficiently on the phytoplankton (Shapiro and Wright 1984, Riemann 1985, Sanders et al. 1989) and the microbial community (Pace et al. 1990, Christoffersen et al. 1993).

These interactions are illustrated by an ecosystem study in Frederiksborg Slotssø (Christoffersen et al. 1993). Large model ecosystems (mesocosms) were stocked with one year old roach (200 m²) or kept free of fish. Changes in biological structure and water chemistry were followed from April to mid September. The proportion of cladocerans decreased (Fig. 10.13) when the planktivorous fish were present. Phytoplankton, bacteria, flagellates, and rotifers were able to proliferate under these conditions and attained high population densities. Interactions within the heterotrophic populations were also observed, such as a decrease in the biomass of heterotrophic nanoflagellates concomitant with an increase in the biomass of rotifers.

The average zooplankton biomass was twice as high in the absence of planktivorous fish and developed asynchronously compared to that in the mesocosms with fish (Fig. 10.13). The mesocosms with fish had zooplankton biomass maxima late in the season (August and September), whereas zooplankton biomass in the absence of fish was high for most of the season. The zooplankton in the mesocosms without fish exhausted the algal biomass during spring so that the growth of macrozooplankton was probably limited by food even when bacteria and flagellates are considered as edible components of the planktonic biomass (Sanders and Porter 1990). Similar consequences of predation by planktivorous fish have frequently been observed in mesocosm experiments (e.g. Lynch and Shapiro 1984, Riemann 1985, Christoffersen et al. 1993) and in whole lake experiments (Benndorf 1990, Gulati et al. 1990, Jeppesen et al. 1997).

In summary, the mesocosm experiment demonstrated two important aspects of the influence of fish predation on zooplankton. Firstly, the dominance of cladocerans induced an intensive grazing impact that had qualitative and quantitative effects on the lower trophic levels. Secondly, a high zooplankton biomass reduced the ability of cyanobacteria to form large colonies. The combined effect of these factors prevented the blooming of cyanobacteria and improved the water quality.

These findings support the concept of predator control in eutrophic freshwater systems (Persson et al. 1988, Jeppesen et al. 1997) and demonstrate that biomanipulation

can be used in lake restoration programs (Gulati et al. 1990). The results also emphasize the inherent ability of natural populations of cladocerans to control pico- and phytoplankton growth and to prevent proliferation of cyanobacteria in eutrophic lakes. The fish reduction in Frederiksborg Slotssø had no long-term effects (Jeppesen et al. 1990). This is attributed to the high nutrient concentrations in the lake at present.

Future perspective
The present studies of planktonic crustacean ecology have indicated that the distribution and feeding activity of copepods and cladocerans may vary considerably on a seasonal and diurnal basis. It has also been demonstrated that especially *Daphnia* are of importance in the regulation of primary production and that they may "break" the microbial loop by directly feeding on picoplankton and protozoans. Large cladocerans are, on the other hand, extremely vulnerable to predation by fish, and they are replaced by small and less effective grazers when fish predation is high.

Current studies of zooplankton ecology at the Freshwater Biological Laboratory focus on several aspects that remain to be elucidated. One subject is the role of macrophytes as refuges for zooplankton. Macrophytes can attract and repel zooplankton and seem to be important shelters when fish predation is high (Whiteside et al. 1985). By protecting vulnerable species of zooplankton during the day-time, macrophytes may contribute to higher grazing rates by zooplankton (Schriver et al. 1995). Another current subject is to study the adverse effect of toxic cyanobacteria on zooplankton (Christoffersen 1996). Little is known today about the sublethal effects of cyanotoxins but it seems that zooplankton populations may fail to become established because of malgrowth or infertility. It is not clear if zooplankton is able to accumulate cyanotoxins and thereby transfer toxins to other components of the food web.

Literature cited

Ahlgren, G., L. Lundstedt, M. Brett, and C. Forsberg. 1990. Lipid composition and food quality of some freshwater phytoplankton for cladoceran zooplankters. Journal of Plankton Research 12: 809-818.

Andersen, J. M., and O. S. Jacobsen. 1979. Production and decomposition of organic matter in eutrophic Frederiksborg Slotssø, Denmark. Archiv für Hydrobiologie 85: 511-542.

Arndt, H., and B. Nixdorf. 1991. Spring clear-water phase in a eutrophic lake: Control by herbivorous zooplankton enhanced by grazing on components of the microbial web. Verhandlungen der Internationale Theoretische und Angewandte für Limnologie 24: 879-883.

Benndorf, J., H. Kneschke, K. Kossatz, and E. Penz. 1984. Manipulation of the food web by stocking with predacious fishes. Internationale Revue der gesamten Hydrobiologie 69: 407-428.

Benndorf, J. 1990. Conditions for effective biomanipulation; conclusions derived from whole-lake experiments in Europe. Hydrobiologia 200/201: 187-204.

Berg, K. 1931. Studies on the genus *Daphnia* O. F. Müller. Videnskabelige meddelser fra dansk naturhistorisk Forening, vol 92, Bianco Lunos Bogtrykkeri A/S, Copenhagen, Denmark.

Berg, K. 1936. Reproduction and depression of the Cladocera illustrated by the weight of the animals. Archiv für Hydrobiologie 30: 438-462

Berg, K., and G. Nygaard. 1929. Studies on the plankton on the lake of Frederiksborg Castle. De Kongelige Danske Videnskabernes Selskab, Række 9. 1. no. 4: 227-316.

Bosselmann, S. 1975. Production of *Eudiaptomus graciloides* in Lake Esrum, 1970. Archiv für Hydrobiologie 76: 43-64

Bosselmann, S., and B. Riemann. 1986. Zooplankton. Pages 199-236 *in* B. Riemann and M. Søndergaard, editors. Carbon dynamics of eutrophic, temperate lakes. Elsevier, Amsterdam, The Netherlands.

Brendelberger, H. 1991. Filter mesh size of cladocerans predicts retention efficiency for bacteria. Limnology and Oceanography 36: 884-894.

Brooks, J. L., and S. I. Dodson. 1965. Predation, body size, and composition of plankton. Science 150: 28-35.

Burns, C. W. 1987. Insights into zooplankton-cyanobacteria interactions derived from enclosure studies. New Zealand Journal of Marine and Freshwater Research 21: 447-482.

Burns, C. W., D. J. Forsyth, , J. F. Haney, M. R. James, W. Lampert, and R. Pridmore. 1989. Coexistence and exclusion of zooplankton by *Anabaena minutissima* var. *attenuata* in Lake Rotongaio, New Zealand. Archive für Hydrobiology Beiheft Ergebnisse der Limnologie 32: 63-82.

Carpenter, S. J., F. F. Kitchell, and J. R. Hodgson. 1985. Cascading trophic interactions and lake productivity. Fish predation and herbivory can regulate lake ecosystems. BioScience 35: 634-639.

Carpenter, S. J., 1988. Complex interactions in lake communities. Springer, New York, USA.

Christoffersen, K. 1988. Effect of food concentration on gut evacuation of *Daphnia pulicaria* and *Daphnia longispina* measured by the fluorescence technique. Verhandlungen der Internationale Vereinigung für Theoretische und Angewandte Limnologie 23: 2050-2055.

Christoffersen, K. 1990. Evaluation of *Chaoborus* predation on natural populations of herbivorous zooplankton in a eutrophic lake. Hydrobiologia 200/201: 458-466.

Christoffersen, K. 1996. Ecological implications of cyanobacterial toxins in aquatic food webs. Phycologia 36 (6 Supplement): 42-50.

Christoffersen, K. 1997. Abundance, size, and growth of heterotrophic nanoflagellates in eutrophic lakes with contrasting *Daphnia* and macrophyte densities. *In* E. Jeppesen, Ma. Søndergaard, Mo. Søndegaard, K. Christoffersen, editors. The structuring role of submerged macrophytes in lakes. Springer-Verlag, New York, USA, in press.

Christoffersen, K., and A. M. Jespersen. 1986. Gut evacuation rates and ingestion rates of *Eudiaptomus graciloides* measured by means of the gut fluorescence method. Journal of Plankton Research 8: 973-983.

Christoffersen, K., B. Riemann, L. R. Hansen, A. Klysner, and H. B. Sørensen. 1990. Qualitative importance of the microbial loop and plankton community structure in a eutrophic lake during a bloom of cyanobacteria. Microbial Ecology 20: 253-272.

Christoffersen, K., B. Riemann, A. Klysner, and M. Søndergaard. 1993. Potential role of fish predation and natural populations of zooplankton in structuring a plankton community in eutrophic lake water. Limnology and Oceanography 38: 561-573.

Cyr, H., and M. L. Pace. 1992. Grazing by zooplankton and its relationship to community structure. Canadian Journal of Fisheries and Aquatic Science 49: 1455-1465.

Cyr, H., and M. L. Pace. 1993. Allometric theory: extrapolation from individuals to communities. Ecology 74: 1234-1245.

Davidowicz, P., Z. M. Gliwicz, and R. D. Gulati. 1988. Can *Daphnia* prevent a blue-green algal bloom in hypertrophic lakes? A laboratory test. Limnologia

(Berlin) 19: 21-26.

De Bernardi, R., and G. Giussani. 1990. Are blue-green algae a suitable food for zooplankton? An overview. Hydrobiologia 200/201: 29-43.

DeMott, W. R. 1989. Optimal foraging theory as a predictor of chemically mediated food selection by suspension feeding copepods. Limnology and Oceanography 34:140-154.

DeMott, W. R., Q. -Z. Zhang, and W. W. Carmichael. 1991. Effects of toxic cyanobacteria on the survival and feeding of a copepod and three species of *Daphnia*. Limnology and Oceanography 36: 1346-1357.

DeMott, W. R., and F. Moxter. 1991. Foraging on cyanobacteria by copepods: responses to chemical defenses and resource abundance. Ecology 72: 1820-1844.

Downing, J. A. 1984. Assessment of secondary production: the fist step. Pages 1-18 *in* J. A. Downing, and F. H. Rigler, editors. Secondary productivity in freshwaters. Blackwell. England.

Fulton, R. S., and H. W. Pearl. 1987. Toxic and inhibitory effects of the blue-green algae *Microcystis aeruginosa* on herbivorous zooplankton. Journal of Plankton Research 9: 837-855.

Hansen, A-M., and B. Santer. 1995. The influence of food resources on the development, survival and reproduction of the two cyclopoid copepods: *Cyclops vicinus* and *Mesocyclops leuckarti*. Journal of Plankton Research 17: 632-646.

Haney, J. F. 1973. An *in situ* examination of the grazing activities of natural zooplankton communities. Archiv für Hydrobiologie 72: 87-132.

Haney, J. F. 1993. Environmental control of diel vertical migration behaviour. Archiv für Hydrobiologie Beiheft Ergebnisse der Limnologie 39: 1-17.

Haney, J. F., and D. J. Hall. 1975. Diel vertical migration and filter-feeding activities of *Daphnia*. Archiv für Hydrobiologie 75: 413-441.

Gilbert, J. J. 1988. Suppression of rotifer populations by *Daphnia*: A review of the evidence, the mechanisms, and the effects on zooplankton community structure. Limnology and Oceanography 33: 1286-1303.

Gliwicz, Z. M. 1968. The use of anaesthetizing substances in studies on the food habits of zooplankton. Ekologia Polska 16: 279-295.

Gliwicz, Z. M. 1990. *Daphnia* growth at different concentrations of blue-green filaments. Archiv für Hydrobiologie 120: 51-65.

Gliwicz, Z. M., and J.Pijanowska. 1989. The role of predation in zooplankton succession. Pages 253-296 *in* U. Sommer, editor. Plankton ecology – Succession in plankton communities. Springer-Verlag, Berlin, Germany.

Gorant, E., G. Prensier, and N.Lair. N. 1994. Specific immunological probes for the identification and tracing of prey in crustacean gut contents. The example of cyanobacteria. Archiv für Hydrobiologie 131: 243-252.

Gulati, R. D., K. Siwertsen, and G. Postema, G. 1982. The zooplankton: its community structure, food and feeding, and role in the ecosystem of Lake Vechten. Hydrobiologia 95: 127-163.

Gulati, R. D., E. H. R. R.Lammens, M. L. Meijer, and E. van Donk. 1990, editors. Biomanipulation – Tool for water management. Hydrobiologia 200/201, Kluwer Academic Publishers, The Netherlands.

Jacobs, J. 1987. Cyclomorphosis in *Daphnia*. Pages 325-352 *in* R. H. Peters and R. de Bernardi, editors. *Daphnia*. Memorie dell'Istituto di Idrobiologia "Dott. Marco De Marchi", Pallanza, Italy.

Jeppesen, E. M. Søndergaard, E. Mortensen, P. Kristensen, B. Riemann, J. P. Jensen, J. P. Müller, O. Sortkjær, J. P. Jensen, K. Christoffersen S. Bosselmann, and E. Dall. 1990. Fish manipulation as a lake restoration tool in shallow, eutrophic temperate lakes 1: cross-analysis of three Danish case-studies. Hydrobiologia 200/201: 205-218.

Jeppesen, E., O. Sortkjær, M. Søndergaard, and M. Erlandsen. 1992. Impact of a trophic cascade on heterotrophic bacterioplankton production in two shallow fish-manipulated lakes. Archive für Hydrobiologie Beiheft Ergebnisse der Limnologie 37: 219-231.

Jeppesen, E., M. Søndergaard, J. P. Jensen, E. Mortensen, and O. Sortkjær. 1996. Fish-induced changes in zooplankton grazing on phytoplankton and bacterioplankton: a long-term study in shallow hypertrophic Lake Søbygaard. Journal of Plankton Research 18: 1605-1625.

Jeppesen, E., J. P. Jensen, M. Søndergaard, T. Lauridsen, L. J. Pedersen, and L. Jensen. 1997. Top-down control on freshwater lakes; the role of fish, submerged macrophytes and water depth. Hydrobiologia *in press*.

Jespersen, A. M., and K. Christoffersen. 1985. *In situ* zooplankton grazing on phytoplankton in freshwater. Master Thesis (in Danish), Freshwater Biological Laboratory, University of Copenhagen, Denmark.

Jespersen, A.-M., K. Christoffersen, and B. Riemann. 1988. Annual carbon fluxes between phyto-, zoo-, and bacterio-plankton in eutrophic Lake Frederiksborg Slotssø, Denmark. Verhandlungen der Internationalen für Limnologie 23: 440-444.

Jónasson, P. M., Lastein, E., and Rebsdorf, A. 1974. Production, insolation, and nutrient budget of eutrophic lake Esrum. Oikos 25: 255-277.

Jónasson, P. M. (1977). Lake Esrum Research 1867-

1977. Pages 67-89 *in* C. Hunding, editor. Danish Limnology. Reviews and Perspectives. Folia Limnologica Scandinavica. Vol. 17. OAB Press, Odense, Denmark.

Jürgens, K. 1994. Impact of *Daphnia* on planktonic microbial food webs – A review. Marine Microbial Food Webs 8:295-324.

Jürgens, K., H. Arndt, and K. O. Rothhaupt. 1994. Zooplankton mediated changes of microbial food web structure. Microbial Ecology 27: 27-42.

Jürgens, K., and G. Stolpe. 1995. Seasonal dynamics of crustacean zooplankton, heterotrophic nanoflagellates and bacteria in a shallow, eutrophic lake. Freshwater Biology 33: 27-38.

Kandel, A., K. Christoffersen, and O. Nybroe. 1993. Filtration rates of *Daphnia cucullata* on *Alcaligenes eutrophus* JMP134 by a fluorescent antibody method. FEMS Microbiology Ecology 12: 1-8.

Krog, A., and K. Berg. 1931. Über die chemische Zusammensetzung des Phytoplanktons aus dem Frederiksborg-Schlossee und Ihre Bedeutung für die Maxima der Cladoceren. Internationale Revue der gesamten Hydrobiologie und Hydrographie 25: 204-218.

Lampert, W., W. Fleckner, H. Rai, and B. E. Taylor. 1986. Phytoplankton control by grazing zooplankton: A study on the spring clear-water phase. Limnology and Oceanography 31: 478-490.

Lampert, W. 1987. Feeding and nutrition in *Daphnia*. Pages 143-192 *in* R. H. Peters and R. de Bernardi, editors. *Daphnia*. Memorie dell'Istituto di Idrobiologia "Dott. Marco De Marchi", Pallanza, Italy.

Lampert, W. 1988. The relationship between zooplankton biomass and grazing: A review. Limnologica (Berlin) 19: 11-20.

Lampert, W. 1993. Ultimative causes for diel vertical migration of zooplankton: New evidence for the predator-avoidance hypothesis. Archive für Hydrobiologie Beiheft Ergebnisse der Limnologie 39: 79-88.

Lampet, W., and B. E. Taylor. 1985. Zooplankton grazing in a eutrophic lake: quantitative and qualitative implications of diel vertical migration. Ecology 66:68-82.

Lange, J. C. 1756. Science of natural waters (in Danish). Copenhagen, Denmark.

Lynch, M., and J. Shapiro. 1981. Predation, enrichment, and phytoplankton community structure. Limnology and Oceanography 26: 86-102.

Mackas, D., and R. Bohrer. 1976. Fluorescence analysis of zooplankton gut contents and an investigation of diel feeding patterns. Journal of experimental Marine Biology and Ecology 25: 77-85.

Michelsen, K., J. Pedersen, K. Christoffersen, and F. Jensen. 1994. Ecological consequences of food partitioning for fish population structure in a eutrophic lake. Hydrobiologia 291: 35-45.

Moss, B. 1988. Ecology of Fresh Waters. Man and Medium. Blackwell, London, England.

Muck, P., and W. Lampert. 1984. An experimental study on the importance of food conditions for the relative abundance of calanoid copepods and cladocerans. I. Comparative feeding studies with *Eudiaptomus gracilis* and *Daphnia longispina*. Archiv für Hydrobiologie Supplement 66: 157-169.

Müller, O. F. 1785. Entomostraca seu insecta testacea qvae in aqvis Daniae et Norvegiae reperit, descripset, et iconibus illustravit. Havniae, Denmark.

Müller, P. E. 1867. Danish Cladocera (in Danish). Naturhistorisk Tidsskrift III: 53-240.

Nilssen, J. P. 1978. On the evolution of life histories of limnetic cyclopoid copepods. Memorie dell'Istituto di Idrobiologia 36:193-214.

Ohman, M. D. 1994. Predation on planktonic protists assessed by immunochemical assays. Pages 731-737 *in* P. F. Kemp, B. F. Sherr and E. B. Sher, editors. Handbook of methods in aquatic microbial ecology. Lewis Publishers, Boca Raton, USA.

Pace, M. L., G. B. McManus, and S. E. G. Findlay. 1990. Plankton community structure determines the fate of bacterial production in a temperate lake. Limnology and Oceanography 35: 795-808.

Persson, L. 1983. Food consumption and the significance of detritus and algae to intraspecific competition in roach (*Rutilus rutilus*) in a shallow eutrophic lake. Oikos 41: 118-125.

Persson, L., S. Andersson, F. Hamrin, and L. Johansson. 1988. Predator regulation and primary production along a eutrophication gradient of temperate lake ecosystems. Pages. 45-65 *in* S. R. Carpenter, editor. Complex Interactions in lake communities. Springer-Verlag, New York, USA.

Peters, R. H. 1983. The ecological implications of body size. Cambridge University Press, Cambridge, England.

Peters, R. H. 1984. Methods for the study of feeding, grazing and assimilation by zooplankton. Pages 336-412 *in* J. A. Downing, and F. H. Rigler, editors. Secondary productivity in freshwaters. Blackwell, England.

Peters, R. H., and J. A. Downing. 1984. Empirical analysis of zooplankton filtering and feeding rates. Limnology and Oceanography 29: 763-784.

Porter, K. C., and R. McDonough. 1984. The energetic cost of response to blue-green algal filaments by cladocerans. Limnology and Oceanography 9: 365-369.

Ramcharan, C. W., and W. G. Sprules. 1988. Ingestion

rates of *Daphnia pulex* as measured by both carbon-14 uptake and gut fluorescence. Verhandlungen Internationale Vereinigung für Theoretische und Angewandte Limnologie 23: 2045-2049.

Riemann, B. 1985. Potential importance of fish predation and zooplankton grazing on natural populations of freshwater bacteria. Applied Environmental Microbiology 50: 187-193.

Riemann, B., K. Christoffersen, H. J. Jensen, J. P. Müller, C. Lindegaard, and S. Bosselmann. 1990. Ecological consequences of a manual reduction of roach and bream in a eutrophic, temperate lake. Hydrobiologia 200/201: 241-250.

Riemann, B., and M. Søndergaard. 1986. Carbon dynamics in eutrophic, temperate lakes. Elsevier, New York, USA.

Riemann, B., and K. Christoffersen. 1993. Microbial trophodynamics in temperate lakes. Marine Microbial Food Webs 7: 69-100.

Rigler, F. H. and J. A. Downing. 1984. The calculation of secondary productivity. Pages 19-58 *in* J. A. Downing, and F. H. Rigler, editors. Secondary productivity in freshwaters. Blackwell, England.

Sanders, R. W., and K. G. Porter. 1990. Bacterivorous flagellates as food resources for the freshwater crustacean zooplankter *Daphnia ambigua*. Limnology and Oceanography 35: 188-191.

Sanders, R. W. K. G. Porter, S. L. Bennett, and A. E. DeBiase. 1989. Seasonal patterns of bacterivory by flagellates, ciliates, rotifers, and cladocerans in a freshwater planktonic community. Limnology and Oceanography 34: 673-687.

Sanders, R. W., D. A. Leeper, C. H. King, and K. G. Porter. 1994. Grazing by rotifers and crustacean zooplankton on nanoplanktonic protists. Hydrobiologia 288: 167-181.

Schoenberg, S. A., and R. E. Carlson. 1984. Direct and indirect effects of zooplankton on phytoplankton in a hypereutrophic lake. Oikos 42: 291-302.

Schriver, P., J. Bøgestrand, E. Jeppesen, and M. Søndergaard. 1995. Impact of submerged macrophytes on fish-zooplankton-phytoplankton interactions: Large-scale enclosure experiments in a shallow eutrophic lake. Freshwater Biology 33: 255-270

Shapiro, J., and D. I. Wright. 1984. Lake restoration by biomanipulation: Round lake, Minnesota – the first two years. Freshwater Biology 14. 371-383.

Sommer, U., Z. M. Gliwicz, W. Lampert, and A. Duncan. 1986. The PEG-model of seasonal succession of planktonic events in fresh waters. Archiv für Hydrobiologie 106: 433-471.

Sterner, R. W. 1989. The role of grazers in phytoplankton succession. Pages 107-169 *in* U. Sommer, editor. Plankton ecology – Succession in plankton communities. Springer-Verlag, Berlin, Germany.

Stockner, J. G., and K. G. Porter. 1988. Microbial food webs in freshwater planktonic ecosystems. Pages 69-84 *in* S. R. Carpenter Editor. Complex interactions in lake communities. Springer Verlag, New York, USA.

Vaqué, D., and M. L. Pace. 1992. Grazing on bacteria by flagellates and cladocerans in lakes of contrasting food-web structure. Journal of Plankton Research 14: 307-321.

Weisse, T., H. Müller, R. M. Pinto-Coelho, A. Schweizer, A., D. Springmann, and, G. Baldringer, 1990. Response of the microbial loop to the phytoplankton spring bloom in a large prealpine lake. Limnology and Oceanography 35: 781-794.

Wickham, S. A., J. J. Gilbert, and U-G. Berninger. 1993. Effects of rotifers and ciliates on the growth and survival of *Daphnia*. Journal of Plankton Research 15: 317-334.

Wesenberg-Lund, C. 1904. Plankton investigations of the Danish lakes I (in Danish). Copenhagen, Denmark.

Wesenberg-Lund, C. 1908. Plankton investigations of the Danish lakes II (in Danish). Copenhagen, Denmark.

Whiteside, M. C., W. L. Doolittle, and C. M. Swindoll. 1985. Zooplankton as food resource for larval fish. Verhandlungen Internationale Vereinigung für Angewandte und Theoretische Limnologie 22: 2523-2526.

Wetzel, R. G. 1975. Limnology. Saunders, Philadelphia, USA.

Zaret, T. M. 1980. Predation and Freshwater Communities. Yale University Press, New Haven, USA.

11 Metabolism and survival of benthic animals short of oxygen

By Kirsten Hamburger, Claus Lindegaard and Peter C. Dall

The profundal zone of eutrophic lakes experiences oxygen deficiency during summer stratification. Profundal macroinvertebrates contain haemoglobin to maintain aerobic metabolism down to a very low oxygen level. Below this oxygen level the animals stop feeding, reduce metabolism and start to degrade large glycogen stores anaerobically. Comparison of the slow decline of glycogen stores in natural populations of *Chironomus anthracinus* and *Potamothrix hammoniensis* in the Lake Esrom with the rapid decline of glycogen in organisms exposed to experimental anaerobiosis suggests that the profundal invertebrates never experience total anoxia for long periods. We hypothesise that wind-induced internal circulation in the hypolimnion of Lake Esrom provides sufficient oxygen for maintaining an almost aerobic metabolism during summer stratification, resulting in one of the highest profundal densities ever recorded.

Macroinvertebrates living in the profundal zone of lakes have adapted metabolically to survive periods with little or no available oxygen. The organisms often contain haemoglobin, allowing them to maintain an almost constant aerobic metabolism even at low oxygen concentrations. Below a certain oxygen concentration, they stop feeding, reduce metabolism and eventually enter a state of dormancy (Hand 1991). Typically, during favourable feeding periods they build up large glycogen stores which are mobilised during anoxic periods.

Anaerobic glycogen conversion is an expensive process. In the Embden Meyerhof pathway (EMP) the end-product is lactate or ethanol. The energetic efficiency in the production of high-energy phosphate components (ATP) is 10-12-fold lower than in complete combustion under aerobic conditions. The facultative anaerobic organisms (e. g.

some midge larvae of the genus *Chironomus* and some oligochaete worms) depend on oxygen, but can survive long-term anaerobiosis utilising glycogen in fermentation processes leading to a variety of end-products. In *Chironomus* larvae, the end-product is ethanol (Wilps and Zebe 1976, Frank 1983). In oligochaete worms (*Tubifex*, *Lumbriculus*), the end-products comprise a variety of organic acids (Putzer et al. 1990, Seuss et al. 1983). This multiple end-product fermentation takes place in the mitochondria and leads to increased efficiency of energy production (ATP), which is about twice as high as in the hololactic fermentation. However, the energy production by glycogen degradation in these animals is still 5-6 times lower during anaerobiosis than under aerobic conditions.

The profundal fauna of Lake Esrom is exposed to severe oxygen deficiency for 2-4 months each year during summer stratification. The few macroinvertebrate species which are able to survive in this environment all show adaptations to meet the periodically unfavourable conditions. This paper focuses on adaptations of energy metabolism and the ability to survive oxygen deficiency, as demonstrated by the two most abundant species of the deep profundal zone, the chironomid *Chironomus anthracinus* Zetterstedt and the oligochaete *Potamothrix hammoniensis* Michaelsen.

The energetic efficiency of ethanolic fermentation in *Chironomus* is unknown, but probably enhanced compared to that of the classic EMP (Wilps and Schöttler 1980). The end-products of fermentation in *P. hammoniensis* are unknown. In view of the uncertainties in energetic efficiency in both species considered here, all calculations of energetic efficiencies are based on the classic glycolytic scheme.

In order to evaluate the importance of these metabolic adaptations, we compare measurements of glycogen conversion in profundal populations with that of organisms exposed to experimental anaerobiosis, and relate the results to ambient oxygen conditions in the profundal zone.

Temperature and oxygen conditions in the profundal zone

The history and present status of the Lake Esrom ecosystem is discussed by Lindegaard et al. (1997). The deep profundal zone (20-22 m depth) has low temperatures and large fluctuations in oxygen concentration during the year. Temperatures in the hypolimnion vary from a minimum of 1-3 °C in February-March to a maximum of 11-13 °C at the time of overturn in the autumn (Fig. 11.1). Temperature in the hypolimnion during summer is usually between 8°C and 10°C, but the actual level primarily depends on solar insolation and wind-induced mixing of the water column. The temperature is almost constant during calm summer months (e.g. 1992), or it rises gradually because of partial deep vertical mixing in windy summers (e.g. 1994). The thermocline is usually established in late May, and the oxygen concentration has declined to less than 1 mg O_2 l^{-1} one to two months later. The period with microxia (< 0.2 mg O_2 l^{-1}) lasts 2-3 months and is abruptly ended by complete vertical mixing (overturn) in late September or early October (Fig. 11.1).

To explain the observed changes in glycogen metabolism of *C. anthracinus* and *P. hammoniensis,* we need to know the oxygen

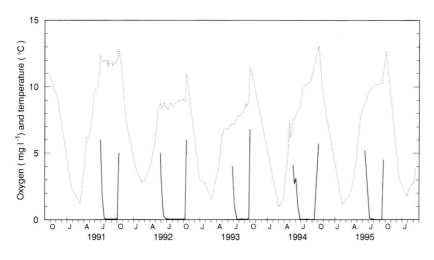

Figure 11.1.
Oxygen concentration (mg $O_2 l^{-1}$) during periods of summer stratification (solid line) and bottom temperature (dotted line) in lake Esrom from September 1990 to March 1996.

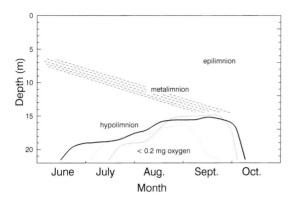

Figure 11.2. Schematic presentation of oxygen conditions in the hypolimnion of Lake Esrom during summer stratification. The curves represent oxygen concentrations of 0.2 mg O_2 l^{-1}. The location of metalimnion and solid line curve show the situation in 1992. Broken and dotted curves show levels in 1991 and 1993 respectively. From Hamburger et al. 1995 (with kind permission from Kluwer Academic Publishers).

conditions in the hypolimnion during summer stratification (Fig. 11.2). Measurements of oxygen concentration in the lower hypolimnion during 1992 showed an increase in the thickness of the bottom layer with less than 0.2 mg O_2 l^{-1} from about 2 m in mid-July to 6 m in mid-August with no subsequent change until the overturn in October. Similar, though less pronounced changes were observed in 1991 and 1993 (Fig. 11.2). This decrease in hypolimnetic oxygen content may be explained by the still deeper location of the metalimnion through the summer. Fig. 11.2 shows schematically that the upper oxygen-rich hypolimnion gradually reduces in volume with time. Thus, the hypolimnion volume in 1992 was gradually reduced by 30% from mid-June to mid-July and by a further 40% during the following month. This increasing reduction of the hypolimnion volume is due to the morphometry of the lake basin and may explain the observed increase in magnitude of the microxic bottom layer. The increase in thick-

ness of the microxic water layer and the simultaneous decrease of the oxygen-rich hypolimnion, apparently deteriorate the oxygen conditions in the water layer adjacent to the sediment, as distinct changes in growth rate and metabolism of the profundal fauna are observed within this period. We therefore distinguish between a first and second phase of the period with microxia, i. e. the period before and after the increase in thickness of the water layer with less than 0.2 mg O_2 l^{-1} (cf. Figs 11.5-11.7).

The macroinvertebrate populations in the deep profundal zone

The harsh environmental conditions in the profundal zone limit the number of species living here. In Lake Esrom seven macroinvertebrates permanently maintain populations in the deepest part from 20 m to 22 m. However, numbers of these few species, particularly of *C. anthracinus* and *P. hammoniensis*, are very high (20,000-40,000 individuals m^{-2}; Jónasson 1972, Lindegaard et al. 1997), making the profundal zone of Lake Esrom one of the most abundant profundal communities ever recorded. The life cycle of *C. anthracinus* is annual or biennial in the deep profundal zone (Jónasson 1972). A new generation starts in April-May, and the larvae grow quickly and reach the 2nd and/or 3rd instar before the oxygen depletion during summer stratification arrests growth. After the autumn overturn, the larvae resume growth and moult immediately to 3rd and/or 4th instar larvae. Growth continues during autumn but stops when the bottom temperature has decreased to about 4 °C. During spring, the mature 4th instar larvae pupate, and the adults emerge and swarm, while the rest of the population of 4th instar larvae remains in the profundal, continues to grow and emerges in the following spring. If this remaining population surpasses 2,000 larvae m^{-2}, it prevents recruitment of a new generation simply by eating the newly-settled eggmasses produced by that part of the population which has emerged as adults (Jónasson 1972).

The population density varied from about 5,000 larvae m^{-2} in 1991/92 to about 10,000 larvae m^{-2} in 1993/94 (Lindegaard et al. 1997, Fig. 4.7). In even years (1992, 1994) only about one third of the population succeeded in swarming, leaving the remaining populations large enough to prevent new settlement. Consequently in the 1990's new recruitment took place in uneven years (1991, 1993, 1995).

The life span of *P. hammoniensis* in the profundal is 5-6 years, and reproduction takes place when the worms are 3-4 years old. Cocoons are found in the bottom sediments from late spring, throughout summer and early autumn (Jónasson and Thorhauge 1972, 1976). The summer stratification probably arrests the development of embryos, as suggested from peaks in density of worms observed just after the autumn overturn (Lindegaard et al. 1997, Fig. 4.8). Growth of *P. hammoniensis* is fastest during spring and autumn and individuals lose weight during summer stratification with very low oxygen concentrations (Thorhauge 1976). Cocoons are eaten by 4th instar larvae of *C. anthracinus*, which is reflected by high densities of *P. hammoniensis* in years with few *C. anthracinus* (Jónasson and Thorhauge 1972, 1976).

Respiratory adaptations

C. anthracinus and *P. hammoniensis* are typical respiratory regulators as originally shown by Berg et al. (1962) and later confirmed by Hamburger et al. (1994) and Nilson (1995). At 10°C the rate of oxygen consumption decreases slightly, though not significantly, between air saturation ($\approx$10 mg O_2 l^{-1}) and a critical point, which is about 3 mg O_2 l^{-1} for *C. anthracinus* and about 1 mg O_2 l^{-1} for *P. hammoniensis* (Fig. 11.3). Below the critical point, the rate of oxygen consumption decreases steeply. A lower threshold concentration for oxygen uptake at about

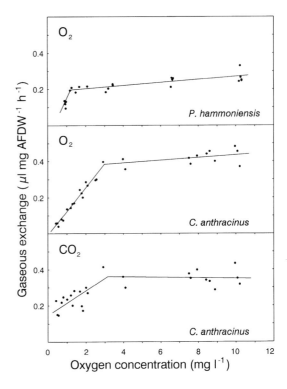

Figure 11.3. The rate of oxygen consumption (μl O_2 mg^{-1} AFDW h^{-1}) at various oxygen concentrations (mg O_2 l^{-1}) of *Potamothrix hammoniensis* and *Chironomus anthracinus* 4th instar larvae, and carbon dioxide production (μl CO_2 mg^{-1} AFDW h^{-1}) of *Chironomus anthracinus* 4th instar larvae. AFDW is ash free dry weight. From Hamburger et al. 1994 (with kind permission from Kluwer Academic Publishers).

0.1 mg O_2 l^{-1} is suggested for *C. anthracinus*. Concurrent with the drop in oxygen consumption, the feeding activity decreases, and below 1 mg O_2 l^{-1} all larvae have empty guts.

The release of carbon dioxide has been measured in *C. anthracinus* as a function of the oxygen concentration. The release of carbon dioxide remains unchanged between saturation and the critical point and decreases below this level. However, below the critical point, the decrease in carbon dioxide release is significantly less than the decrease in oxygen consumption, indicating a change of energy metabolism. Between saturation and

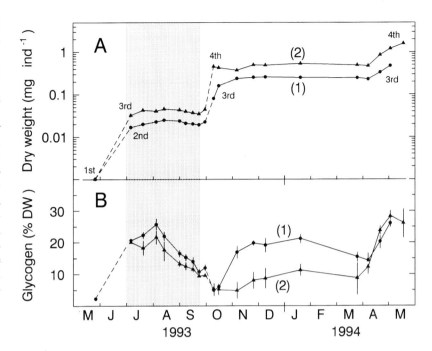

Figure 11.4. Seasonal changes in dry weight (A) and glycogen concentration (% of DW) (B) of two subpopulations of the 1993-cohort of *Chironomus anthracinus* in 1st to 3rd instar (curve 1) and 1st to 4th instar (curve 2). Transitions from one instar to the next are indicated by dashes. The shaded area delimits the period during stratification with oxygen concentrations below 1 mg O_2 l^{-1} in the bottom water. Vertical bars represent the standard deviation.

the critical point, the respiratory quotient (RQ = vol CO_2/vol O_2) is between 0.8 and 0.95, which indicates aerobic metabolism only. Below the critical point, the respiratory quotient increases gradually to 3.4 at 0.5 mg O_2 l^{-1}, suggesting an increase of anaerobic metabolism. At 3 mg O_2 l^{-1}, the total energy production (in kJ mg $AFDW^{-1}$ h^{-1}) constitutes 92% of the energy production at air saturation, of which the anaerobic part is 3%. At 0.5 mg O_2 l^{-1}, total energy production has declined to 20% of which the anaerobic contribution amounts to 8%. Thus, in *C. anthracinus,* anaerobic metabolism is low and cannot compensate for the decrease in aerobic energy production at low environmental oxygen concentration (Hamburger et al. 1994).

Seasonal fluctuations in glycogen content
Glycogen concentrations vary extensively at different larval stages of *C. anthracinus* (Fig. 11.4). The glycogen concentration increases during periods of intensive growth and decreases during periods with no growth or weight loss. The glycogen concentration of *C.*

anthracinus increased, for example, from 2% of dry weight in May 1993 in 1st instar larvae to 20% in July 1993 when the larvae were in the 2nd or 3rd instar (Hamburger et al. 1996).

The glycogen store is partially utilised during summer stratification with pronounced oxygen deficiency (Fig. 11.5). The dry weight of 2nd and 3rd instar larvae increases slightly during the first phase of the microxic period but decreases in the second phase, whereas the glycogen concentration is almost constant in the first phase and decreases in the second phase. Apparently, during the first phase, the oxygen concentration of the bottom water is still sufficiently high to support a certain, though greatly reduced, growth of the larvae. The change from growth to loss of weight and utilisation of the glycogen reserves in the second phase suggests a decrease of the oxygen concentration and a switch from fully aerobic to partly anaerobic metabolism. At the autumn overturn, the glycogen concentration has decreased to about 10% of dry weight in both instars. Similar, though less pronounced, changes in growth rate and glycogen concen-

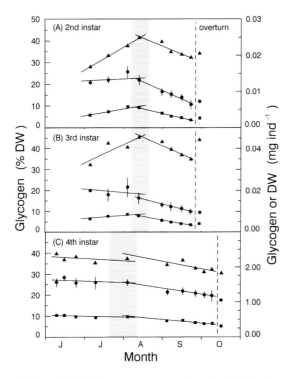

Figure 11.5. Glycogen concentration (% of DW) (●), dry weight content (▲), and glycogen content (■) of *Chironomus anthracinus* during summer stratification in 1992 (4th instar larvae) and 1993 (2nd and 3rd instar larvae). Shaded areas show the transition period between the first and second phase of microxia. Broken lines indicate the time of autumn overturn. Vertical bars represent the standard deviation. From Hamburger et al. 1995 (with kind permission from Kluwer Academic Publishers).

tration are observed in the large, almost full-grown 4th instar larvae.

The moulting from 2nd to 3rd or from 3rd to 4th instar takes place immediately after the autumn overturn and causes a further loss of the glycogen reserves to about 5% of dry weight (Fig. 11.4). After an increase of the glycogen concentration in the autumn and a slight decrease during winter, the glycogen concentration increases rapidly in spring accompanied by high larval growth rate and a bloom of diatoms. The glycogen concentration reaches a maximum of 25-30% of dry weight by early May. At this time all larvae are in 4th instar and part of the population pupates and emerges.

During each larval instar, the glycogen concentration increases with increasing weight of the larvae. The higher concentration of glycogen in 3rd compared to 4th instar larvae in autumn, can therefore be explained by the time the larvae have spent in the instar (Fig. 11.4). Thus, the 3rd instar larvae are almost full-grown and have a glycogen concentration close to maximum in November, well before growth ceases towards winter. The 4th instar larvae, however, are in the first stages of the instar at this time, and they have significantly lower glycogen concentrations.

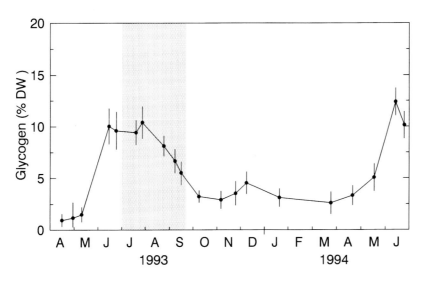

Figure 11.6. Seasonal changes in glycogen concentration (% of DW) of *Potamothrix hammoniensis* in 1993/1994. The shaded area indicates the period during summer stratification with less than 1 mg O_2 l^{-1} in the bottom water. Vertical bars indicate the standard deviation.

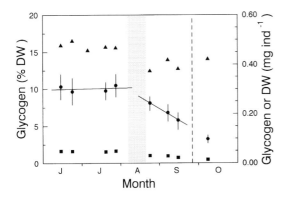

Figure 11.7. Glycogen concentration (% of DW) (●), dry weight content (▲), and glycogen content (■) of *Potamothrix hammoniensis* during summer stratification in 1993. The shaded area indicates the transition period between the first and second phase. Vertical bars indicate the standard deviation.

The glycogen concentration of *P. hammoniensis* varies from a few per cent up to 10% of dry weight during the year, regardless of the size of the organisms ranging between 0.1 and 2 mg DW (Nilson 1995). From early April to mid-May the glycogen concentration is only 1-2% (Fig. 11.6). During the following month it increases steeply to 10% by mid-June. The mean dry weight is constant during the first phase but lower during the second phase of the microxic period (Fig.

11.7). The glycogen concentration is constant in the first phase but decreases in the second phase. Although *P. hammoniensis* is able to maintain unchanged oxidative metabolism down to lower oxygen concentrations than *C. anthracinus*, the changes in glycogen concentration during the transition between the first and second phase of the microxic period indicate a switch from aerobic to partly anaerobic metabolism. The glycogen concentration remains low from autumn to early spring, and it starts to increase in mid-April, but the maximum level is not observed until June (Fig. 11.6).

Survival and glycogen conversion during experimental anaerobiosis

Individuals of the two species were exposed to deoxygenated water in the laboratory in closed bottles at 10°C (*C. anthracinus*) and 12°C (*P. hammoniensis*), which are typical hypolimnion temperatures during summer stratification. Before the experiments, the water was bubbled with nitrogen to produce a value of less than 0.15 mg O_2 l^{-1}. Due to respiration by the enclosed animals and bacteria, total or functional anoxia was achieved within a few hours.

Experiments were carried out at different seasons with *C. anthracinus* of varying size, instar and glycogen content. Small 2nd and

Figure 11.8. Survival (left) and glycogen concentration (% of DW) (right) of *Chironomus anthracinus* with time of exposure to experimental anoxia. (✳) 2nd instar larvae, from second phase 1993; (■) 3rd instar larvae, from March 1994; (●) 4th instar larvae, from March 1994; (▲) 4th instar larvae, from first phase 1992.

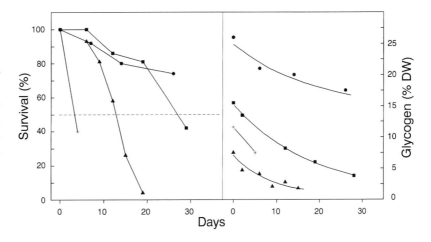

3rd instar larvae containing relatively high glycogen concentrations were sampled during the second phase of the microxic period in 1993, and large 4th instar larvae in their second year of growth during the first phase in 1992 (cf. Fig. 11.5). In March 1994, experiments were carried out with large 3rd instar larvae and small newly-moulted 4th instar larvae, with high and low glycogen concentrations respectively (cf. Fig. 11.4). The results show that survival time (i. e. survival of 50% of the individuals) depends on larval size and glycogen concentration at the start of the anoxic treatment (Fig. 11.8). The shortest survival time (4 days) was observed in 2nd and small 3rd instar larvae. The longest survival time of more than 26 days was found in large 4th instar larvae; after seven weeks 10% of these larvae were still alive. Intermediate survival times of 27 and 12.5 days respectively were found in 3rd and small 4th instar larvae. The glycogen concentration decreased in all larvae throughout the experimental period, and the decline was faster in the beginning than later during anaerobiosis. The larvae moved little while kept in anoxia, and their colour changed from bright red to bluish red. After transfer to aerated water they almost immediately performed normal movements, and their colour changed to bright red. We therefore suppose that *C. anthracinus* enters a state of dormancy, where the normal metabolic functions are kept on a low level. This type of dormancy is called quiescence because it is induced by environmental factors, and the larvae react with normal behaviour immediately after the supply of oxygen-rich water.

The importance of the initial content of glycogen for the survival time is most obvious in 3rd and small 4th instar larvae of approximately the same size but with highly different glycogen concentration. The 3rd instar larvae initially containing 15.5% glycogen lived more than twice as long as 4th instar larvae with initially 8% glycogen.

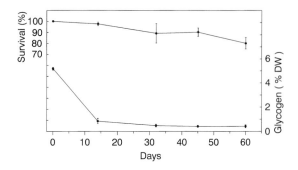

Figure 11.9. Survival (upper curve) and glycogen concentration (% of DW) (lower curve) with the time of exposure of *Potamothrix hammoniensis* to experimental anoxia. Vertical bars indicate the standard deviation.

Within the experimental period, the glycogen concentration of the 4th instar larvae declined to 2-3%, which is probably close to the lower limit for survival of *C. anthracinus*.

Small and large specimens of *P. hammoniensis* survived equally well during anoxia. The survival time exceeded 60 days (Fig. 11.9). Within the first two weeks, the glycogen concentration decreased from 5% to about 1% of dry weight, then decreased slowly during the following two weeks to about 0.5% and finally stayed constant at this level throughout the remaining part of the experiment. During the experiment, the animals clumped together and did not move, but they resumed movement immediately after transfer to aerated water (Nilson 1995). Thus, *P. hammoniensis* enters a state of dormancy (quiescence) during anoxia with an extremely reduced metabolic rate.

Are the profundal organisms in Lake Esrom exposed to anoxia?

The rate of glycogen conversion in field populations during the second phase of the microxic period is compared to that of organisms exposed to anoxic conditions in the

Table 11.1. Rate of glycogen conversion in field populations during the second microxic phase of summer stratification (A) and in anoxic laboratory experiments (B).

Instar	Dry weight mg	Glycogen conversion (μg glycogen mg^{-1} DW d^{-1})	Interval days	Year
A: Field populations				
C. anthracinus				
2	0.022	3.8	0-15	1993
3	0.040	3.8	0-15	1993
4	2.045	1.5	0-15	1992
P. hammoniensis				
	0.5	1.1	0-15	1993
B: Laboratory experiments				
C. anthracinus				
3	0.22	5.4	0-15	
4	0.32	3.6	0-15	
4	2.00	3.5	0-15	
P. hammoniensis				
	0.3	3.1	0-14	

laboratory (Table 11.1). In *C. anthracinus* larvae, the rate of glycogen conversion decreases with increase in larval size, but the rate is independent of size in *P. hammoniensis*. Animals exposed to experimental anaerobiosis show a 2-3 times higher conversion rate than animals from the lake. It is therefore possible that the profundal organisms in the lake rarely experience complete anoxia.

Hamburger et al. (1995) argued that the energy metabolism of *C. anthracinus* during the second phase of the microxic period was mainly aerobic. An estimation of the aerobic and anaerobic metabolism from the decrease in dry weight and glycogen content during the second phase (cf. Fig. 11.5) indicated that total metabolism made up from 13% to 20% of the energy metabolism at air saturation, and that only 5% of this metabolism was anaerobic. Relating this estimate to the measured respiratory metabolism at different oxygen concentrations (cf. Fig. 11.3) led to the conclusion that in order to maintain aerobic

metabolism of this magnitude, the bottom water must contain about 0.5 mg O_2 l^{-1}.

This discrepancy between measured and estimated oxygen concentrations is difficult to understand. The most likely hypothesis is that oxygen is constantly transferred at a very low rate from the upper oxygen-rich to the lower microxic part of the hypolimnion and utilised at the same rate by the bottom sediment and the associated fauna and bacteria. Wind-induced circulation in the hypolimnion of stratified lakes is well documented (e.g. Imboden et al. 1983, Monismith 1985), but has not been studied in Lake Esrom. However, the increase in temperature, for example in 1993, from 7.2°C in June to 8.5°C by the end of September, indicates transfer of relatively warm, oxygen-rich water from the upper hypolimnion to the lower oxygen-poor hypolimnion. The switch from a totally aerobic metabolism during the first phase to a partially anaerobic metabolism during the second phase may be explained by a reduced transfer of oxygen from a decreasing volume

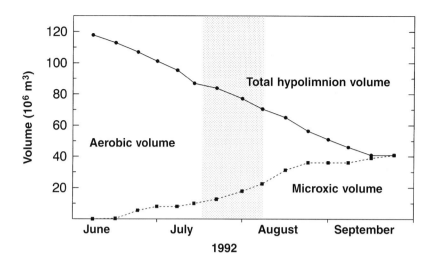

Figure 11.10. Diagram showing the increase in microxic volume and the concurrent decrease in hypolimnic aerobic volume in Lake Esrom during the stratification period in 1992. The shaded area shows the transition period between the first and second phase of microxia.

of oxygen-rich hypolimnion to a microxic hypolimnion of increasing volume (Fig. 11.10).

Survival in Lake Esrom

The laboratory experiments showed that the mortality in *C. anthracinus* increases strongly at glycogen concentrations below 2-3% (Fig. 11.8). In the following estimation of the theoretical maximum survival time of *C. anthracinus* in Lake Esrom, it is assumed that glycogen reserves can be exploited down to 2.5%, and that the individuals die when the concentrations drop below this level. From the rate of decrease in glycogen concentration during summer stratification in 1993 it is estimated that 2nd instar larvae would have been able to survive oxygen deficiency during a second phase of the microxic period for two and a half months and 3rd instar larvae for three months; thus, survival would have occurred if the overturn had been postponed until the end of October. The 4th instar larvae would theoretically be able to survive, as much as six months longer than the time that was spent with oxygen deficiency in the lake during summer stratification in 1992.

During experimental anoxia *P. hammoniensis* utilises glycogen reserves down to 0.5% of dry weight and thereafter enters a state of dormancy, the length of which is unknown but exceeds at least 60 days. At the time of overturn in 1993, the glycogen reserves of *P. hammoniensis* was 3% and would theoretically reach the level of 0.5% one month later. *P. hammoniensis* is therefore able to survive true anoxic conditions better than *C. anthracinus*, which is confirmed by the distribution pattern of these two species in many eutrophicated lakes. This difference in ability to survive anoxic conditions is useful when monitoring long term changes in profundal conditions (Lindegaard et al. 1993, 1997).

The duration of the stratification period is shown for the years 1991 to 1995 (Fig. 11.11). The earliest record of less than 1 mg O_2 l^{-1} in the hypolimnion was by mid-May in 1992. The autumn overturn occurred in mid-October that year, and the period with oxygen deficiency thus lasted four months, out of which the second phase constituted about two months. A prolongation of the stratification to the end of October is not unrealistic, as it was observed as late as 25 October in 1955 (Jónasson 1972). The longest second phase is therefore theoretically about 100 days long (Fig. 11.11). This period is probably cri-

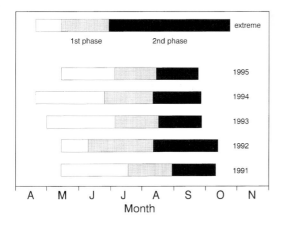

Figure 11.11. The duration of stratification in Lake Esrom during 1991-1995. White and dark bars show oxygen concentrations above the sediment higher and lower than 1 mg O_2 l^{-1} respectively. The top bar shows the theoretically longest duration of stratification and thereby the worst possible condition for profundal life. The first and second phases of microxia are indicated schematically.

tical for the small first year *C. anthracinus* larvae, but not for the large second year larvae. It is, therefore, not surprising that *C. anthracinus* in Lake Esrom survived a prolongation of the microxic period from a few days in the 1930's to several months at present. In fact we have not been able to measure any enhanced mortality during summer stratification in the 1990's (Lindegaard et al. 1997).

These estimates of theoretical survival times for *C. anthracinus* and *P. hammoniensis* in Lake Esrom, however, presuppose that 95% of the reduced metabolism during microxia continues to be aerobic, and the survival therefore depends on a more or less continuous supply of oxygen caused by internal circulation in the hypolimnion. That oxygen is a key factor in maintaining large populations of *C. anthracinus* is supported by the fact that in other eutrophic lakes exposed to eutrophication, population abundance and

vertical distribution of *C. anthracinus* are strongly decreased due to a prolongation of the anaerobic period. The 31 m deep Lake Hald (Jutland, Denmark) before 1950 possessed a high density of *C. anthracinus* all over the profundal zone. Due to heavy eutrophication, the population rapidly decreased in density and was in 1980 restricted to the 10-20 m depth zone. After an artificial oxygenation of the hypolimnion was established in 1984, the *C. anthracinus* population increased dramatically and repopulated the deep profundal areas.

We believe differences in wind fetch and lake morphometry of the two lakes to be decisive in determining internal circulation in the hypolimnion and thereby indirectly influencing the microxic conditions. Lake Hald is a small (2 km^2), relatively deep lake (31 m) with a typical V-shaped depth profile resulting in a profundal zone of moderate size preventing circulation to the deepest part of the hypolimnion. Lake Esrom is 17 km^2, 21 m deep with a large U-shaped profundal zone providing excellent conditions for wind induced hypolimnetic circulation. We therefore suggest that lake morphometry and the initiation and location of the metalimnion are important factors in controlling the microxic regime in hypolimnions. Indirectly, these factors are decisive for the survival and vertical distribution of *C. anthracinus*.

Concluding remarks

Our results demonstrate that three adaptations to survive profundal oxygen depletion during the summer stratification are successfully combined by *Chironomus anthracinus* and *Potamothrix hammoniensis* in Lake Esrom: (1) the ability to maintain an almost unchanged oxygen uptake down to 10-25% of air saturation, which enables them to eat and grow for one or two months after establishment of the thermocline; (2) during anoxia they enter a type of dormancy with reduced metabolism; and (3) during anoxia they provide the necessary energy by anaerobic degradation of glycogen.

Growth during the first part of stratification when the oxygen content decreases is especially important for the new-settled 1st instar larvae of *C. anthracinus*, because they must reach 2nd instar and increase their glycogen content from 2% to about 20% of dry weight before the start of the microxic period. Similarly, *P. hammoniensis* increases the glycogen content to about 10% of dry weight prior to the microxic period.

The combination of dormancy and anaerobic degradation of glycogen to about 2% of dry weight enables *C. anthracinus* to survive experimental anoxia for 20 to 40 days depending on the larval instar. *P. hammoniensis* exploits its glycogen down to 0.5% of dry weight during 15 days of experimental anoxia and subsequently enters a type of dormancy with extremely reduced metabolism. It survives at least 60 days exposed to experimental anaerobiosis. We therefore consider *P. hammoniensis* to be better adapted to oxygen depletion than *C. anthracinus*.

The microxic period in Lake Esrom varies from two to four months and clearly exceeds the period which *C. anthracinus* is able to survive in anoxia. Neither of the two species exploit their glycogen down to the lower limit in Lake Esrom, and anaerobic metabolism accounts for only 5% of the total metabolism during the late phase of microxia. We therefore hypothesise that the profundal animals in Lake Esrom rarely experience complete anoxia, and that wind-induced circulations provide sufficient oxygen for maintaining an almost aerobic, although reduced, metabolism during summer stratification.

We conclude, that *C. anthracinus* is unable to survive the stratification period in lakes where (1) the period prior to the microxic conditions is too short for the larvae to build up glycogen stores large enough for surviving later anoxic conditions; (2) the anoxic period exceeds 20-40 days; (3) the lake morphometry does not allow wind-induced internal circulations providing the profundal species with sufficient oxygen to maintain an almost aerobic metabolism during summer stratification.

Literature Cited

Berg, K., P. M. Jónasson, and K. W. Ockelmann. 1962. The respiration of some animals from the profundal zone of a lake. Hydrobiologia 19:1-39.

Frank, C. 1983: Ecology, production and anaerobic metabolism of *Chironomus plumosus* L. larvae in a shallow lake. II. Anaerobic metabolism. Archiv für Hydrobiologie 96:354-362.

Hamburger, K., P. C. Dall, and C. Lindegaard. 1994. Energy metabolism of *Chironomus anthracinus* (Diptera: Chironomidae) from the profundal zone of Lake Esrom, as a function of body size, temperature and oxygen concentration. Hydrobiologia 294: 43-50.

Hamburger, K., P. C. Dall, and C. Lindegaard. 1995. Effects of oxygen deficiency on survival and glycogen content of *Chironomus anthracinus* (Diptera, Chironomidae) under laboratory and field conditions. Hydrobiologia 297:187-200.

Hamburger, K., C. Lindegaard, and P. C. Dall. 1996.

The role of glycogen during the ontogenesis of *Chironomus anthracinus* (Chironomidae, Diptera). Hydrobiologia 318:51-59.

Hand, S. C. 1991. Metabolic dormancy in aquatic invertebrates. Advances in Comparative and Environmental Physiology 8: 1-49.

Imboden, D. M., U. Lemmin, J. Joller, and M. Schurter. 1983. Mixing processes in lakes: mechanisms and ecological relevance. Schweizerische Zeitschrift für Hydrobiologie 45: 11-44.

Jónasson, P. M. 1972. Ecology and production of the profundal benthos in relation to phytoplankton in Lake Esrom. Oikos Supplementum 14:1-148.

Jónasson, P. M., and F. Thorhauge. 1972. Life cycle of *Potamothrix hammoniensis* (Tubificidae) in profundal of a eutrophic lake. Oikos 23:151-158.

Jónasson, P. M., and F. Thorhauge. 1976. Population dynamics of *Potamothrix hammoniensis* in the profundal of Lake Esrom with special reference to environmental and competitive factors. Oikos 27:193-203.

Lindegaard, C., P. C. Dall, and S. B. Hansen. 1993. Natural and imposed variability in the profundal fauna of Lake Esrom, Denmark. Verhandlungen der Internationale Vereinigung für Theoretische und Angewandte Limnologie 25:576-581.

Lindegaard, C., P. C. Dall, and P. M Jónasson. 1997. Long-term patterns of the profundal fauna in Lake Esrom. Chapter 4 *in* K. Sand-Jensen and O. Pedersen, editors. Freshwater Biology. Priorities and Developments in Danish Limnology Gad, Copenhagen, Denmark.

Monismith, S. G. 1985. Wind-forced motions in stratified lakes and their effect in mixed-layer shear. Limnology and Oceanography 30: 771-783.

Nilson, I. B. 1995. *Potamothrix hammoniensis* (Tubificidae, Oligochaeta) in Lake Esrom. A study of energy production and survival during microxia. MS-thesis, Freshwater Biological Laboratory, University of Copenhagen, Denmark.

Putzer, V. M., A. De Zwaan, and W. Wieser. 1990. Anaerobic energy metabolism in the oligochaete *Lumbriculus variegatus* Müller. Journal of Comparative Physiology B 159: 707-715.

Seuss, J., E. Hipp, and K. H. Hoffmann. 1983. Oxygen consumption, glycogen content and the accumulation of metabolites in *Tubifex* during aerobe-anaerobe shift and under progressing anoxic shift. Comparative Biochemistry and Physiology 75 A: 557-562.

Thorhauge, F. 1976. Growth and life cycle of *Potamothrix hammoniensis* (Tubificidae, Oligochaeta) in the profundal of eutrophic Lake Esrom. A field and laboratory study. Archiv für Hydrobiologie 78: 71-85.

Wilps, H., and U. Schöttler. 1980. In vito studies on the anaerobic formation of ethanol by the larvae of *Chironomus thummi* (Diptera). Comparative Biochemistry and Physiology 67 B: 239-242.

Wilps, H., and E. Zebe. 1976. The end-products of anaerobic carbohydrate metabolism in the larvae of *Chironomus thummi*. Journal of Comparative Physiology 112: 263-272.

12 The nature of water transport in aquatic plants

By Ole Pedersen

Evaporation of water from leaf surfaces drives the transpiration stream in terrestrial plants. The transpiration stream carries major nutrients from the roots to the leaves, where new biomass is formed. Transpiration cannot take place in submerged plants, which may therefore face problems in supplying the apical shoots with sufficient nutrients derived from the sediment. It is shown here that submerged plants, nevertheless, transport water from root to shoot. This water transport probably serves as an important vehicle for inorganic nutrients and phytohormones, both derived from the roots. It thereby ensures optimal plant growth in the absence of a transpiration stream.

The transpiration of terrestrial plants serves several important functions. It replenishes water lost by evaporation from the leaves and thereby maintains the water potential in the tissue. Evaporation of water from the leaf surface serves to cool the leaf and eliminates the disruption of enzymatic processes that can be caused by excessive heat. The transpiration stream through the xylem serves to translocate inorganic nutrients taken up in the root and phytohormones produced by the root to the sites of active growth in the shoot. The stomata are by far the most important pathway for water loss in terrestrial plants. About 90% of the water is lost via the stomatal pathway. The remaining 10% are lost through the cuticle and lenticels (Salisbury and Ross 1992). The stomata act as the major site of water loss because they have to be open to the surrounding environment to allow CO_2 to diffuse into the leaf. Thus, the terrestrial plants face a dynamic balance between CO_2 acquisition and water loss. In contrast, floating plants and amphibious plants, which generally grow in water-saturated soil, never experience a water deficit, though they produce numerous stomata providing access to atmospheric CO_2 (Gessner 1959, 1956, Sculthorpe 1967). The truly submerged plants lack the transpiration stream and consequently may suffer from nutrient deficiency if they have no transport mechanism taking the place of the transpiration stream (Bowes 1987).

In the following, we will consider how the vascular tissue of submerged plants is adapted to an environment with no transpiration. The water transport which, nevertheless, takes place in submerged plants is probably driven by root pressure, and a control mechanism is proposed to explain how the water is channelled to sites of active growth, thereby providing an efficient transport system for inorganic nutrients and phytohormones. Finally, I challenge the allegation that the functional hydraulic performance of submerged plants is inferior to that of terrestrial plants.

The challenge for aquatic plants

In an environment where transpiration is absent, communication between root and shoot is a fundamental challenge to aquatic plants. In the water columns of lakes, the availability of dissolved phosphorus, iron and manganese is usually low. It is often below detection limits due to low solubility in the oxygenated water and also due to efficient competition from phytoplankton which is able to strip the water of nutrients. The major re-mineralisation takes place in the sediment, providing a steady supply of ammonia and phosphorus and a low redox potential. This assures that PO_4^{3-}, Fe^{2+} and Mn^{2+} are accessible in soluble forms. Due to a thin cuticle, most aquatic plants can absorb nutrients over the entire plant surface. (Arber 1920, Denny 1980). Most experiments have shown that aquatic plants predominantly grow using the sediment as the main source of nitrogen and phosphorus (Carignan and Kalff 1980, Barko and Smart 1981, Chambers et al. 1989). In the case of phosphorus, 90-100% of the uptake can be derived from the sediment (Carignan and Kalff 1980, Barko and Smart 1981). Consequently, we need to search for a mechanism and transport route to account for nutrient translocation from the roots to the distant apical shoot meristems. Moreover, we may look for a type of mechanism also present in terrestrial herbaceous plants, because the aquatic rooted plants have preserved much of the vascular system of tracheids and vessels present in their terrestrial ancestors (Schenck 1886, Arber 1920, Gessner 1959, Sculthorpe 1967, Hutchinson 1974, Hostrup and Wiegleb 1991). In the following, I consider how the water is pressed through the vascular system of aquatic plants when the driving force of transpiration is absent.

The driving force for non-transpirational water transport

Back in 1938 Crafts and Broyer presented a model to account for *root pressure*. Root pressure causes water to move up through the stalk of terrestrial herbaceous plants, and during periods of low transpiration the water is pressed out through the apical leaf openings to produce drops of guttation. A slightly modified version of the Crafts and Broyer model is still widely accepted. According to Crafts and Broyer (1938), ions can travel from the surrounding medium in either the symplastic or apoplastic compartment in the root until they meet the endodermis. Here the Casparian band usually blocks the apoplastic

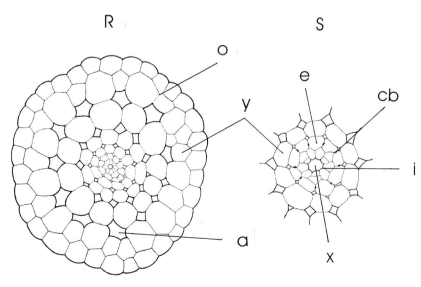

Figure 12.1. Cross-section of an entire *Vallisneria spiralis* root, R and the central stele, S (Schenck, 1886). Inner, i and outer, o, apoplast (the cell wall continuum), the symplast, y (the interconnected protoplasts and their plasmodesmata), the endodermis, e, with the Casparian band, cb, the xylem lacuna, x and air-filled lacunae, a.

Figure 12.2. Guttation is the formation of small droplets of water at the leaf tips. Root pressure propels water up through the plant where it is exuded from special pores in the leaves called hydathodes. The hydathodes are located at the rim of leaves and are often numerous. Guttation builds up when the water transport exceeds the evaporation from the leaves, which is usually only the case during night time or early in the morning when the atmosphere is highly humid (see the terrestrial *Alchemilla*, top). Guttation is also observed in some aquatic plants (see *Lobelia dortmanna*, bottom) where it has been used to quantify the water transport.

pathway (Fig. 12.1). At this point the ions have to enter the symplast by means of energy requiring ion pumps. The symplast runs continuously via plamodesmata from the endodermis cells to other living cells within the stele. Once located in the symplast, the ions are pumped into the inner apoplast. This build-up of high ion concentrations leads to an associated passive flow of water into the apoplast forming a hydrostatic overpressure. Because of the differential permeability of the endodermis, the root acts as an osmotic system, and the resulting hydrostatic pressure can only equilibrate by moving quantities of water up through the stem.

The precise nature of root pressure in aquatic plants has not been worked out in detail. Frankly, only one literature value from a very preliminary experiment has been published. In an intact root system of *Myriophyllum spicatum*, I measured a root pressure of 30 kPa. This is enough to press water up through plants more than 2m tall (Pedersen 1994). There are other indications that root pressure operates in submerged aquatic plants. By measuring the guttation rate (i.e. the formation of water drops from leaf hydathodes, Fig. 12.2) in *Sparganium emersum*, I have shown that the driving force for guttation is located in the root (Pedersen 1993). This experiment was done by cooling the root (Fig. 12.3, Pedersen 1993) or by introducing an ATPase inhibitor to the root cells. Both treatments resulted in a significant drop in guttation rate, thus demonstrating that the driving force resides in the root and most probably is present as root pressure.

Additional driving forces may operate besides root pressure. Especially, *Münch counter flow* may be important in submerged aquatic plants (Münch 1930). Münch flow is still the most commonly accepted mechanism for explaining how pressurised mass flow is generated in the phloem. According to this theory, sugars and amino acids are loaded into the phloem sap at the source (photosyn-

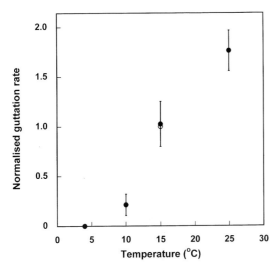

Figure 12.3. Temperature dependence of the guttation rate in the submerged form of *Sparganium emersum*. The guttation rate is normalised to the rate at 15°C. The plant was kept in a two-compartment system where the temperature around the root and the shoot could be controlled independently. Guttation stops when the temperature around the root is adjusted to 4°C, whereas it is increased 1.8-fold when the temperature is raised from 15 to 25°C. The experiments confirm that the driving force for guttation is restricted to the root.

thetically active leaves), and water follows by down-hill diffusion from the nearby xylem and generates hydrostatic pressure. This pressure equilibrates by mass flow through the phloem, and the sugars and amino acids are withdrawn from the phloem at the sink sites (roots and shoot meristems). During Münch counter flow, the water re-enters the xylem and thereby generates a closed circuit. Therefore, the occurrence and importance of Münch counter flow is difficult to track since the water is retained within the plant rather than guttated through the apical openings. Currently we do not know how important this transport mode is compared to root pressure. Regardless of the driving force *per se,* the water has to travel up through the plant within the vascular bundles of tracheids and

vessels. These cells are pipes of very small dimensions, and the flow through the vascular bundles therefore obeys the laws of capillary flow (Nobel 1991).

The physics of capillary flow

Fluid flow in capillaries is strongly controlled by the law of physics. The Hagen Poiseuille equation is usually applied to describe the *discharge of water* (Q) of an ideal capillary as the function of the *pressure gradient* (dP/dl), the *radius* (r) and the *dynamic viscosity* (η) of the fluid involved. If the vascular system contains *n* capillaries, the discharge can be expressed as (Gibson et al. 1984, Zimmermann 1983),

$$Q = \frac{dP}{dl} \frac{\pi \sum_{i}^{n} r_i^4}{8\eta} \quad (12.1)$$

The most important aspect to notice from eq. 12.1 is that the water discharge scales to the fourth power of the capillary radius. Thus, vessel dimensions are extremely important in determining the resistance and flow capacity. Eq. 12.1 allows us to estimate the discharge of a given vascular system from measurements of the pressure gradient and the number and dimensions of the vessel. If one wishes to characterise a system of capillaries in terms of transport efficiency, the parameter *hydraulic conductance per unit length* (K_h) can be applied (Gibson et al. 1984),

$$Q = K_h \frac{dP}{dl} \quad (12.2)$$

and by combining eq.12.1 and 12.2, the hydraulic conductance per unit length can be written as

$$K_h = \frac{\pi \sum_{i}^{n} r_i^4}{8\eta} \quad (12.3)$$

We are now able to work with the hydraulic properties of plants in theory and experimentally. At the theoretical level, the hydraulic conductance per unit length can be estimated on the basis of microscopic measurements of vessel dimensions (eq. 12.3). At the experimental level, the hydraulic conductance per unit length can be measured in a pressure bomb (Pedersen et al. 1997) where, for example, a piece of stem (l) is mounted, pressure (P) is applied and the following discharge (Q) of water is measured (eq. 12.2). Usually, one will find a discrepancy between the two estimates in terms of a larger estimated K_h (eq. 12.3) as compared to the measured K_h (eq. 12.2). This difference can be explained by the fact that not all vessels take an active part in water transport; a problem which is primarily associated with heartwood-forming species (Zimmermann 1983). Equation 1-3 are only strictly valid for ideal capillary systems, and the vascular elements of angiosperms are not ideal at all. Nevertheless, the three equations are indispensable tools when trying to evaluate the importance of vessel dimensions for the water transport capacity.

The anatomy of the vascular pathway

Around the turn of the century, Arber (1920) intensively studied the anatomy of submerged vascular plants. Because of the limited availability of micro-techniques, she was led to conclude that the functional performance of the vascular system in aquatic plants was much reduced compared to terrestrial plants. Her major argument was the striking lack of lignification of the xylem in aquatic plants. The lignification was interpreted as a necessity for a functional xylem pathway. As we shall see later, this may not impede the function of the xylem if it operates under a positive pressure. Arber (1920) also described the xylem lacunae which may be of general importance in the adaptation of the vascular tissue to a submerged life form (Fig. 12.1). Xylem lacunae seem to be common in submerged vascular plants. They form huge continuous pipes in the central stele with large diameters, so that resistance to water transport is significantly lowered and the hydraulic conductance many-fold increased (Fig. 12.4). An extraordinarily low resistance to water flow is probably important for submerged aquatic plants, which

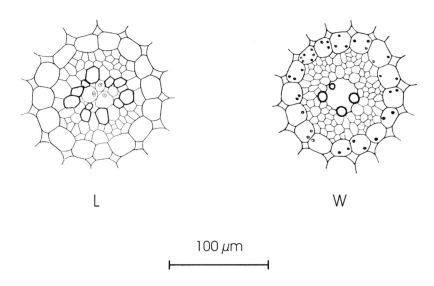

L   W

100 μm

Figure 12.4. The central stele of *Callitriche stagnalis* in the land form (L) and the water form (W) (Schenck, 1886). In the water form, cortex cells has been resorbed and one large xylem lacuna is formed in place of the 11 xylem elements present in the land form. The 11 xylem elements provide an estimated hydraulic conductance (eq. 12.3) of $1.5 \cdot 10^{-17}$ m^4 Pa^{-1} s^{-1} whereas the single xylem lacuna of the water form supports a hydraulic conductance 9-fold higher ($1.3 \cdot 10^{-16}$ m^4 Pa^{-1} s^{-1}).

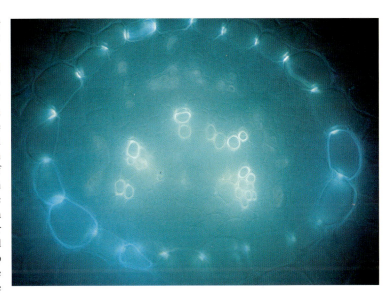

Figure 12.5. The central vascular cylinder of a *Lobelia dortmanna* leaf stained with berberine hemisulphate reacting with lignin and waxes. The five groups of lignified xylem elements are distinctly stained as well as the Casparian band (see also Fig. 12.1). The presence of a Casparian band in stalks and leaves is quite rare in terrestrial plants. The presence of a continuous Casparian band in roots, stalks and leaves of aquatic plants probably provides a hydraulic insulation to the outer apoplast, ensuring that water and nutrients are carried all the way up to the distal hydathode before superfluous water escapes to the surroundings.

have to invest energy to press up water through the stalk. In contrast, water transport is almost energy-neutral in transpiring terrestrial plants.

Once water has entered the roots, and sufficient root pressure has been generated, the water is forced up through the stem and out into the leaves. This positive pressure has probably required further adaptations of the vascular bundles in aquatic plants. Currently, we have only preliminary evidence of a Casparian band in the endodermis of both the stalk and the leaves. The Casparian band, a strip of fatty substances and lignin covering the anticlinal walls of the endodermis (Fig. 12.1), is usually accepted as a hydraulic insulator between the outer and the inner apoplast of the roots in most angiosperms. When water is taken up in the roots, it travels through the apoplast until it reaches the Casparian band. Here, it enters the symplast of the endodermis cells, and the water later follows ions released from the inner side of the endodermis to the inner apoplast. The Casparian band hinders water movements through the apoplastic pathway (Nagahashi et al. 1974) and thereby allows the build-up of hydrostatic pressure within the inner apo-

plast, root pressure. In terrestrial plants, root pressure is not the prime driving force for acropetal water transport, and the Casparian band usually only lines the endodermis cell walls in the root (Raven et al. 1982). In most aquatic angiosperms tested so far, however, the Casparian band is found in the stalks, the petioles and even in the leaves (Fig. 12.5). It is logical to interpret the presence of a continuous Casparian band in these aquatic plants as an important adaptation to root pressure, which could then be the prime driving force for water transport. The Casparian band acts as an efficient hydraulic insulator and probably hinders significant radial water flow from high pressure areas within the central vascular cylinder and out through the stalk to the surrounding environment. Instead, the water is retained within the inner apoplast. Here it probably serves to bring up the nutrients and hormones it carries from the root to the apical meristems.

Measurements of mass transport of water

Water transport in submerged plants has previously been measured by qualitative and quantitative techniques. Experiments performed a hundred years ago with dyes and potometers suggested that acropetal water transport took place in various submerged plants (Unger 1862, Sauvageau 1891, Thut 1932, Wilson 1947). However, the results were ambiguous, and no consensus was reached until Höhn and Ax (1961) published more convincing measurements of water transport in the amphibious *Nomaphila stricta* under submerged conditions. They used a very sensitive double-potometer. The root and the shoot parts were hydraulically separated in two chambers. They observed a change in the volume in both compartments, but they were not able to deduce, which proportion was due to genuine water transport or to changes in the volume of the internal air-filled lacunae.

Tritiated water has been used as a suitable qualitative and quantitative tracer for water transport in submerged plants (Pedersen and Sand-Jensen 1993). We tested nine species of aquatic plants comprising different growth forms in a two-compartment system with ^{3}HHO labelled water in the root chamber. Eight species transported water from the root to the shoot and out into the surrounding shoot chamber at a greater rate than could be accounted for by pure diffusion. Analysis of the plants showed that the tritiated water was channelled to the most growth-active sites in the shoot, namely new leaves and adventitious shoots formed during incubation (Pedersen and Sand-Jensen 1993).

Tritiated water was also used to measure the amount of water transported by root pressure and by transpiration (Pedersen and Sand-Jensen 1997). The amphibious plant *Mentha aquatica* maintained a rate of water transport under submerged conditions at 14% of the transpiration rate in air. Additional studies of nutrient concentrations in amphibious plants, grown with air contact or totally submerged, showed no significant difference in nitrogen and phosphorus content. Moreover, because no difference in growth rate was observed, we concluded that the trans-

piration stream *per se* was not necessary for maintaining maximum growth rates in *Mentha aquatica* (Pedersen and Sand-Jensen 1997).

In 1899 von Minden observed guttation (Fig. 12.2) in a variety of submerged and amphibious plants. These observations were adapted to include guttation in quantitative measurements of water transport in submerged plants (Pedersen 1993). This technique can provide a conservative measure for the acropetal long-distance transport of water which takes place in the submerged plants, but the technique is confined to species with functional leaf hydathodes. With the leaf tips held in a small, enclosed, highly-humid atmosphere, guttation drops emerge and can be collected with an accuracy of within 0.01 µl by means of microglass capillaries. Determination of guttation rates has supported data obtained by application of tritiated water and has yielded transport rates up to 0.75 ml H_2O g^{-1} leaf DW h^{-1}, though average rates are usually in the range of 50 to 250 µl H_2O g^{-1} leaf DW h^{-1} (Pedersen 1993). The distribution

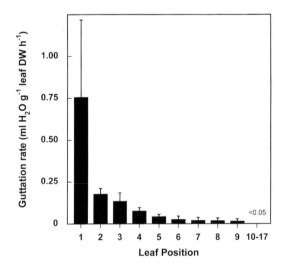

Figure 12.6. Guttation from leaves of *Lobelia dortmanna*. The young leaves are most actively guttating, and the entire pool of tissue water is flushed up to 2.5 times d^{-1}.

pattern, with the main flow channelled to the young leaves, is supported by guttation measurements (Fig. 12.6, Pedersen 1993). All plants tested so far show a more active guttation in the young leaves – an observation which also holds for many terrestrial plants (Takeda et al. 1991).

The measurements of water discharge rates allow us to calculate the flow velocity of water in the vascular system. This is important for judging the effectiveness of the communication system when root-produced phytohormones exert their information on distant apical tissues. Water moves in capillaries in a manner which can be best described as parabolic front flow, according to Zimmermann (1983)

$$v_{max} = \frac{dP}{dl} \frac{r^2}{4\eta} \qquad (12.4)$$

where v_{max} is the maximum flow velocity and the other variables appear from eq. 12.1 to 12.3. From the knowledge of vessel dimensions, measurements of volume-based discharge rates, and calculated pressure gradients (eq. 12.1), I estimated that the water moves with a velocity of up to 1 m h^{-1} in, for example, *Lobelia dortmanna* (Pedersen 1993), which is within the upper-end of flow velocities measured in the phloem of terrestrial plants (Biddulph and Cory 1957, Cataldo et al. 1972). Thus, this flow provides a sufficiently fast transport system, ensuring efficient communication between root and shoot in submerged plants – and in the absence of transpiration.

Control of the through-flow of water

The guttation results inspired us to examine the anatomical details of the hydathodes because they provide a potential site for through-flow control. Hostrup and Wiegleb (1991) were the first to present microscopic documentation suggesting that the leaf tip may act as an important site for flow regula-

tion. In *Littorella uniflora*, the leaf tips underwent considerable anatomical modification in aerial leaves compared to the leaf tip of normal, submerged leaves. The submerged leaves have the vascular tissue running all the way out to the apical opening, which is without epidermis (Hostrup and Wiegleb 1991). Consequently, there is a direct contact from the xylem elements to the surrounding water. The scenario is completely different in leaves with air contact. Here, the leaf tip is intact and there is no indication of a hydathode or water pore (Hostrup and Wiegleb 1991). The implications of these differences in structure are that the terrestrial forms need strict control of water loss, which is an inevitable consequence when functional stomata develop in the aerial leaves. The submerged forms, on the other hand, do not face any risk of critical water loss to the surroundings, but the exposed apical openings instead allow excessive water from the xylem elements to escape to the surroundings as submerged guttation.

We may now consider whether this escape of water to the surroundings is subject to any physiological control? Apparently it is not, though Mortlock (1952) ingeniously noticed clogging of the hydathodes in *Ranunculus fluitans*. He observed that older *Ranunculus* leaves were clogged by what he referred to as brown gum. He suggested that the clogging might "be of some value in diverting the slight flow in the xylem, with the solutes it probably contains, to the region of most active growth" (Mortlock 1952). Occlusion of leaf hydathodes has also been reported in terrestrial plants. Takeda et al. (1991) demonstrated that occlusion of old leaf hydathodes by a waxy substance hindered guttation, and that guttation resumed when the wax was removed by gently scraping the leaf tips. Therefore, clogging of the apical openings may provide a very simple and effective way of controlling the through-flow of water in submerged leaves because the water will fol-

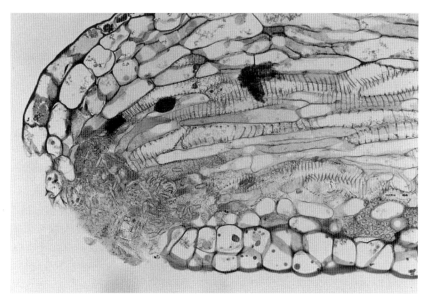

Figure 12.7. The leaf tip of an old *Sparganium emersum* leaf. The hydathode is completely clogged by a matrix of micro-organisms forming a high resistance to through-flow of water.

low the route of less resistance through the leaves.

We set out to test whether such a system regulated water flow through the leaves of the submerged form of *Sparganium emersum* by examining the hydraulic performance among leaves of different age (Pedersen et al. 1997). Experiments with leaf segments in the pressure bomb showed that the hydraulic conductance per unit length (K_h, eq. 12.3) was much larger in old leaves than in young leaves. However, the pattern changed completely when we used leaves with intact leaf tips. The hydraulic conductance per unit length was then significantly higher in the young leaves. This indicates that the leaf tip can form a major resistance to the flow in old *Sparganium emersum* leaves (Pedersen et al. 1997). Microscopic analysis of the leaf tips of *Sparganium emersum* revealed an apical opening much similar to the one common among species of Alismataceae and Potamogetonaceae (von Minden 1899, Weinrowsky 1898, 1899). The apical opening of *Sparganium emersum* leaves underwent substantial development from young to old leaves. However, we only observed minor

filling of the opening with $CaCO_3$ and an unidentified substance, compared to the much more conspicuous colonisation of micro-organisms in the hydathode reflecting senescence (Fig. 12.7). We believe that the dense matrix of micro-organisms may severely impede through-flow of water in the apical opening, whereas the $CaCO_3$ and the unidentified material are of minor importance to the total flow resistance because these substances only block a minority of vessels (Pedersen et al. 1997). Overall, these experiments show that the hydraulic conductance per unit length is greatest in young leaves in accordance with the findings in the guttation experiments (Pedersen et al. 1997). The control of hydraulic conductance, on the other hand, cannot be characterised as active because the plant can hardly control its own colonisation by the microorganism, unless hitherto unidentified organic polymers are excreted from the hydathode to attract the micro-organisms.

The leaf-specific conductivity (LSC)
Most of the submerged plants tested so far do transport water up through the stalk in the

absence of transpiration. I shall finally consider how their hydraulic performance compares to that of terrestrial plants.

Conceptual understanding of the hydraulic performance of plants improved significantly when Huber (1928) introduced what later became the *Huber value*. The Huber value is by definition the cross-sectional xylem area at a given point of the stem, divided by the fresh weight of leaves supplied acropetally to the cross-sectional point. This ratio was of great value, because for the first time it allowed comparisons of hydraulic properties among species across the entire plant kingdom.

To avoid the complications associated with the calculation of the Huber value, it was later modified by Zimmermann (1978) and presented as the leaf-specific conductivity (LSC). LSC is defined as K_h divided by the leaf fresh weight supplied, though the more recent preference is to express K_h relative to the leaf area supported (*A*) (Zimmermann 1983, Gibson et al. 1984, Tyree and Sperry 1988, Yang and Tyree 1993),

$$LSC = \frac{K_h}{A} \qquad (12.5)$$

Eq. 12.5 illustrates that LSC can be based on estimates (microscopy of cross sections, eq. 12.3) as well as direct determinations (pressure-driven flow measurements, eq. 12.2) of hydraulic conductance. Thus, the latter allows estimates of LSC devoid of the disturbing contribution of non-functional xylem elements.

We have recently compiled terrestrial LSC data from the literature and calculated the LSC for the submergent form of *Sparganium emersum* (Pedersen et al. 1997). The data obtained for *Sparganium emersum* do not support the traditional belief in poor functional performance of the vascular tissue in aquatic plants. So far only one species has been tested, and more work is required to

establish a general overview. The submergent form of *Sparganium emersum,* however, is an obligate hydrophyte, and we have no reason to believe that this species should perform better, or worse, than other aquatic plants (Pedersen et al. 1997).

Rooted aquatic plants form a very diverse group but with a common, specialised anatomy and physiology, which are requisite for successful growth in water. I have shown here that submerged plants possess an efficient and fast water transport system which maintains communication between root and shoot. The water moves nutrients and phytohormones up through the stalk and out into the leaves. Here the nutrients are utilised for new biomass formation, and the hormones exert their effect on the tissue. The water is mainly channelled to sites of active growth. This is ensured by controlling the flow resistance of the leaf hydathodes, allowing the excess water to escape to the surroundings by means of mass flow. The effectiveness of the transport system is not systematically lower than in terrestrial herbaceous plants and this indicates that the aquatic plants, have preserved the functionality of the vascular system found in their terrestrial ancestors.

Unclarified questions, important for our general understanding of water transport in aquatic plants, still remain. The coupling between the functional structure of the vascular pathway and the underlying physiology is of special interest. We have here a rare situation in biology where the significance of evolutionary adaptations of the vascular system can be tested instead of remaining subject to speculation alone. Fluid movements in capillary vessels are subject to strong physical control, and the influence of even minor modifications of the vascular system can be tested in terms of the effectiveness of acropetal water transport. In addition, amphibious plants constitute a pool of experimental material which may reveal important information on the basic needs in both environments – air and water – and they allow us to effectively separate transpirational and non-transpirational water transport to evaluate the importance of both transport mechanisms.

Hence, anatomy, the laws of physics and the experimental method join to create a delimited scenario in which the hypotheses are testable.

Literature cited

Arber, A. 1920. Water plants: A study of aquatic angiosperms. University Press, Cambridge, England.

Barko, J. W., and R. M. Smart. 1981. Sediment-based nutrition of submerged macrophytes. Aquatic Botany 10: 339-352.

Biddulph, O., and R. Cory. 1957. An analysis of translocation in the phloem of the bean plant using THO, ^{32}P, and ^{14}C. Plant Physiology 32: 608-619.

Bowes, G. 1987. Aquatic plant photosynthesis: strategies that enhance carbon gain. Pages 79-98 *in* R.M.M. Crawford, editor. Plant life in aquatic and amphibious habitats. Blackwell Scientific Publishers, Oxford, England.

Carignan, R., and J. Kalff. 1980. Phosphorous source for aquatic weed: water or sediments? Science 207: 987-989.

Cataldo, D. A., A. L. Christy, and C. L. Coulson. 1972. Solution flow in the phloem. Plant Physiology 49: 690-695.

Chambers, P.A., E. E. Prepas, M. L. Bothwell, and H. R. Hamilton. 1989. Roots versus shoots in nutrient uptake by aquatic macrophytes in flowing waters. Canadian Journal of Fisheries and Aquatic Sciences 46: 435-439.

Crafts, A. S.,, and T. C. Broyer. 1939. Migration of salts and water into xylem of the roots of higher plants. American Journal of Botany 25: 529-535.

Denny, P. 1980. Solute movement in submerged angiosperms. Biological Reviews Cambridge Philosophical Society 55: 65-92.

Gessner, F. 1956. Der Wasserhaushalt der Hydrophyten und Helophyten. Pages 854-901 *in* O. Stocker, editor. Handbuch der Pflanzenphysiologien. Springer Verlag, Berlin, Germany.

Gessner, F. 1959. Hydrobotanik I. VEB Deutscher Verlag der Wissenschaften, Berlin, Germany.

Gibson, A. C., H. W. Calkin, and P. S. Nobel. 1984. Xylem anatomy, water flow, and hydraulic conductance in the fern *Cyrtomium falcatum*. American Journal of Botany 71: 564-574.

Hostrup O., and G. Wiegleb. 1991. Anatomy of leaves of submerged and emergent forms of *Littorella uniflora* (L.) Ascherson. Aquatic Botany 39: 195-209.

Höhn, K., and W. Ax. 1961. Untersuchungen über Wasserbewegung und Wachstum submerser Pflanzen. Beiträge zur Biologie der Pflanzen. 36: 273-298.

Huber, B. 1928. Weitere quantitative Untersuchungen über das Wasserleitungssystem der Pflanzen. Jahrbuch der Wissenschaftlichen Botanik 67: 877-959.

Hutchinson, G. E. 1975. A treatise on limnology. John Wiley and Sons Ltd, New York, USA.

von Minden, M. 1899. Beiträge zur anatomischen und physiologischen Kenntnis Wasser-sezernierender Organe. Bibliotheca Botanica, Heft 46.

Mortlock, C. 1952. The structure and development of the hydathodes of *Ranunculus fluitans* Lam. New Phytologist 51: 129-138.

Münch, E. 1930. Die Stoffbewegungen in der Pflanze. Verlag von Gustaf Fischer, Jena, Germany.

Nagahashi, W., W. W. Thomson, and R. T. Leonard. 1974. The Casparian Strip as a barrier to the movement of lanthanum in corn roots. Science 183: 670-671.

Nobel, P. S. 1991. Physicochemical and environmental plant physiology. Academic Press, San Diego, USA.

Pedersen, O. 1993. Long-distance water transport in aquatic plants. Plant Physiology 103: 1369-1375.

Pedersen, O. 1994. Acropetal water transport in submerged plants. Botanica Acta 107: 61-65.

Pedersen, O., and K. Sand-Jensen. 1993. Water transport in submerged macrophytes. Aquatic Botany 44: 385-406.

Pedersen, O., and K. Sand-Jensen. 1997. Transpiration does not control growth and nutrient supply in the amphibious plant, *Mentha aquatica*. Plant Cell and Environment 20: 117-123.

Pedersen, O., L. B. Jørgensen, and K. Sand-Jensen. 1997. Through-flow of water in leaves of a submerged plant is influenced by the apical opening. Planta, *in press.*

Raven, P. H., R. F. Evert, and H. Curtis. 1982. Biology of plants. Worth Publishers Inc., New York, USA.

Salisbury, F. B., and C. W. Ross. 1992. Plant Physiology, 4th ed. Wadsworth Publishing Company, Belmont, USA.

Sauvageau, C. 1891. Sur les feuilles de quelques monocotylédones aquatiques. Annales des Sciences Naturelles VII. série Botanique 13: 103-296.

Schenck, H. 1886. Vergleichende Anatomie der submersen Gewächse. Bibliotheca Botanica, Heft 1.

Sculthorpe, C. D. 1967. The biology of aquatic vascular plants. Edward Arnold Ltd., London, England.

Takeda, F., M. E. Wisniewski, and D. M. Glenn. 1991. Occlusion of water pores prevents guttation in older strawberry leaves. Journal of the American Society of Horticultural Sciences 116: 1122-1125.

Thut, H. F. 1932. The movement of water through some submerged water plants. American Journal of Botany 19: 693-709.

Unger, F. 1862. Beiträge zur Anatomie und Physiologie der Pflanzen. IX. Neue Untersuchungen über die Transspiration der Gewächse. Sitzungsberichte der Wiener Akademie der Wissenschaften 44: 181-217.

Weinrowsky, P. 1898. Untersuchungen über die Scheitelöffnungen bei Wasserpflanzen. Dissertation, Berlin, Germany.

Weinrowsky, P. 1899. Untersuchungen über die Scheitelöffnung bei Wasserpflanzen. Beiträge zur Wissenshaftlichen Botanik (Fünfstück) 3: 205-247.

Wilson, K. 1947. Water movement in submerged aquatic plants, with special reference to cut shoots of *Ranunculus fluitans*. Annals of Botany new series 11: 91-122.

Zimmermann, M. H. 1978. Hydraulic architecture of some diffuse porous trees. Canadian Journal of Botany 56: 2286-2295.

Zimmermann, M. H. 1983. Xylem structure and the ascent of sap. Springer Verlag, Berlin, Germany.

13 Macroinvertebrate communities in Danish streams: the effect of riparian forest cover

By Dean Jacobsen and Nikolai Friberg

It is unquestionable that terrestrial leaf litter is an important food source for macroinvertebrates in forested headwater streams of the temperate deciduous biome. But is invertebrate species richness and community structure determined by forest cover and food sources? We demonstrate that the invertebrate fauna of small Danish streams is unaffected by the degree of forest cover. Instead, stream size itself appears to be the main determinant. This pattern is probably due to the patchiness of forested and open reaches along most Danish streams, and to the high feeding plasticity of stream invertebrates.

After the last pleistocene glaciation some 12,000 years ago, Denmark became covered by mixed deciduous forest. The clearing for agricultural purposes in north-western Europe including Denmark intensified 2,000 years ago (Higler 1993). For 10,000 years, therefore, Danish headwater streams have been shaded by forest (Higler 1993, Petersen et al. 1995), and during this period the streams have been colonized by aquatic invertebrates. Consequently, forest streams are the original habitat of many invertebrates. Today, only about 4% of the country is covered by deciduous forest (8% is covered by coniferous plantations), and the number of original forest stream habitats has therefore greatly diminished so that they occur only in small isolated forest patches. Within the next 100 years, the area of forest in Denmark is to be doubled, one expectation being an overall increase in biological diversity. With respect to streams, however, the knowledge on invertebrate richness in relation to riparian vegetation type is very limited.

The main food source for macroinvertebrates in forested headwater streams was, and still is, coarse detritus of allochthonous origin (e.g. Minshall 1967, Fisher and Likens 1973, Iversen et al. 1982). Unshaded streams typically have small inputs of allochthonous litter but substantial autochthonous production of microalgae and/or rooted macrophytes. This is also common in large streams that are little influenced by riparian vegetation. One of the main features of the River Continuum Concept (Vannote et al. 1980) is that the relative proportions of functional feeding groups (*sensu* Cummins 1973) will vary in a predictable way in pristine river systems from headwaters to large rivers, following the decrease of riparian influence and the change of available food sources. Several studies have even found that the species composition of riparian forest may influence the structure of the shredder community (Cummins et al. 1989, Stout et al. 1993, Friberg 1997). According to the River Continuum Concept, forested headwater streams should have a high proportion of shredders feeding on coarse detritus, whereas scrapers feeding on benthic and epiphytic algae should take over in streams with little riparian influence. An unstated implication is that rooted macrophytes are not suitable food sources for shredders. Therefore, unshaded headwater streams with growth of macrophytes and algae do not fit into the River Continuum Concept very well.

Our previous studies suggest that shredders have considerable feeding plasticity. In preference experiments, invertebrates normally regarded as shredder-detritivores can feed on fresh filamentous green algae and rooted macrophytes in addition to coarse allochthonous detritus (Friberg and Jacobsen 1994, Jacobsen and Friberg 1995). Shredders may also consume fresh macrophytes in natural streams (Jacobsen and Sand-Jensen 1992), and feeding on macrophytes by trichopteran shredders in particular may be more common than previously recognized (Newman 1991, Jacobsen 1993).

The restricted longitudinal distributions of macroinvertebrates within the same stream system, from forested headwaters to unshaded large rivers, obscure the influence of different food sources, because small streams and large rivers are simply widely different habitats with respect to temperature regime, current velocity and sediment type. Therefore, if physical features are as important for the stream fauna as food sources (Statzner and Higler 1986), the fauna of small, unshaded, macrophyte-rich streams may be more similar to the fauna of small forest streams than to the fauna of large open streams, though the available food sources of unshaded small and large streams may resemble each other.

The purpose of this paper is to examine the influence of riparian forest, and therefore available food sources, on the composition and diversity of macroinvertebrate communities in Danish streams. The study addresses three basic questions: 1) Is diversity, either on a local or a regional scale, different in for-

est and unshaded streams? 2) Is there a systematic difference in community structure between forest streams and open streams? 3) Does the composition of functional feeding groups differ between forested and open streams? To address the three questions we compiled and analysed faunal composition data for small forested, small unshaded and large unshaded streams. A possible fourth category, large forest streams, does not exist in Denmark.

Characteristics of the streams and the study – The streams

Primary forests no longer exist in Denmark. About 200 years ago the area of forest was smaller than ever before, covering only 2% of the country. Hence, the open streams included here have been more or less unshaded for centuries. At times, however, they may have had some scattered trees (typically alder) along their banks. Some of the localities regarded as forest streams in this study may have become surrounded by trees within the last century. Recent afforestation is, however, mainly due to an increase in coniferous plantations (Friberg 1996). Streams draining coniferous plantations in Denmark are often severely acidified and house a species-poor invertebrate fauna (Friberg 1997). However, none of the forest streams included here were in coniferous plantations, so the streams have probably had the current status for centuries. We therefore assume that a time span of centuries has been sufficiently long for the invertebrate communities to be structured by prevailing environmental conditions and available food sources in the two stream types. Macroinvertebrate communities have been shown to change considerably during a 10-year period accompanying a succession in riparian vegetation from grassland to young forest (Haefner and Wallace 1981).

The study included three groups of 12 lowland streams in Jutland, Denmark. The first group comprised small streams (width < 1.5

m; depth < 30 cm) running through, and completely covered by, deciduous forests of beech (*Fagus sylvatica* L.), ash (*Fraxinus excelsior* L.), and alder (*Alnus glutinosa* L.). The small forest streams had no growth of aquatic macrophytes. The second group consisted of small unshaded streams of similar size to the forest streams, but with growth of aquatic macrophytes, mainly species of *Callitriche, Berula, Myosotis,* and *Veronica.* The third group comprised large streams 7 to 20 m wide and typically 0.5 to 1 m deep. These streams were mainly unshaded, with scattered trees along the banks and aquatic vegetation dominated by species of *Sparganium, Batrachium, Elodea, Callitriche* and *Potamogeton.* The maximum distance between any of the streams within the three groups was 106-148 km, and the minimum distance was 2-9 km. None of the 36 streams showed significant signs of organic or chemical pollution. The data used in the following analysis were collected by local river authorities as part of regional surveys (Vejle County 1988a, 1988b, 1989, 1991, 1992, Viborg County 1991a, 1991b, Ribe County 1994, Ringkjøbing County 1994).

Food sources

The amounts of coarse allochthonous detritus, macrophyte biomass, and periphyton biomass were not measured in this study. However, literature values representative of the studied forest streams report mean pools of coarse particulate organic matter of 200-900 g DW m^{-2}, and mean benthic algal biomasses of 20-70 mg chl. m^{-2} (Friberg 1997). Mean biomasses of macrophytes in small open streams often range from 50 to 100 g DW m^{-2} (Andersen and Andersen 1991, Sode 1996), but mean biomasses of benthic microalgae are extremely variable, and are usually several-fold higher than in shaded streams (Iversen et al. 1990). In addition, macrophytes in open streams may serve as substrata for epiphytic microalgae, and biomasses

reach 13-29 mg chl. m^{-2} stream bottom during summer (Sand-Jensen et al. 1989a). Unfortunately, we have found no measurements of coarse detrital biomass in small open Danish streams. Although aquatic macrophytes increase the retention capacity of small streams and trap coarse detritus, we assume that the amount of allochthonous leaf litter is considerably smaller in open streams than in forest streams, a contention supported by general observations. Though some of our small open streams may have had forested reaches upstream, the retention capacity of many natural forest streams is high, and little coarse detritus is normally exported downstream (Cummins et al. 1989). Hence, amounts, origin and types of food resources available to invertebrate detritivores and primary consumers differ considerably in small forest and open streams.

The sampling

Invertebrate samples were collected by standardized 'kicksampling' using a 25 x 25 cm nylon handnet (mesh size: 0.5 mm) placed on the stream bottom. A sample was obtained in the following way: three transects across a 20 m stream reach were chosen, and at each transect the stream bottom in front of the net was disturbed by kicking twice in the substrate at four positions: next to the bank, at 25%, at 50% and at 75% of the stream width (Kirkegaard et al. 1992). Although this method is not strictly quantitative, it allows comparisons to be made among samples. All 36 streams were sampled between November and May. Invertebrates were assigned to functional feeding groups according to Merritt and Cummins (1996). To calculate relative biomasses of functional feeding groups, approximate weights of major taxonomic groups were obtained from Mortensen and Simonsen (1983), Iversen (1988), and Friberg (unpublished data).

Table 13.1. Total number of species within invertebrate groups collected in the 12 streams of each type. Mean number of species per stream are given in parentheses. Other non-insects include Nematoda and Nematomorpha. Other Diptera include Tipulinae, Ptychopteridae, Dixidae, Thaumaleidae, Ceratopogonidae, Stratiomyidae, Empididae, Tabanidae and Athericidae. Other insects include Odonata, Megaloptera and Lepidoptera. An asterix denotes that the group was not identified to species level.

	Small forest		Small open		Large open		Total
Non-insects							
Turbellaria	2	(0.9)	4	(1.1)	5	(1.1)	7
Oligochaeta	5	(1.4)	4	(1.6)	7	(2.8)	9
Hirudinea	1	(0.1)	3	(0.5)	8	(2.6)	8
Bivalvia	1	(0.5)	1	(0.8)	3	(2.6)	3
Gastropoda	2	(0.6)	2	(0.7)	11	(1.5)	11
Hydracarina	1	(1.0)	1	(0.8)	1	(0.9)	1[*]
Crustacea	4	(1.7)	4	(1.4)	7	(2.8)	8
Other non-insects	1	(0.1)	2	(0.3)	2	(0.1)	2[*]
Insects							
Plecoptera	14	(5.8)	12	(3.6)	16	(3.5)	21
Ephemeroptera	3	(1.2)	9	(1.7)	22	(7.1)	22
Hemiptera	1	(0.1)	1	(0.3)	9	(1.7)	9
Coleoptera	16	(3.7)	22	(3.3)	22	(4.3)	42
Trichoptera	23	(8.2)	31	(8.3)	40	(10.3)	59
Diptera							
Chironomidae	45	(10.8)	51	(11.7)	64	(16.0)	86
Psychodidae	22	(3.2)	16	(2.0)	7	(0.8)	24
Simuliidae	7	(3.1)	8	(3.0)	9	(2.8)	13
Limoniinae	11	(2.8)	9	(2.3)	6	(1.8)	15
Other Diptera	15	(5.3)	15	(2.5)	13	(3.8)	26
Other insects	–		1	(1.0)	7	(0.9)	7
Total	175	(50.3)	196	(46.3)	259	(67.7)	373
		(31-77)		(33-76)		(37-96)	

Species richness

In the present study 373 species of macroinvertebrates were identified from the 36 streams (Table 13.1). However, the number of species obtained depends largely on the method and effort of sampling and identification. The 1,044 species of invertebrates recorded from one single foreign stream, the Breitenbach in Germany, clearly illustrate this fact (Zwick unpublished, cited by Allan 1995). For our purpose, standardized samples serve to examine general patterns in species richness among streams.

Species richness of stream invertebrate communities is thought to be influenced by many variables such as flow and substrate stability (Stanford and Ward 1983; Death and Winterbourn 1994), temperature (Ward 1994, Jacobsen et al. 1997), stream age (Malmqvist et al. 1991, Milner 1994) and pH (Hildrew et al. 1984, Friberg et al. 1997a). Large streams normally also house more species than small streams because of their higher habitat diversity, larger habitat area, and a greater drainage basin from where recruitment may occur (Brönmark et al. 1984; Minshall et al. 1985b). Our large streams were significantly more species-rich than the other two stream

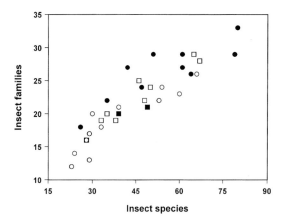

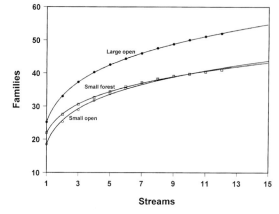

Figure 13.1. Family richness in relation to species richness of aquatic insects for 36 Danish streams. Open circles denote small open streams, squares denote small forest streams, and filled circles denote large streams.

Figure 13.2. Family accumulation curves for each of the three stream types. The power function $y=c(x-x_0)^z$ was fitted to the data. All three regression coefficients: $r^2 = 0.99$. Symbols as in Fig. 13.1. See text for further explanation.

types (P < 0.05; Mann-Whitney U test), while species richness of the two small stream types was not significantly different. The presence of aquatic macrophytes in open streams may compensate for the absence of habitats associated with woody debris and coarse detritus in forest streams, although some studies on streams without macrophytes have also found no effect of forest cover on species richness (Hawkins et al. 1982, Reed et al. 1994, Schultz 1997). Diversity of food sources is also thought to influence species richness (Covich 1988), but evidence to support this contention is scarce. The higher species richness in large streams was mainly attributable to the presence of more leeches (Hirudinea), snails (Gastropoda), mayflies (Ephemeroptera), bugs (Hemiptera), caddisflies (Trichoptera) and midges (Chironomidae). On the other hand, Psychodidae and Limoniinae were more diverse in small streams. No significant differences in species richness within any of the major taxonomic groups given in Table 13.1 were observed between the small forest and small open streams.

The mean number of species in the small forest streams was, however, slightly higher than in the small open streams, although the opposite was true for the total number of species found in the two stream types. A given forest stream, therefore, houses a higher proportion of the total species pool inhabiting forest streams than is the case for open macrophyte-rich streams. This finding suggests that the taxonomic composition of forest streams is less variable than that of open streams (i.e., lower beta diversity), and that the total species pool of forest streams (the gamma diversity) is smaller than that of open streams. To further examine this pattern, and recognizing that species richness increases with the size of the sample and the size of the sampled area, species accumulation curves for insects were drawn with each stream regarded as a locality. Insects contributed about 90% of total species richness, and almost all insects were identified to species. There were only weak correlations between species richness of single insect orders and total species richness, whereas the richness of aquatic insect families was clearly related

Table 13.2. Mean abundance of invertebrate groups in the three stream groups. Percentages of totals are given in parentheses. Other non-insects include Nematoda and Nematomorpha. Other Diptera include Tipulinae, Psychodidae, Ptychopteridae, Dixidae, Thaumaleidae, Ceratopogonidae, Stratiomyidae, Empididae, Tabanidae and Athericidae. Other insects include Odonata, Megaloptera and Lepidoptera.

	Small forest		Small open		Large open	
Non-insects						
Turbellaria	15	(0.8)	9	(0.4)	8	(0.2)
Oligochaeta	13	(0.7)	23	(1.0)	242	(7.4)
Hirudinea	0.1	(0.0)	1	(0.0)	11	(0.3)
Bivalvia	2	(0.1)	4	(0.2)	31	(0.9)
Gastropoda	13	(0.7)	89	(3.7)	41	(1.3)
Hydracarina	14	(0.7)	17	(0.7)	116	(3.6)
Crustacea	437	(22.9)	836	(35.5)	276	(8.5)
Other non-insects	0.2	(0.0)	0.5	(0.0)	3	(0.1)
Insects						
Plecoptera	402	(21.1)	316	(13.4)	50	(1.5)
Ephemeroptera	268	(14.0)	212	(9.0)	333	(10.2)
Hemiptera	2	(0.1)	1	(0.0)	12	(0.4)
Coleoptera	86	(4.5)	80	(3.4)	59	(1.8)
Trichoptera	101	(5.3)	122	(5.2)	381	(11.7)
Diptera						
Chironomidae	110	(5.8)	215	(9.1)	1241	(38.0)
Simuliidae	293	(15.3)	305	(13.0)	299	(9.2)
Limoniinae	32	(1.7)	33	(1.4)	15	(0.5)
Other Diptera	60	(3.1)	29	(1.2)	67	(2.1)
Other insects	–		0.2	(0.0)	2	(0.0)
Total	1909		2353		3265	
	(629-4332)		(80-4848)		(457-11659)	

to species richness (Fig. 13.1). We therefore performed the analysis on the family level. However, because of the very high taxonomic and trophic diversity of the two diptera families Chironomidae and Tipulidae, these were included as subfamilies. Fourtyone families (and subfamilies) of insects were found in the small forest streams, 41 were found in the small open streams and 52 were found in the large open streams. The order of plotting the 12 localities was randomized 500 times using a Jackknifing procedure, and a power function, $y = c(x-x_0)^z$ with zero offset, was fitted to the mean data (Fig. 13.2). The main parameter of interest is the exponent z, which is the rate of increase in family rich-

ness with stream number (the slope in the alternative linear log-log plot). The z-value can be regarded as a measure of beta-diversity (Rozenzweig 1995). The small forest streams had the lowest z-value of 0.19, and the small open streams the highest z-value of 0.22. The curve suggests that the difference in species richness between small and large streams will increase as more streams are sampled. Though differences in the z-values were slight, the accumulation curve for the small open streams rises above the curve for the forest streams at about 10 sampled localities. The localities in this study were chosen so that the 12 streams in each group were distributed within similarly sized geographical areas and were separated by the same mean distance. Therefore, we do not believe that the lower beta and gamma diversity in forest streams relative to open streams is due to regional differences in species richness and dispersal. Forest stream habitats may be more similar to each other than macrophyte-rich open streams, and so have less inter-stream faunal variability.

Community structure – Invertebrate density

Invertebrate density and production are often higher in open streams than in forest streams (e.g. Behmer and Hawkins 1986, Iversen 1988, Friberg et al. 1997b), probably because of higher food availability and quality from microalgae and rooted macrophytes, but also because macrophytes increase the surface area as substrata for invertebrates (Iversen et al. 1985; Sand-Jensen 1997). In this study, the overall density of invertebrates was not significantly different in our three stream types (P > 0.05; Kruskal-Wallis test) (Table 13.2), largely because the high mean abundance for the large streams was due to one very large sample. However, the two open stream types tended to have higher densities than the forest streams.

The higher abundances of invertebrates in the small open streams compared to the forest streams was mainly due to very high densities of the amphipod *Gammarus pulex* in several of them. No significant differences in abundance of major taxonomic groups were found between small forest and small open streams (Table 13.2), primarily because there was considerable inter-stream variability in the abundance of specific invertebrate groups. Nevertheless, Chironomidae were significantly more abundant in large streams (38% of individuals) than small streams (P < 0.05; Mann-Whitney U test). Oligochaeta were also more common in large than in small streams (P < 0.002) whereas Plecoptera were significantly more abundant in small

forest streams than in large streams (P < 0.002). It is noteworthy that even though the density of plecopterans was 6-8 times higher in the small streams than in the large streams, the number of species collected in the three stream types was about the same. Likewise, mayflies made the same contribution to the overall density of invertebrates in the three stream types, although the large streams were much richer in species.

Common species

No common species in either the small forest streams or the small open streams (i.e., species occurring in more than half of the localities) was completely absent from the other stream type. Hence, there appeared to be little overall difference in species composition between small forest and small open streams. Forest streams often have lower and more constant temperatures than streams fully exposed to sunlight, and temperature is an important structuring parameter for invertebrate communities (Jacobsen et al. 1997). However, springbrooks are common in Denmark and were included in this study, so the variability in ground water inflows may have been more important than forest cover in determining temperature stability. Because inflows of ground water vary irrespectively of forest cover, small forest patches do not seem to serve as refugia for stream invertebrates preferring a particular temperature regime.

The most common invertebrate species occurring in most of the 12 streams of each type are listed in Table 13.3. The forest streams had 13 species occurring in 10 or more of the 12 streams, and three species were found in all streams. Eight species occurred in 10 or more of the small open streams. Hence, the forest streams were more similar to each other in faunal composition than the small open streams, a result that is consistent with the family accumulation curves and Table 13.1.

Table 13.3. The most common macroinvertebrate species in three Danish stream types.

	Small forest	Small open	Large open
Species present in:			
All streams	*Gammarus pulex* *Plectrocnemia conspersa* *Brillia modesta*	*Dicranota sp.*	Limnephilidae *Baetis rhodani*
11 out of 12 streams	Hydracarina *Nemoura flexuosa* *Leuctra nigra* *Elodes minuta* *Sericostoma personatum* *Dicranota sp.*	*Gammarus pulex* *Baetis rhodani* *Brillia modesta* *Micropsectra sp.*	Tubificidae Hydracarina *Asellus aquaticus* *Gammarus pulex* *Micropsectra sp.*
10 out of 12 streams	*Dugesia gonocephala* *Baetis rhodani* *Amphinemura stanfussi* *Eleophila sp.*	Tubificidae *Nemoura cinerea* Limnephilidae	*Erpobdella octoculata* *Elmis aenea* *Hydropsyche pellucidula* *Prodiamesa olivacea* *Pisidium sp.*

Gammarus pulex (Crustacea), *Baetis rhodani* (Ephemeroptera), *Amphinemura stanfussi* (Plecoptera) and *Odagmia spinosa* (Simuliidae) were, in decreasing order, the four most abundant species in small forest and small open streams (Fig. 13.3). The four most abundant species in large streams were *Micropsectra sp.* (Chironomidae), *Gammarus pulex*, Tubificidae indet. and *Rheotanytarsus sp.* (Chironomidae). In habitats char-

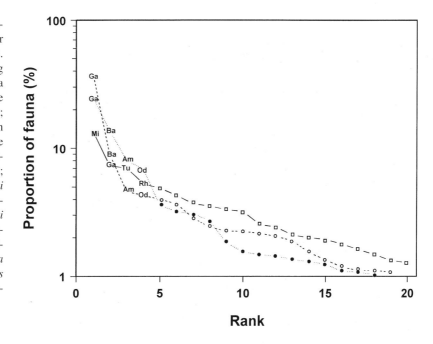

Figure 13.3. Mean rank-abundance diagrams for the three stream types. Only species contributing >1 % of the total fauna are included. Dotted line = small forest streams; dashed line = small open streams; full line = large open streams. Ga=*Gammarus pulex* (Crustacea); Ba=*Baetis rhodani* (Ephemeroptera); Am=*Amphinemoura stanfussi* (Plecoptera); Od=*Odagmia spinosa* (Simuliidae); Mi=*Micropsectra sp.*; Rh=*Rheotanytarsus sp.* (Chironomidae); Tu=Tubificidae.

acterised by high resource availability, pronounced dominance by few species might be expected, and Hawkins et al. 1982 found different rank-abundance curves for open and shaded stream sites with higher dominance in open sites. We found that the forms of the mean rank-abundance curves were not significantly different among our three stream types (P < 0.05; Kolmogorov-Smirnov test for goodness of fit). Therefore, the higher evenness and lower dominance by few species seen from the curve for the large streams was a consequence of their greater number of species.

Overall faunal similarity

Similarity in overall faunal structure among the 36 streams was analysed using the non-parametric multivariate technique provided in the computer software package PRIMER (Clarke and Warwick 1994). The ordination (multidimensional scaling) based on Bray-Curtis similarities of faunal composition among all 36 streams shown in Fig. 13.4 utilised species composition as well as the abun-

dance of each species. It is evident that the small forest and small open streams did not form separate groups, indicating that there was no overall difference in their communities. The large open streams, on the other hand, formed a separate group.

The Danish landscape is a mosaic of open areas and small forest patches, and riparian vegetation is also patchily distributed along small streams. The behaviour of many adult stream insects is to fly upstream to oviposit (Madsen et al. 1973, Svensson 1974, Müller 1982), and this behaviour probably occurs irrespective of riparian vegetation type. Because the small streams are short (a few km) this migration behaviour probably diminishes the differences in species composition and the richness of aquatic insects in small headwater streams, with or without a forested riparian zone. If, on the other hand, forested and open streams had been fully separated, as is often the case elsewhere, differences in species richness and taxonomic composition may have been more evident. In summary, the riparian vegetation type of small streams does not seem to be crucial in structuring the invertebrate community. Instead, stream size appears to be more important, a conclusion in accordance with the findings of Brönmark et al. (1984).

Functional feeding groups

Studies of functional feeding groups in relation to riparian vegetation cover have produced divergent results. Much of this inconsistency arises because streams around the world differ greatly with respect to hydrological stability, retention of coarse organic material and riparian vegetation characteristics. But probably also because invertebrate species are often misplaced into functional feeding groups, and food sources are underestimated. A factor further complicating such studies is the occasional opposite conclusions reached by studies examining invertebrate abundance (numbers) and others examining

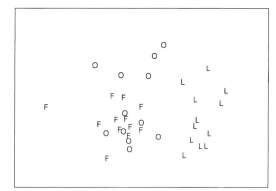

Figure 13.4. Ordination plot of the 36 stream macroinvertebrate communities using multidimensional scaling. To downweigh very abundant species, the data were 4th root transformed before analysis. The small forest (F) and the small open (O) streams do not form separate groups, whereas the large open streams (L) form a distinct group to the right of the diagram.

biomass. Some studies have found that shredder biomass or abundance is positively, and scraper biomass or abundance negatively related to riparian forest cover (e.g. Haefner and Wallace 1981, Dudgeon 1989, Reed et al. 1994). Other studies have not observed these relationships (e.g. Winterbourn et al. 1981, Hawkins et al. 1982, Malicky 1990). Our work adds to the last category of studies.

We chose to examine mainly the invertebrate biomass to facilitate comparison with the River Continuum Concept. In the Danish streams, the estimated relative biomasses of functional feeding groups were almost the same in small forest and small open streams (Fig. 13.5). Shredders formed the dominant functional feeding group in small streams, irrespective of riparian vegetation type, and comprised 46% (forest streams) and 49% (open streams) of the total biomass of invertebrates. In terms of absolute biomass, more shredders were found in small open streams than in forest streams. In the large streams, shredders accounted for only 15% of the biomass. Collector-gatherers were important in all three stream types and made up 31% – 35% of the biomass. Filterers were sparsely represented in the two small stream types (< 3%) but were quite important in the large streams (15%). Scrapers were not common in any of the stream types (5 – 7% of total biomass).

The proportions of the four functional feeding groups in the small forest streams and the large open streams are in accord with the predictions of the River Continuum Concept, although the proportion of scrapers in the open stream types is lower than expected. Our main interest, however, is the dominance of shredders in small open streams in addition to the forest streams. That shredder dominance is unrelated to coarse detrital inputs from terrestrial vegetation appears to be in contrast to the original River Continuum Concept, as we understand it. However, following much critique and discussion (see

Small forest streams

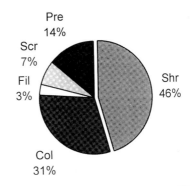

Small open streams

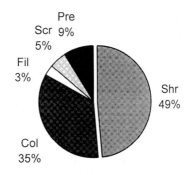

Large open streams

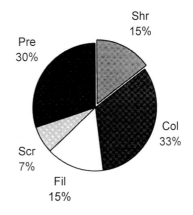

Figure 13.5. Mean biomass (%) contributed by five functional feeding groups in three Danish stream types. Shr=shredders; Col=collectors; Fil=filteres; Scr=scrapers; Pre=predators.

references in Statzner and Higler 1985), the original concept has been modified (Minshall et al. 1985a) to such a degree that apparent exceptions could be explained and accommodated by the concept.

Nevertheless, we have demonstrated here that it is not solely the inputs (or pool) of terrestrial leaf litter that result in shredder dominance in small lowland streams in Denmark. Rather, high shredder biomass seems to be an intrinsic property of these small streams in general. This pattern may arise if inputs of allochthonous coarse detritus to the small open streams are underestimated. More importantly, however, we believe that food sources of autochthonous origin such as microalgae and rooted aquatic macrophytes, to some degree, may substitute for terrestrial leaf litter as food for shredders. Our knowledge on the feeding habits of stream invertebrates is still very incomplete. Thus, the functional feeding group concept may be too rigid to be realistic.

Feeding plasticity

We propose that the similarity of invertebrate communities in small forest and small open streams is, in part, due to a high degree of opportunistic feeding among stream invertebrates, especially nominal shredders. In other words, we believe that the more or less similar and dominant shredders are consuming different foods in the two small stream types. These arguments are in accordance with those presented by Hawkins et al. (1982), Rounick et al. (1982) and Malicky (1990). In preference tests, species normally regarded as shredder-detritivores have been shown to choose fresh aquatic macrophytes or filamentous algae before for example beech leaves (Friberg and Jacobsen 1994; Jacobsen and Friberg 1995). However, some terrestrial leaf species are generally preferred to others; leaves that are nitrogen-rich and quickly processed, such as *Alnus* (alder) and *Faxinus* (ash), are often preferred to slowly degrad-

able, nutrient-poor leaves of *Fagus* (beech), *Quercus* (oak) and *Picea* (spruce) (e.g. Iversen 1974, Webster and Benfield 1986, Cummins et al. 1989, Friberg and Jacobsen 1994). The forest streams in this study represented streams with inputs of 'slow', 'fast' and a mixture of both leaf types. Hence, shredders in beech forest streams, the most common deciduous forest type in Denmark today, apparently live on food they do not prefer, and a food that is not particularly nutrient-rich compared to fresh macrophytes and algae (Friberg and Jacobsen 1994, Jacobsen and Friberg 1995). Furthermore, growth experiments have confirmed that some trichopteran shredder-detritivores are able to utilise and grow on a diet of fresh macrophytes or algae (Jacobsen and Sand-Jensen 1994, Friberg and Jacobsen 1996). However, some aquatic macrophytes seem to be more attractive as food for shredders than others (Otto and Svenson 1981, Jacobsen and Sand-Jensen 1992, Jacobsen 1994). Nevertheless, we would not expect shredders to be confined to Danish forest streams solely for reasons of food quality.

Several studies have shown that shredder biomass, even in forested headwater streams, may be food limited (Richardson 1991, Dobson and Hildrew 1992, Friberg 1997). If we assume that the pool of coarse detritus is lower in open streams than in forest streams, as might be expected, it is therefore somewhat surprising that 'shredder' biomass was even higher in the small open than in the forest streams, at least if we follow the traditional belief that shredders consume coarse detritus. As suggested above, this finding might be explained by high feeding plasticity, as demonstrated for *Gammarus pulex* for example. In the present study we consistently regarded *Gammarus* as 50% shredder and 50% collector-gatherer, but this amphipod, so dominant in Danish streams, is known to feed on almost any organic material (Marchant 1981, Gee 1988, Pedersen 1990, Friberg and Jacobsen 1994). Consequently, *Gammarus* may be feeding as a shredder in the forest streams and as collector-gatherer or scraper in the small open streams. Malicky (1990) also found that *Gammarus* and limnephilid caddis larvae changed from shredding to scraping habits depending on food availability, and we believe that the same may be true for many plecopteran and trichopteran larvae. Similar considerations may also explain the unexpectedly low proportion of scrapers in the open streams.

Concluding remarks

We found no systematic difference in either species richness, community structure or pro-portions of functional feeding groups of invertebrates between small forested and small open streams. The similarity in species richness and species composition may, in part, be due to the close proximity of forested and open reaches along nearly all Danish streams. Consequently, in terms of inverte-brate species richness, little will probably be gained by an increase in the forest cover of Denmark. Riparian vegetation and the type of available food sources does not seem to be so crucial in determining the proportion of func-tional feeding groups. Stream size itself appears to be more important in structuring invertebrate communities in small Danish lowland streams.

Litterature cited

Allan, J.D. 1995. Stream Ecology – Structure and function of running waters. Chapman & Hall, London, England.

Andersen, T.L., and K. Andersen. 1991. Ø-dynamik hos *Callitriche cophocarpa*. Masters thesis. Botanical Institute, University of Aarhus, Denmark.

Behmar, D.J., and C.P. Hawkins. 1986. Effects of over-head canopy on macroinvertebrate production in a Utah stream. Freshwater Biology 16: 287-300.

Brönmark, C., J. Herrmann, B. Malmquist, C. Otto, and P. Sjöström. 1984. Animal community structure as a function of stream size. Hydrobiologia 112: 73-79.

Clarke, K.J., and R.M. Warwick. 1994. Changes in marine communities. An approach to statistical analysis and interpretation. Plymouth Marine Laboratory, England.

Covich, A.P. 1988. Geographical and historical comparisons of neotropical streams: biotic diversity and detrital processing in highly variable habitats. Journal of the North American Benthological Society 7: 361-386.

Cummins, K.W. 1973. Trophic relations of aquatic insects. Annual Review of Entomology 18: 183-206.

Cummins, K.W., M.A. Wilzbach, D.M. Gates, J.B. Perry, and W.B. Taliaferro. 1989. Shredders and riparian vegetation. Bioscience 39: 24-30.

Death, R.G., and Winterbourn, M.J. 1994. Environmental stability and community persistence: a multivariate perspective. Journal of the North American Benthological Society 13: 125-139.

Dobson, M., and A.G. Hildrew. 1992. A test of resource limitation among shredding detritivores in low order streams in southern England. Journal of Animal Ecology 61: 69-77.

Dudgeon, D. 1989. The influence of riparian vegetation on the functional organization of four Hong Kong stream communities. Hydrobiologia 179: 183-194.

Fisher, S.G., and G.E. Likens. 1973. Energy flow in Bear Brook, New Hampshire: an integrative approach to stream ecosystem metabolism. Ecological Monographs 43: 421-439.

Friberg, N. 1996. Biological structure of forest streams and effects of afforestation. National Environmental Research Institute, Silkeborg, Denmark.

Friberg, N. 1997. Benthic invertebrate communities in six Danish forest streams: Impact of forest type on structure and function. Ecography 20: 19-28.

Friberg N., and D. Jacobsen. 1994. Feeding plasticity in two detritivore shredders. Freshwater Biology 32: 133-142.

Friberg, N., and D. Jacobsen. 1996. Growth variability of the detritivore-shredder *Sericostoma personatum*. Pages 87-95 *in* N. Friberg. Biological structure of forest streams and effects of afforestation. National Environmental Research Institute, Silkeborg, Denmark.

Friberg, N., A. Rebsdorf, and S.E. Larsen. 1997a. Effects of afforestation on acidity and invertebrates in Danish streams and implications for freshwater communities in Denmark. Water, Air and Soil Pollution, in press.

Friberg, N., M. Winterbourn, K.A. Shearer, and S.E.

Larsen. 1997b. Benthic communities of forest streams in the South Island, New Zealand: effects of forest type and location. Archiv für Hydrobiologie, 138: 289-306.

Gee, J.H.R. 1988. Population dynamics and morphometrics of *Gammarus pulex* L.: evidence of seasonal food limitation in a freshwater detritivore. Freshwater Biology 19: 333-343.

Haefner, J.D., and J.B. Wallace. 1981. Shifts in aquatic insect populations in a first-order southern Appalachian stream following a decade of old field succession. Canadian Journal of Fisheries and Aquatic Sciences 38: 353-359.

Hawkins, C.P., M.L. Murphy, and N.H. Anderson. 1982. Effects of canopy, substrate composition, and gradient on the structure of macroinvertebrate communities in Cascade Range streams of Oregon. Ecology 63: 1840-1856.

Higler, L.W.G. 1993. The riparian community of northwest European lowland streams. Freshwater Biology 29: 229-241.

Hildrew, A.G., C.R. Townsend, J. Francis, and K. Finch. 1984. Cellulolytic decomposition in streams of contrasting pH and its relationship with invertebrate community structure. Freshwater Biology 14: 323-328.

Iversen, T.M. 1974. Ingestion and growth in *Sericostoma personatum* (Trichoptera) in relation to nitrogen content of ingested leaves. Oikos 25: 278-282.

Iversen, T.M. 1988. Secondary production and trophic relationships in a spring invertebrate community. Limnology and Oceanography 33: 582-592.

Iversen, T.M., J. Thorup, T. Hansen, J. Lodal, and J. Olsen. 1985. Quantitative estimates and community structure of invertebrates in a macrophyterich stream. Archiv für Hydrobiologie 102: 291-301.

Iversen, T.M., J. Thorup, and J. Skriver. 1982. Inputs and transformation of allochthonous particulate organic matter in a headwater stream. Holarctic Ecology 5: 10-19.

Iversen, T.M., N. Thyssen, K. Kjeldsen, P. Lund-Thomsen, J. Thorup, N.B. Jensen, C.L. Pedersen, T. Winding, and L.P. Nielsen. 1990. Biologisk struktur i små vandløb. NPo-forskning fra Miljøstyrelsen C7. Miljøstyrelsen, Copenhagen, Denmark.

Jacobsen, D. 1993. Trichopteran larvae as consumers of submerged angiosperms in running waters. Oikos 67: 379-383.

Jacobsen, D. 1994. Food preference of the caddis larva *Anabolia nervosa* feeding on aquatic macrophytes. Verhandlungen der Internationale Vereinigung für Theoretische und Angewandte Limnologie 25: 2478-2481.

Jacobsen, D., and N. Friberg. 1995. Food preference of the trichopteran larva *Anabolia nervosa* from two streams with different food availability. Hydrobiologia 308: 139-144.

Jacobsen, D., and K. Sand-Jensen. 1992. Herbivory of invertebrates on submerged macrophytes from Danish freshwaters. Freshwater Biology 28: 301-308.

Jacobsen, D., and K. Sand-Jensen. 1994. Growth and energetics of a trichopteran larva feeding on fresh submerged and terrestrial plants. Oecologia 97: 412-418.

Jacobsen, D., R. Schultz, and A. Encalada. 1997. Structure and diversity of stream insect assemblages: the influence of temperature with latitude and altitude. Freshwater Biology, in press.

Kirkegaard, J., P. Wiberg-Larsen, J. Jensen, T.M. Iversen, and E. Mortensen. 1992. Biologisk bedømmelse af vandløbskvalitet. Metode til anvendelse på vandløbsstationer i Vandmiljøplanens overvågningsprogram. Teknisk anvisning fra Danmarks Miljøundersøgelser nr. 5. National Environmental Research Institute, Silkeborg, Denmark.

Madsen, B.L. J. Bengtson, and I. Butz. 1973. Observations on upstream migrations by imagines of some Plecoptera and Ephemeroptera. Limnology and Oceanography 18: 678-681.

Malicky, H. 1990. Feeding tests with caddis larvae (Insecta: Trichoptera) and amphipods (Crustacea: Amphipoda) on *Platanus orientalis* (Platanacea) and other leaf litter. Hydrobiologia 206: 163-173.

Malmqvist, B., S. Rundle, C. Brönmark, and A. Erlandsson. 1991. Invertebrate colonization of a new, man-made stream in southern Sweden. Freshwater Biology 26: 307-324.

Marchant, R. 1981. The ecology of *Gammarus* in running water. Pages 225-249 *in* M.A. Lock and D.D. Williams, editors. Perspectives in running water ecology. Plenum Press, New York, USA.

Minshall, G.W. 1967. Role of allochthonous detritus in the trophic structure of a woodland springbrook community. Ecology 48: 139-149.

Minshall, G.W., K.W. Cummins, R.C. Petersen, C.E. Cushing, D.A. Bruns, J.R. Sedell, and R.L. Vannote. 1985a. Developments in stream ecosystem theory. Canadian Journal of Fisheries and Aquatic Sciences 42: 1045-1055.

Minshall, G.W., R.C. Petersen, and C.F. Nimz. 1985b. Species richness in streams of different size from the same drainage basin. The American Naturalist 125: 16-38.

Merritt, R.W., and K.W. Cummins. (editors) 1996. An introduction to the aquatic insects of North America. Third edition. Kendall/Hunt Publishing Company, Dubuque, USA.

Milner, A. 1994. Colonization and succession of invertebrate communities in a new stream in Glacier Bay National Park, Alaska. Freshwater Biology 32: 387-400.

Mortensen, E., and J.L. Simonsen. 1983. Production estimates of the benthic invertebrate community in a small Danish stream. Hydrobiologia 102: 155-162.

Müller, K. 1982. The colonization cycle of freshwater insects. Oecologia 52: 202-207.

Newman, R.M. 1991. Herbivory and detritivory on freshwater macrophytes by invertebrates: a review. Journal of the North American Benthological Society 10: 89-114.

Otto, C., and B.S. Svensson. 1981. How do macrophytes growing in or close to water reduce their consumption by aquatic herbivores? Hydrobiologia 78: 107-112.

Pedersen, C.L. 1990. Populationsforhold og ernæringsbiologi for *Gammarus pulex* L. i Skærbæk. Masters thesis. Freshwater Biological Laboratory, University of Copenhagen, Denmark.

Petersen, R.C., G.M. Gíslason, and L.B.M. Vought. 1995. Rivers of the Nordic countries. Pages 295-341 *in* C.E. Cushing, K.W. Cummins, and G.W. Minshall, editors. Ecosystems of the world 22: River and stream ecosystems. Elsevier, Amsterdam, The Netherlands.

Reed, J.L., I.C. Cambell, and P.C.E. Bailey. 1994. The relationship between invertebrate assemblages and available food at forest and pasture sites in three southeastern Australian streams. Freshwater Biology 32: 641-650.

Ribe County. 1994. Smådyrsfaunaen i udvalgte vandløb i Ribe Amt i 1987-1992. Report from Ribe County, Denmark.

Richardson, J.S. 1991. Seasonal food limitation of detritivores in a montane stream: an experimental test. Ecology 72: 873-887.

Ringkjøbing County. 1994. 70 faste vandløbsstationer. Beskrivelse af forureningsgrad og smådyrsfauna 1992 og 1993 samt udviklingen de sidste ca. 5 år. Bilagsdel. Report from Ringkjøbing County, Denmark.

Rosenzweig, M.L. 1995. Species diversity in space and time. Cambrige University Press, Cambrige, England.

Rounick, J.S., M.J. Winterbourn, and G.L. Lyon. 1982. Differential utilization of allochthonous and autochthonous inputs by aquatic invertebrates in some New Zealand streams: a stable carbon isotope study. Oikos 39: 191-198.

Sand-Jensen, K. 1997. Macrophytes as biological engineers in the ecology of Danish streams. Chapter 6 *in* K. Sand-Jensen, and O. Pedersen, editors. Freshwater Biology – Priorities and Development in Danish Research. Gad, Copenhagen, Denmark.

Sand-Jensen, K., D. Borg, and E. Jeppesen. 1989a. Biomass and oxygen dynamics of the epiphyte community in a Danish lowland stream. Freshwater Biology 22: 431-443.

Schultz, R. 1997. Biologisk struktur i ecuadorianske lavlandsvandløb med forskellig grad af riparisk skygning. Masters thesis. Freshwater Biological Laboratory, University of Copenhagen, Denmark.

Sode, A. 1996. Effekt af strømrendeskæring på vandføringsevne og sammensætning og mængde af grøde i Ringe Å. Fyns Amt, Natur- og Miljøafdelingen. Unpublished report from Fyns County, Denmark.

Stanford, J.A., and J.V. Ward. 1983. Insect diversity as a function of environmental variability and disturbance in stream ecosystems. Pages 265-278 *in* J.R. Barnes and G.W. Minshall, editors. Stream Ecology – Application and testing of general ecological theory. Plenum Press, New York, USA.

Statzner, B., and B. Higler. 1985. Questions and comments on the River Continuum Concept. Canadian Journal of Fisheries and Aquatic Sciences 42: 1038-1044.

Statzner, B., and B. Higler. 1986. Stream hydraulics as a major determinant of benthic invertebrate zonation patterns. Freshwater Biology 16: 127-139.

Stout, B.M., E.F. Benfield, and J.R. Webster. 1993. Effects of a forest disturbance on shredder production in southern Appalachian headwater streams. Freshwater Biology 29: 59-69.

Svensson, B.W. 1974. Population movements of adult Trichoptera in a South Swedish stream. Oikos 25: 157-175.

Vannote, R.L., G.W. Minshall, K.W. Cummins, J.R. Sedell, and C.E. Cushing. 1980. The River Continuum Concept. Canadian Journal of Fisheries and Aquatic Sciences 37: 130-137.

Vejle County. 1988a. Smådyrsfaunaen og forureningsgraden i Grejs Å med tilløb – 1987. Forvaltningen for Teknik og Miljø. Report from Vejle County, Denmark.

Vejle County. 1988b. Smådyrsfaunaen og forureningsgraden i Mattrup Å med tilløb – 1987. Forvaltningen for Teknik og Miljø. Report from Vejle County, Denmark.

Vejle County. 1989. Smådyrsfaunaen og forureningstilstanden i Gudenåsystemet, 1988. Udvalget for Teknik og Miljø. Report from Vejle County, Denmark.

Vejle County. 1991. Smådyr og vandkvalitet i Vejle Å og dens tilløb 1990. Hovedrapport. Forvaltningen for Teknik og Miljø. Report from Vejle County, Denmark.

Vejle County. 1992. Smådyr og forureningstilstand i

Kolding Å og dens tilløb 1991-92. Bilagsdel. Forvaltningen for Teknik og Miljø. Report from Vejle County, Denmark.

Viborg County. 1991a. Smådyr og forureningsgrad i Skovbæk, Vinge Møllebæk, Gemmehøje Bæk, Kildebæk ved Vansø, Kildeområde ved Subæk-mølle, Vestergård Bæk og Harrestruplund Bæk. Nr. 106 i miljøserien. Forvaltningen for Teknik og Miljø. Report from Viborg County, Denmark.

Viborg County. 1991b. Smådyrslivet i kilderne ved Hald Sø. Nr. 107 i miljøserien. Forvaltningen for Tek-

nik og Miljø. Report from Viborg County, Denmark.

Ward, J.V. 1994. Ecology of alpine streams. Freshwater Biology 32: 277-294.

Webster, J.R., and E.F. Benfield. 1986. Vascular plant breakdown in freshwater ecosystems. Annual Review of Ecology and Systematics 17: 567-594.

Winterbourn, M.J., J.S. Rounick, and B. Cowie. 1981. Are New Zealand stream ecosystems really different? New Zealand Journal of Marine and Freshwater Research 15: 321-328.

14 Publications from the Freshwater Biological Laboratory

Publications 1894 – 1910

Wesenberg-Lund, C. 1894. Vore hjuldyr. – Naturen og Mennesket 12: 96-120.

Wesenberg-Lund, C. 1894. Grønland ferskvandsentomostraca. I. Phyllopoda, Branchiopoda et Cladocera. – Videnskabelige Meddelelser Dansk Naturhistorisk Forening København 56: 82-143.

Wesenberg-Lund, C. 1895. Lidt om vore økonomisk vigtige krebsdyr. – Dansk Fiskeritidsskrift 4: 206-285.

Wesenberg-Lund, C. 1895. Livet i overfladen af vore ferske vande og disses farve. – Naturen og Mennesket 14: 333-347.

Wesenberg-Lund, C. 1895. Om forekomsten af *Cordylophora lacustris* i danske ferskvande. – Videnskabelige Meddelelser Dansk Naturhistorisk Forening 57: 169-174.

Wesenberg-Lund, C. 1895. Fersk- og saltvandsentomostraca (Østgrønlandske expedition). – Meddelelser fra Grønland 19: 133-143.

Wesenberg-Lund, C. 1896. Biologiske undersøgelser over ferskvandsorganismer. – Videnskabelige Meddelelser Dansk Naturhistorisk Forening 58: 105-168.

Wesenberg-Lund, C. 1896. Om ferskvandsfaunaens kitin og kisellevninger i tørvelagene. – Meddelelser fra Dansk Geologisk Forening 3: 51-84.

Wesenberg-Lund, C. 1896. Biologiske studier over ferskvandsbryozoer. – Videnskabelige Meddelelser Dansk Naturhistorisk Forening 58: 252-363.

Wesenberg-Lund, C. 1898. Om biologiske ferskvandsstationer og deres opgaver. – Biologisk Selskabs Forhandlinger: 48-51.

Wesenberg-Lund, C. 1898. Über dänische Rotiferen und über die Fortpflanzungsverhältnisse der Rotiferen (Vorl. Mitt.). – Zoologischer Anzeiger 21: 200-211.

Wesenberg-Lund, C. 1899. Danmarks Rotifera I. Grundtrækkene i Rotiferernes økologi, morfologi og systematik. – Videnskabelige Meddelelser Dansk Naturhistorisk Forening 61: 1-145.

Wesenberg-Lund, C. 1899. Danmarks Insektverden. Den danske Stat. – Danmarks Natur, Frem: 573- 752.

Wesenberg-Lund, C. 1900. Von dem Abhängigkeitsverhältnis zwischen dem Bau der Planktonorganismen und dem spezifischen Gewicht des Süsswassers. – Biologischen Centralblatt 20: 606-619, 644-656.

Wesenberg-Lund, C. 1901. Studier over søkalk, bønnemalm og søgytje i danske indsøer. – Meddelelser fra Dansk Geologisk Forening 7: 1-180.

Wesenberg-Lund, C. 1902. Sur l'existance d'une faune relicte dans le lac de Furesö. – Oversigt over Videnskabernes Selskabs Forhandlinger 6: 257-303.

Wesenberg-Lund, C. 1903. Sur les *Ægagropila Sauteri* de lac de Sorø. – Oversigt over Videnskabernes Selskabs Forhandlinger 2: 167-204.

Ditlevsen, A. 1904. Studien an Oligochæten. – Zeitschrift für wissenschaftliche Zoologie 77: 398-480.

Jensen, S. 1904. Biologiske og systematiske undersøgelser over ferskvands-ostracoder. – Videnskabelige Meddelelser Dansk Naturhistorisk Forening 66: 1-79

Wesenberg-Lund, C. 1904. Om en nulevende i vore søer indelukket marin arktisk istidsfauna. – Geografisk Tidsskrift 17: 229-238.

Wesenberg-Lund, C. 1904. Planktoninvestigations of the Danish lakes. Special Part. 1: 1-276. Copenhagen.

Wesenberg-Lund, C. and Warming, E. 1904. Bidrag til vadernes, sandenes og marskens naturhistorie. – Det Kongelige Danske Videnskabers Selskabs Skrifter 7 R. 2: 1-56.

Wesenberg-Lund, C. 1904. Om svævet i vore ferske vande. – Ferskvandsfiskeribladet 2: 154-185.

Jensen, S. 1905. Faunistisk fortegnelse over de danske ferskvandscopepoder. – Videnskabelige Meddelelser Dansk Naturhistorisk Forening 67: 111-125.

Wesenberg-Lund, C. 1905. A comparative study of the lakes of Scotland and Denmark. – Proceedings of the Royal Society of Edinburgh 25: 401-448.

Brinkmann, A. 1906. Studier over Danmarks rhabdocoele og acoele turbellarier. – Videnskabelige Meddelelser Dansk Naturhistorisk Forening 68: 1-160.

Bøving, A. 1906. Bidrag til kundskaben om Donaciin-Larvernes naturhistorie. – Dissert. København. 263 pp.

Ostenfeld, C. H. and Wesenberg-Lund, C. 1906. A regular fortnightly exploration of the plankton of the two Icelandic lakes: Thingvallavatn and Mývatn. – Proceedings of the Royal Society of Edinburgh 25: 1092-1176.

Ostenfeld, C. H. and Wesenberg-Lund, C. 1906. Über Süsswasserplankton. – Prometheus 17: 785- 790, 801-804, 817-820.

Wesenberg-Lund, C. 1906. Om kvartærgeologernes stilling til begrebet biologisk variation. – Meddelelser fra Dansk Geologisk Forening 12: 65-69.

Wesenberg-Lund, C.1906. Om naturforholdene i skotske og danske søer. – Geografisk Tidsskrift 18: 15-26.

Wesenberg-Lund, C. 1906. Umformungen des Erdbodens. Beziehungen zwischen Dammerde, Marsch, Wiesenland und Schlamm. – Prometheus 16: 561.

Wesenberg-Lund, C. 1907. On the occurrence of *Frederi-cellla sultana* Blumenb. and *Paludicella Ehrenbergii* van Bened in Greenland. – Meddelelser fra Grønland 34: 64-75.

Wesenberg-Lund, C. 1908. Plankton investigations of the Danish lakes. General Part. – Gyldendalske Boghandel, Copenhagen. 389 pp.

Wesenberg-Lund, C. 1908. Die littoralen Tiergesellschaften unserer grösseren Seen. – Internationale Revue der gesamten Hydrobiologie 1: 574-609.

Wesenberg-Lund, C. 1908. Über "pelagische" Ernährung der Uferschwalben. – Internationale Revue der gesamten Hydrobiologie 1: 510-512.

Wesenberg-Lund, C. 1908. Wasseramseln (*Cinclus aquaticus*) am Ufer des Fursees. – Internationale Revue der gesamten Hydrobiologie: 1: 512.

Wesenberg-Lund, C. 1908. *Culex-Mochlonyx-Corethra*, eine Anpassungreihe. – Internationale Revue der gesamten Hydrobiologie 1: 513-516.

Wesenberg-Lund, C. 1908. Über tropfende Laichmassen. – Internationale Revue der gesamten Hydrobiologie 1: 869-871.

Wesenberg-Lund, C. 1908. Den biologiske Station i Lunz. – Særtryk af Berlingske Tidende. 30 pp.

Boysen-Jensen, P. 1909. Über Steinkorrosion an den Ufern von Furesee. – Internationale Revue der gesamten Hydrobiologie 2: 163-173.

Petersen, H. E. 1909. Studier over Ferskvands-Phycomyceter. – Botanisk Tidsskrift 29: 345-440.

Wesenberg-Lund, C. 1909. Beiträge zur Kenntnis des Lebenszyklus der Zoochlorellen. – Internationale Revue der gesamten Hydrobiologie 2: 153-162.

Wesenberg-Lund, C. 1909. Über pelagische Eier, Dauerzustände und Larvenstadien der pelagischen Region des Süsswassers. – Internationale Revue der gesamten Hydrobiologie 2: 424-448.

Wesenberg-Lund, C. 1909. Über die praktische Bedeutung der jährlichen Variationen in der Viscosität des Wassers. – Internationale Revue der gesamten Hydrobiologie 2: 231-233.

Wesenberg-Lund, C. 1909. Grundtrækkene i ferskvandsplanktonets biologi og geografi. – Ymer: 90-133.

Wesenberg-Lund, C. 1909. Om limnologiens betydning for kvartærgeologien. – Geologiska Föreningen Stockholm 31: 449-470.

Bøving, A. G. 1910. Natural history of the larvae of Donaciinae. – Internationale Revue der Gesamten Hydrobiologie Biologisches Suppl. 1, 3: 1-108.

Ussing, H. 1910. Beiträge zur Biologie der Wasserwanze: *Aphelocheirus Montandoni* Horvath. – Internationale Revue der gesamten Hydrobiologie 3: 115-121.

Wesenberg-Lund, C. 1910. Grundzüge der Biologie und Geographie des Süsswasser-planktons nebst Bemerkungen über Hauptprobleme zukünftiger limnologischer Forschungen. – Internationale Revue der gesamten Hydrobiologie Biologisches Suppl. 1, 3: 1-44.

Wesenberg-Lund, C. 1910. Summary of our knowledge regarding various limnological problems. Pages 374-438 *in* J. Murray and F. P. Pullar, editors. A bathymetic survey of the freshwater locks of Scotland. II.

Wesenberg-Lund, C. 1910. Über die Biologie von *Glyphotaelius punctatolineatus* Retz. nebst Bemerkungen über das freilebende Puppenstadium der Wasserinsekten. – Internationale Revue der gesamten Hydrobiologie 3: 93-114.

Wesenberg-Lund, C. 1910. Über eine eventuelle Brutpflege bei *Gordius aquaticus* L. – Internationale Revue der gesamten Hydrobiologie 3: 122-127.

Wesenberg-Lund, C. 1910. Über die Süsswasserbiologischen Forschungen in Dänemark. – Internationale Revue der gesamten Hydrobiologie 3: 128-135.

Publications 1911 – 1920

Wesenberg-Lund, C. 1911. Über die Respirationsverhältnisse bei unter dem Eise überwinternden luftatmenden Wasserinsekten, besonders der Wasserkäufer und Wasserwanzen. – Internationale Revue der gesamten Hydrobiologie 3: 467-486.

Wesenberg-Lund, C. 1911. Über die Biologie der *Phryganea grandis* und über die Mechanik ihres Gehäusebaues. – Internationale Revue der gesamten Hydrobiologie 4: 65-90.

Wesenberg-Lund, C. 1911. Erweiterung des dänischen Süsswasserlaboratorium. – Internationale Revue der gesamten Hydrobiologie 4: 234-235.

Wesenberg-Lund, C. 1911. Om nogle ejendommelige temperaturforhold i de baltiske søers littoralregion og deres betydning. – Biologiske Arbejder tilegnede Warming: 87-106.

Wesenberg-Lund, C. 1911. Biologische Studien über netzspinnende, campodeoide Trichopterenlarven. – Internationale Revue der gesamten Hydrobiologie Biologisches Suppl. 3, 4: 1-64.

Brønsted, J. N. and Wesenberg-Lund, C. 1912. Chemisch-physikalische Untersuchungen der dänischen Gewässer. – Internationale Revue der gesamten Hydrobiologie 4: 251-290, 437-492.

Bøving, A. 1912. Studies relating to the anatomy, the biological adaptations on the mechanism of ovipositor in the various genera of Dytiscidae. – Internationale Revue der gesamten Hydrobiologie Biologisches Suppl. 5: 1-28.

Ferdinandsen, C. and Winge, Ø. 1912. Kobberdammene i Aldershvile Skov ved Bagsværd, med en indledende oversigt af C. Wesenberg-Lund. – Botanisk Tidsskrift 33: 1-44.

Wesenberg-Lund, C. 1912. Über einige eigentümliche Temperaturverhältnisse in der Litoralregion der baltischen Seen und deren Bedeutung. – Internationale Revue der gesamten Hydrobiologie 5: 287-306.

Wesenberg-Lund, C. 1912. Über die geographische Verbreitung der zwei Geschlechter von *Stratiotes aloides*. – Internationale Revue der gesamten Hydrobiologie 5: 307-316.

Wesenberg-Lund, C. 1912. Biologische Studien über Dytisciden. – Internationale Revue der gesamten Hydrobiologie Biologisches Suppl. 5: 1-129.

Wesenberg-Lund, C. 1913. Fortpflanzungsverhältnisse: Paarung und Eiablage der Süsswasserinsekten. – Fortschritte der Naturfachliche Forschung (Abderhalden) 8: 162-286.

Wesenberg-Lund, C. 1913. Wohnungen und Gehäusebau der Süsswasserinsekten. – Fortschritte der Naturfachliche Forschung (Abderhalden) 9: 55-132.

Wesenberg-Lund, C. 1914. Odonaten-Studien. – Internationale Revue der gesamten Hydrobiologie 6: 155-228, 373-422.

Wesenberg-Lund, C. 1914. Bidrag til nogle myggeslægters biologi. Særlig *Mochlonyx* og *Corethras*. – Mindeskrift for J. Steenstrup 34: 1-24.

Ege, R. 1915. On the respiratory function of the air stores carried by some aquatic insect (Corixidæ, Dytiscidæ and Notonectidæ). – Zeitschrift für allgemeine Physiologie 17: 81-124.

Ege, R. 1915. On the respiratory conditions of the larva and pupa of Donaciae. – Videnskabelige Meddelelser Dansk Naturhistorisk Forening 66: 183-195.

Wesenberg-Lund, C. 1915. Insektlivet i ferske vande. – Gyldendal, København. 524 pp.

Wesenberg-Lund, C. 1915-16. V. Bergsøe: Nybearbejdet af C.Wesenberg-Lund : Fra Mark og Skov. Vol. I: 1-596. Vol. II: 1-569. – Gyldendal, København.

Bardenfleth, K S. and Ege, R. 1916. On the anatomy and physiology of the air sacs of the larva of *Corethra plumicornis*. – Videnskabelige Meddelelser Dansk Naturhistorisk Forening 67: 25-42.

Wesenberg-Lund, C. 1917. Furesøstudier. – Det Kongelige Danske Videnskabers Selskabs Skrifter 8 R. 3: 1-208.

Wesenberg-Lund, C. 1917. Myreløvelarvens biologi. – Naturens Verden 1: 486-506.

Wesenberg-Lund, C. 1918. Anatomical description of the larva of *Mansonia Richardii* (Ficalbi) found in Danish freshwaters. – Videnskabelige Meddelelser Dansk Naturhistorisk Forening 69: 277-328.

Wesenberg-Lund, C. 1918. Om planktonolien og dens eventuelle betydning i fedtstoffattige tider. – Naturens Verden 2: 9-19.

Wesenberg-Lund, C. 1919. Contributions to the knowledge of the postembryonal development of the *Hydracarina*. – Videnskabelige Meddelelser Dansk Naturhistorisk Forening 70: 5-57.

Wesenberg-Lund, C. 1919. Bidrag til stikmyggenes biologi. – Naturens Verden 3: 1-26, 49-67, 150- 179, 312-320.

Wesenberg-Lund, C. 1919. Danmarks Fauna. Pages 170-208 *in* D. Bruun, editor. Danmark, Land og Folk. Danmark.

Wesenberg-Lund, C. 1920. The pupa-stage of the mosquitoes. – Festschrift für Zschokke 23, Basel: 1- 17.

Wesenberg-Lund, C. 1920. Nyere undersøgelser over vandrende Pinde og Blade. – Naturens Verden 4: 481-495.

Lundblad, O. 1920. Süsswasseracarinen aus Dänemark. – Det Kongelige Danske Videnskabers Selskabs Skrifter 8 R. 6: 135-258.

Wesenberg-Lund, C. 1920-21. Contributions to the biology of the Danish Culicidæ. – Det Kongelige Danske Videnskabers Selskabs Skrifter 8 R. 7: 1-208.

Publications 1921 – 1930

Wesenberg-Lund, C. 1921. Vore smaavande og deres fremtidsskæbne. – Naturens Verden 5: 256-264.

Wesenberg-Lund, C. 1921. Den lollandske feber før og nu. – Hjemstavnskursus: 1-14.

Wesenberg-Lund, C. 1921. Undersøgelser over danske malariamyg og dansk malaria. – Nordisk Hygienisk Tidsskrift: 229-247.

Wesenberg-Lund, C. 1921. Sur les causes du changement intervenu dans le mode de nourriture de *l'Anopheles maculipennis*. – Comptes rendue des séances de la Societe de Biologique 55: 383-385.

Wesenberg-Lund, C. 1921. Les Anophélinés du Danemark el les Fièvres paludéennes. – Comptes rendue des séances de la Societe de Biologique 55: 386.

Wesenberg-Lund, C. 1922. Contributions to the biology of Danish Batrachia. – Internationale Revue der gesamten Hydrobiologie 10: 24-30, 209-232, 321-361.

Wesenberg-Lund, C. 1922. Fra Sø og Aa. – Gyldendal, København. 137 pp.

Wesenberg-Lund, C. 1923. Contributions to the biology of the Rotifera. Part I. The males of the Rotifera. – Det Kongelige Danske Videnskabers Selskabs Skrifter 8 R. 4: 192-345.

Wesenberg-Lund, C. 1924. Bidrag til nogle danske padders biologi. – Naturens Verden 8: 12-30, 50- 61.

Micoletzky, H. 1925. Die freilebenden Süsswasser- und Morrnematoden Dänemarks. – Det Kongelige Danske Videnskabers Selskabs Skrifter 8 R. 10: 57-310.

Wesenberg-Lund, C. 1925. Den forstlige behandling af vore skoves dyreverden. – Dansk Skovforenings Tidsskrift: 31-75.

Wesenberg-Lund, C. 1925. Contributions to the biology of *Zoothamnium geniculatum* A. – Det Kongelige Danske Videnskabers Selskabs Skrifter 8 R. 10: 1-53.

Wesenberg-Lund, C. 1925. Om myg og myggeplage. – Schultz Forlag, København. 62 pp.

Wesenberg-Lund, C. 1925. Om forureningen af vore ferske vande. – Ferskvandsfiskeribladet 23: 157- 163.

Berg, K. 1926. Lidt om biernes sansefysiologi og "sprog". – Naturens Verden 10: 97-105.

Lakjer, T. 1926. Studien über die Trigeminus-versorgte Kaumuskulatur der Sauropsiden. Nach seinem Tode, herausgegeben von Prof. A. Luther, Helsingfors, und Professor C. Wesenberg-Lund, Kopenhagen: 1-154.

Lundblad, O. 1926. Zur Kenntnis der Quellenhydracarinen auf Møns Klint. – Det Kongelige Danske Videnskabernes Selskabs Biologiske Meddelelser 6, 1: 1-102.

Wesenberg-Lund, C. 1926. Contributions to the biology and morphology of the genus *Daphnia* with some remarks on heredity. – Det Kongelige Danske Videnskabers Selskabs Skrifter 8 R. 11: 91-250.

Wesenberg-Lund, C. 1926. Some features of the avifauna of Denmark and its present life-conditions. – Verhandlungen den VI Internationale Ornitholog Kongres. Kopenhagen: 414-425.

Wesenberg-Lund, C. 1926. Fuglefredningsarbejdet 1925. – Naturens Verden 10: 1-15.

Lakjer, T. 1927. Studien über die Gaumenregion bei Sauriern im Vergleich mit Anamniern und primitiven Sauropsiden. Herausgegeben von Professor A. Luther und Professor C. Wesenberg-Lund. – Zoologische Jahrbuch 49: 57-356.

Wesenberg-Lund, C. 1927. Peter Érasmus Müller: Mindetale i Videnskabernes Selskab. – Det Kongelige Danske Videnskabers Selskabs Oversigter: 1-24.

Wesenberg-Lund, C. 1927. Bondelandets fauna. – Gyldendal, København. 89 pp.

Berg, K. 1928. Prof. Dr. Phil. August Krogh. – Arosia 7: 18-21.

Wesenberg-Lund, C. 1928. Rotifera. Pages 7-120 *in* W. Kükenthal, editor. Handbuch der Zoologie, 2.

Wesenberg-Lund, C. 1928. Fuglefredningsarbejdet 1927-1928. – Danmarks Naturfredningsforenings Årsskrift 1927/1928: 95-114.

Berg, K. 1929. Faunistic and biological studies of Danish Cladocera. – Videnskabelige Meddelelser Dansk Naturhistorisk Forening 88: 31-111.

Berg, K. 1929. Nyere undersøgelser over honningbiens biologi. – Naturens Verden 13: 455-462.

Berg, K. 1929. Ecological studies on the zooplankton in the Lake of Frederiksborg Castle. – Report of the 18. Scandinavian Naturalist Congress, Copenhagen. 5 pp.

Berg, K. and Nygaard, G. 1929. Studies on the plankton in the Lake of Frederiksborg Castle. – Det Kongelige Danske Videnskabers Selskabs Skrifter 9 R. 1: 222-316.

Wesenberg-Lund, C. 1929. Insekter og Edderkopper. Pages 1-269 *in* A. Brehm, editor. Dyrenes Liv. Gyldendal, København.

Lundblad, O. 1930. Die Hydracarinen der Insel Bornholm. – Det Kongelige Danske Videnskabernes Selskabs Biologiske Meddelelser 8, 7: 1-96.

Wesenberg-Lund, C. 1930. Contributions to the biology of the Rotifera Part II. The periodicity and sexual periods. – Det Kongelige Danske Videnskabers Selskabs Skrifter 9 R. 2: 1-230.

Publications 1931 – 1940

Berg, K. 1931. Studies on the genus *Daphnia*. – Videnskabelige Meddelelser Dansk Naturhistorisk Forening 92: 1-222.

Krogh, A. and Berg, K. 1931. Über die chemische Zusammensetzung des Phytoplanktons aus dem Frederiksborg-Schlosssee und ihre Bedeutung für die Maxima der Cladoceren. – Internationale Revue der gesamten Hydrobiologie 25: 204-218.

Wesenberg-Lund, C. 1931. Familiebegrebet biologisk belyst. – Nordisk Tidsskrift 7: 1-20.

Wesenberg-Lund, C. 1931. Contributions to the development of the Trematoda Digenea Part 1. The biology of *Leucochloridium paradoxum*. – Det Kongelige Danske Videnskabers Selskabs Skrifter 9 R. 4: 90-142.

Berg, K. 1932. Les Cladocères et leur reproduction. – Bulletin Francais de la Pisciculture: 1-28.

Berg, K. 1932. Ist das Alter der Latenzeier der Daphnien ein geschlechtsbestimmender Faktor. – Archiv für Hydrobiologie 24: 497-508.

Hermes, G. 1932. Studien über die Konstanz histologischer Elemente IV. Die Männchen von *Hydatina senta* Ehrbg., *Rhinops vitrea* Hudson und *Asplachna priodonta* Gosse. – Zeitschrift für Wissenschaftliche Zoologie 141: 581-725.

Wesenberg-Lund, C. 1932. Om forekomsten af udviklingsstadier til blod-parasiterende ikter i danske ferskvande. – Naturens Verden 16: 433-446.

Berg, K. 1933. Note on *Macrothrix hirsuticornis* Norman & Brady, with description of the male. – Videnskabelige Meddelelser Dansk Naturhistorisk Forening 97: 11-24.

Wesenberg-Lund, C. 1933. Naturstudium og religiøs følelse. – Nordisk Tidsskrift 9: 33-54.

Berg, K. 1934. Cyclic reproduction, sex determination and depression in the Cladocera. – Biological Reviews 9: 139-174.

Berg, K. 1934. Die Geschlechtsbestimmung und die Depression bei den Cladoceren. – Verhandlungen Internationale Vereinigung für Theoretische und Angewandte Limnologie 6: 314-318.

Wesenberg-Lund, C. 1934. Contributions to the development of the Trematoda Digenea Part II. The biology of the freshwater Cercariae in Danish freshwaters. – Det Kongelige Danske Videnskabers Selskabs Skrifter 9 R. 5: 1-223.

Wesenberg-Lund, C. 1934. Nyere undersøgelser over malariamyg. – Naturens Verden 18: 241-254.

Wesenberg-Lund, C. 1934. Nyere undersøgelser over leverikten. – Maanedsskrift for Dyrlæger 46: 145-155.

Berg, K. 1935. Ochrida Søen og dens relikt fauna. – Naturens Verden 19: 49-64.

Wesenberg-Lund, C. 1935. Nyere undersøgelser over ikternes forvandling. – Naturens Verden 19: 32- 77.

Wesenberg-Lund, C. 1935. Johan Erik Vesti Boas. Mindetale i Videnskabernes Selskab. – Det Kongelige Danske Videnskabers Selskabs Oversigter: 1-28.

Wesenberg-Lund, C. 1935. Et 75 aars minde, Sven Lovén. – Nordisk Tidsskrift 11: 177-193.

Berg, K. 1936. Reproduction and depression in the Cladocera illustrated by the weight of the animals. – Archiv für Hydrobiologie 30: 438-462.

Berg, K. 1937. Generationswechsel der Cladoceren und die Depressionshypothese. – Zoologische Jahrbücher. Abteilung für Allgemeine Zoologie und Physiologie der Tiere 57: 373-374.

Berg, K. 1937. Contributions to the biology of *Corethra* Meigen. – Det Kongelige Danske Videnskabers Selskabs Biologiske Meddelelser 13, 11: 1-101.

Wesenberg-Lund, C. 1937. Naturstudium og livsopfattelse. – Nordisk Tidsskrift 3: 491-506.

Wesenberg-Lund, C. 1937. Ferskvands-faunaen biologisk belyst, I-II. – Gyldendal, København. 833 pp.

Berg, K. 1938. Studies on the bottom animals of Esrom Lake. – Det Kongelige Danske Videnskabers Selskabs Skrifter 9 R. 8: 1-255.

Nygaard, G. 1938. Hydrobiologische Studien über dänische Teiche und Seen 1. – Archiv für Hydrobiologie 32: 523-692.

Kaiser, E. W. 1939. Et nyt fund af vandtægen *Aphelocheirus* i Danmark. Review of lecture in Dansk Naturhistorisk Forening. – Naturhistorisk Tidende 3: 152-153.

Wesenberg-Lund, C. 1939. Biologie der Süsswassertiere. Wirbellose Tiere. Deutsche Ausgabe besorgt von O. Storch. – Verlag Springer. Wien. 817 pp.

Wesenberg-Lund, C. 1939. Kulturskov, fauna og æsthetik. – Dansk Skovforenings Tidsskrift: 181-206.

Wesenberg-Lund, C. 1939. Faunaens kaar i de danske skove. – Danmarks Naturfredningsforenings Årsskrift 1938/39: 61-76.

Wesenberg-Lund, C. 1939. Lidt om de nordsjællandske søer. – Frederiksborg Amts Historiske Samfunds Aarbog 2: 1-6.

Nielsen, A. 1940. Jyske aaer. – Danmark 1: 855-860.

Nielsen, A. 1940. Om aarsagerne til den rheophile faunas rigdom i Himmerland. Review of lecture in Dansk Naturhistorisk Forening. – Naturhistorisk Tidende 5: 47-48.

Wesenberg-Lund, C. 1940. Gribsø, en dansk dystroph sø. – Naturens Verden 24: 19-34.

Wesenberg-Lund, C. 1940. Malaria som verdens faktor. – Nordisk Tidsskrift 16: 101-118.

Wesenberg-Lund, C. 1940. Det Ferskvandsbiologiske Laboratorium gennem 40 aar. En redegørelse for min virksomhed. – Copenhagen. (With a list of publications from the laboratory up to 1939). 109 pp.

Publications 1941 – 1950

Berg, K. 1941. Contributions to the biology of the aquatic moth *Acentropus niveus* (Oliv). – Videnskabelige Meddelelser Dansk Naturhistorisk Forening 105: 59-139.

Wesenberg-Lund, C. 1941. Om fangtragt-dannende sanddyr. – Naturens Verden 25: 97-110.

Bennike, S. A. B. 1942. De danske iglers biologi. Review of prize paper. – Naturhistorisk Tidende 6: 87.

Nielsen, A. 1942. P. Esben-Petersen. 18. December 1869 – 2. April 1942. – Naturhistorisk Tidende 6: 158.

Nielsen, A. 1942. Über die Entwicklung und Biologie der Trichopteren. Mit besonderer Berücksichtigung der Quelltrichopteren Himmerlands. – Archiv für Hydrobiologie Suppl. 17: 255-631.

Nielsen, A. 1942. Review of thesis. – Naturhistorisk Tidende 6: 90-91.

Wesenberg-Lund, C. 1942. Smaa iagttagelser over dansk fugleliv. – Naturens Verden 26: 433-456.

Bennike, S. A. B. 1943. Contributions to the ecology and biology of the Danish fresh-water leeches *(Hirudinea)*. – Folia Limnologica Scandinavica 2: 1-109.

Berg, K. 1943. Physiographical studies on the River Susaa. – Folia Limnologica Scandinavica 1: 1- 174.

Clemens-Petersen, I. 1943. Om Gribsøs bundfauna og dennes afhængighed af økologiske faktorer. – Naturhistorisk Tidende 7: 56-57.

Nielsen, A. 1943. *Apatidea auricula* Forsslund from a Norwegian mountain lake. Description of the imago and notes on the biology. – Entomologiske Meddelelser 23: 18-30.

Nielsen, A. 1943. Postembryonale Entwicklung und Biologie der rheophilen Köcherfliege *Oligoplectrum maculatum* Fourcroy. – Det Kongelige Danske Videnskabers Selskab Biologiske Meddelelser 19, 2: 1-86.

Nielsen, A. 1943. Trichopterologische Notizen. – Videnskabelige Meddelelser Dansk Naturhistorisk Forening 107: 105-120.

Nielsen, A. 1943. Nogle bemærkninger om *Eubria palustris* Germar, en postglacial varmerelikt. Review of lecture in Dansk Naturhistorisk Forening. – Naturhistorisk Tidende 7: 168.

Wesenberg-Lund, C. 1943. Biologie der Süsswasserinsekten. – Gyldendal og Springer, Kopenhagen, Berlin, Wien. 682 pp.

Wesenberg-Lund, C. 1943. Bemerkungen über die Biologie der Chironomiden. – Entomologiske Meddelelser 23: 179-203.

Wesenberg-Lund, C. 1943. P. Esben-Petersen. 18. December 1869 – 2. April 1942. – Videnskabelige Meddelelser Dansk Naturhistorisk Forening 106: 5-11.

Wesenberg-Lund, C. 1943. Sødragene omkring Mølleaaen. – Danmark 3: 489-496.

Jespersen, P. H. 1944. Om vandedderkoppens (*Argyroneta aquatica* Cl.) overvintring i tomme sneglehuse. – Flora og Fauna 10: 1-11.

Nygaard, G. 1944. Om sødybdens indvirkning på planktonproduktionen. – Naturhistorisk Tidende 8: 34-37.

Wesenberg-Lund, C. 1944. Små iagttagelser over dansk fugleliv. – Naturens Verden 28: 145-160.

Berg, K. 1945. Svend Aage Boisen Bennike. 7 December 1918 – 24. November 1944. – Videnskabelige Meddelelser Dansk Naturhistorisk Forening 108: 1-5.

Nygaard, G. 1945. Dansk Planteplankton. En flora over de vigtigste ferskvandsformer. – Gyldendal, Copenhagen.

Wesenberg-Lund, C. 1945. Fra Sø og Aa. – Gyldendal, Copenhagen. 350 pp.

Wesenberg-Lund, C. 1946. Fra det usynliges verden. – "Mærkelige Dyr". Copenhagen: 48-65.

Jørgensen, E. G. 1947. Algevegetationen i Madum Sø. – Botanisk Tidsskrift 48: 141-155.

Bennike, S. A. B. and Berg, K. 1948. Oligochaeta, freshwater bristle-bearing worms (in River Susaa). – Folia Limnologica Scandinavica 4: 40-54.

Berg, K. (ed.) 1948. Biological studies on the River Susaa. – Folia Limnologica Scandinavica 4: 1-318.

Jónasson, P. M. 1948. Quantitative studies of the bottom fauna (in River Susaa). – Folia Limnologica Scandinavica 4: 204-287.

Jørgensen, E. G. 1948. Diatom communities in some Danish lakes and ponds. – Konglige Danske Videnskabers Selskabs Biologiske Skrifter 5: 1-140.

Keiding, J. 1948. Arachnida. Acarina, Mites (in River Susaa). – Folia Limnologica Scandinavica 4: 79-108.

Nielsen, A. 1948. Trichoptera, Caddis Flies. With description of a new species of *Hydroptila* (in River Susaa). – Folia Limnologica Scandinavica 4: 123-144.

Nielsen, A. 1948. Undersøgelser over vaarfluernes biologi. Familien Hydroptilidae. Review of Lecture in Dansk Naturhistorisk Forening. – Naturhistorisk Tidende 12: 19-41.

Nielsen, A. 1948. Postembryonic development and biology of the Hydroptilidae. A contribution to the phylogeny of the Caddis Flies and to the question of the origin of the case-building instinct. – Det Kongelige Danske Videnskabers Selskabs Biologiske Skrifter 5, 1: 1-298.

Berg, K. 1949. Personalia. C. Wesenberg-Lund. – Hydrobiologia 1: 322-324.

Berg, K. 1949. Notes on *Cercaria splendens* Szidat. – Videnskabelige Meddelelser Dansk Naturhistorisk Forening 111: 263-270.

Berg, K. 1949. Remarks on some Danish river studies. – Verhandlungen Internationale Vereinigung für Theoretische und Angewandte Limnologie 10: 76-79.

Hansen, K. 1949. Ispresning i Tystrup Sø og Esrum Sø vinteren 1946-47. – Geografisk Tidsskrift 49: 67-72.

Jónasson, P. M. 1949. A quantitative river study. – Verhandlungen Internationale Vereinigung für Theoretische und Angewandte Limnologie 10: 232-234.

Kaiser, E. W. (ed.). 1949. Fysisk, kemisk, hydrometrisk og biologisk undersøgelse af Mølleåen fra Lyngby Sø til Øresund 1946 og 1947. – Dansk Ingeniørforening, Spildevandskomitéen. Skrift 1: 1-39.

Nygaard, G. 1949. A simple micromanipulator. – Science 110: 165-166.

Nygaard, G. 1949. Hydrobiological studies on some Danish ponds and lakes. Part II: The Quotient Hypothesis and some new or little known phytoplankton organisms. – Det Kongelige Danske Videnskabers Selskabs Biologiske Skrifter 7, 1: 1-293.

Berg, K. 1950. De ferske Vande. Pages 161-192 *in* F. W. Bræstrup, G. Thorson and E. Wesenberg- Lund, editors. Vort Lands Dyreliv III. Gyldendal, Copenhagen, Denmark.

Berg, K. 1950. News about the Freshwater-Biological Laboratory, University of Copenhagen. – Hydrobiologia 2: 386.

Berg, K. 1950. Den fredede ejendom Strødam 1925-1950. – Dansk Naturfredningsforenings Aarsskrift 1949-50.

Berg, K. 1950. Personalia. August Krogh in memoriam. – Hydrobiologia 2: 380-382.

Fjerdingstad, E. 1950. The microflora of the River Mølleaa. With special reference to the relation of the benthal algae to pollution. – Folia Limnologica Scandinavica 5: 1-123.

Hansen, K. 1950. The geology and bottom deposits of Lake Tystrup Sø, Zealand. – Danmarks geologiske undersøgelser 2. ser. 76: 1-52.

Nielsen, A. 1950. The torrential invertebrate fauna. – Oikos 2: 176-196.

Nielsen, A. 1950. Insekter (Insecta). Odonata, Ephemeroptera. Plecoptera, Orthoptera, Dermaptera, Neuroptera-Mecoptera, Trichoptera, Hydrachnellae. Pages 225-234, 239-255, 272-284, 427-429 *in* F. W. Bræstrup, G. Thorson and E. Wesenberg-Lund, editors. Vort Lands Dyreliv II. Gyldendal, Copenhagen.

Nielsen, A. 1950. Notes on the genus *Apatidea* MacLachlan. With descriptions of two new and possibly endemic species from the springs of Himmerland. – Entomologiske Meddelelser 25: 384-404.

Nielsen, A. 1950. On the zoogeography of springs. – Hydrobiologia 2: 313-321.

Publications 1951 – 1960

Berg, K. 1951. Notes on some large Danish springs. – Hydrobiologia 3: 72-78.

Berg, K. 1951. On the respiration of some molluscs from running and stagnant water. – Annales de Biologie 27: 561-567.

Berg, K. 1951. The content of limnology demonstrated by F. A. Forel and August Thienemann on the shore of Lake Geneva. Some historical remarks. – Verhandlungen Internationale Vereinigung für Theoretische und Angewandte Limnologie 11: 41-57.

Johansen, H. og Nielsen, B. 1951. Ornithologisk arbejde på Strødam-reservatet. – Dansk Ornithologisk Forenings Tidsskrift 45: 205-216.

Johnson, D. S. 1951. *Ceriodaphnia setosa* Matile, en for Danmark ny Dafnia art. – Flora og Fauna 57: 48.

Kaiser, E. W. (ed.). 1951. Biologiske, biokemiske, bakteriologiske samt hydrometriske undersøgelser af Pøleåen 1946 og 1947. – Dansk Ingeniørforening, Spildevandskomitéen. Skrift 3: 1-53.

Nielsen, A. 1951. *Hydroptila occulta* Eaton, new to the Danish fauna. With descriptions of the specific characters. – Entomologiske Meddelelser 26: 122-129.

Nielsen, A. 1951. Contributions to the metamorphosis and biology of the genus *Atrichopogon* Kieffer (Diptera, Ceratopogonidae) with remarks on the evolution and taxonomy of the genus. – Det Kongelige Danske Videnskabers Selskabs Biologiske Skrifter 6, 6: 1- 95.

Nielsen, A. 1951. Spring fauna and speciation. – Verhandlungen Internationale Vereinigung für Theoretische und Angewandte Limnologie 11: 261-263.

Nielsen, A. 1951. Is dorsoventral flattening of the body an adaptation to torrential life? – Verhandlungen Internationale Vereinigung für Theoretische und Angewandte Limnologie 11: 264-267.

Nygaard, G. 1951. How to make permanent fluid mounts of planktonorganisms. – Hydrobiologia 3: 282-289.

Walshe, B. M. 1951. The feeding habits of certain chironomid larvae (Subfamily Tendipedinae). – Proceedings of the Zoological Society of London 121: 63-79.

Walshe, B. M. 1951. The function of haemoglobin in relation to filter feeding in leaf- mining chironomid larvae. – The Journal of Experimental Biology 28: 57-61.

Walshe, B. M. 1951. Observation on the biology and behaviour of larvae of the midge *Rheotanytarsus*. – The Journal of the Quekett microscopical Club ser. 4, 3: 171-178.

Berg, K. 1952. On the oxygen consumption of Ancylidae (Gastropoda) from an ecological point of view. – Hydrobiologia 4: 225-267.

Hemmingsen, A. M. 1952. The oviposition of some crane-fly species (Tipulidae) from different types of localities. – Videnskabelige Meddelelser Dansk Naturhistorisk Forening 114: 365-430.

Røen, U. 1952. Nye fund af ferskvandscopepoder fra Danmark. – Flora og Fauna 58: 1-4.

Røen, U. 1952. On some Euphyllopoda from North China. – Videnskabelige Meddelelser Dansk Naturhistorisk Forening 114: 203-215.

Wesenberg-Lund, C. 1952. De danske søer og dammes dyriske plankton. – Munksgaard, Copenhagen. 182 pp.

Berg, K. 1953. Respirationsforsøg med snegle fra rindende og stillestående vand. Review of lecture in Dansk Naturhistorisk Forening. – Naturhistorisk Tidende 17: 34-35.

Berg, K. 1953. The problem of respiratory acclimatization illustrated by experiments with *Ancylus fluviatilis* (Gastropoda). – Hydrobiologia 5: 331-350.

Green, J. 1953. Records of epibionts on freshwater organisms in Denmark. – Videnskabelige Meddelelser Dansk Naturhistorisk Forening 115: 177-180.

Hynes, H. B. N. 1953. The Plecoptera of some small streams near Silkeborg, Jutland. – Entomologiske Meddelelser 26: 489-494.

Holmen, K. and Mathiesen, H. 1953. *Luzula Wahlenbergii* in Greenland. – Botanisk Tidsskrift 49: 233- 238.

Røen, U. 1953. Ferskvandscopepodernes biologi i mindre søer. Review of prize paper. – Naturhistorisk Tidende 17: 231.

Berg, K. 1954. Københavns Universitets Ferskvandsbiologiske Laboratorium. – Ferskvandsfiskeribladet 52: 113-118.

Dahl, J. 1954. *Orthocladius naumanni* Brundin (Dipt. Chiron.), new to Denmark, with description of the female. – Entomologiske Meddelelser 26: 617-623.

Dunn, D. R. 1954. Notes on the bottom fauna of twelve Danish lakes. – Videnskabelige Meddelelser Dansk Naturhistorisk Forening 116: 251-268.

Foged, N. 1954. On the diatom flora of some Funen lakes. – Folia Limnologica Scandinavica 6: 1-76.

Heyningen, H. E. Van. 1954. A study on the food of some Daphniidae. – Freshwater Biological Laboratory. Hillerød. 14 pp.

Jónasson, P. M. 1954. An improved funnel trap for capturing emerging aquatic insects, with some preliminary results. – Oikos 5: 179-188.

Macan, T. T. 1954. The Corixidae (Hemipt.) of some Danish lakes. – Hydrobiologia 6: 44-69.

Nygaard, G. 1954. A new diatom species and two new varieties from plankton in lake Taserssuatsiaq. – Meddelelser om Grønland 148, 1: 311-313.

Ockelmann, W. 1954. On the interrelationship and the zoogeography of northern species of *Yoldia*, s.str. (Mollusca, Fam. Ledidae). – Meddelelser om Grønland 107, 7: 1-33.

Røen, U. 1954. Om småkrebsfaunaen i en jydsk, oligotroph sø, Langsø i Grane Plantage med et fund af *Eucyclops lilljeborgi* (G. O. Sars) ny for Danmark. – Flora og Fauna 60: 11-18.

Berg, K. 1955. Ecological remarks on the bottom fauna of a Danish humic acid lake, Store Gribsø. – Verhandlungen Internationale Vereinigung für Theoretische und Angewandte Limnologie 12: 569-576.

Berg, K. 1955. C. Wesenberg-Lund. Tale i Videnskabernes Selskabs møde d. 9. marts 1956. – Oversigt over Det Det Kongelige Danske Videnskabers Selskabs Virksomhed 1955-56. 13 pp.

Jónasson, P. M. 1955. The efficiency of sieving techniques for sampling freshwater bottom fauna. – Oikos 6: 183-207.

Olsen, S. 1955. Lake Lyngby Sø. Limnological studies on a culturally influenced lake. – Folia Limnologica Scandinavica 7: 1-152.

Nygaard, G. 1955. On the productivity of five Danish waters. – Verhandlungen Internationale Vereinigung für Theoretische und Angewandte Limnologie 12: 123-133.

Røen, U. 1955. On the number of eggs in some free-living freshwater copepods. – Verhandlungen Internationale Vereinigung für Theoretische und Angewandte Limnologie 12: 447-454.

Berg, K. and Clemens-Petersen, I. 1956. Studies on the humic, acid Lake Gribsø. With contributions by K. Hansen, P. Wolthers, and G. Nygaard. – Folia Limnologica Scandinavica 8: 1-273.

Berg, K. 1956. Professor C. Wesenberg-Lund. – Hydrobiologia 8: 191-192.

Hansen, K. 1956. The profundal bottom deposits of Gribsø. – Folia Limnologica Scandinavica 8: 16- 25.

Mathiesen, H. and Nielsen, J. 1956. Botaniske undersøgelser i Randers Fjord og Grund Fjord. – Botanisk Tidsskrift 53: 1-34.

Nygaard, G. 1956. Ancient and recent flora of diatoms and Chrysophyceae in Lake Gribsø. – Folia Limnologica Scandinavica 8: 32-94.

Røen, U. 1956. Results from the Danish expedition to the French Cameroons 1949-50. XI. Entomostraca. – Bulletin de l'I.F.A.N.T. XVIII, sér. A no. 3: 913-925.

Wolthers, P. 1956. Pollen-analytical studies of the profundal dygyttja. – Folia Limnologica Scandinavica 8: 25-32.

Berg, K. 1957. Professor Carl Jørgen Wesenberg-Lund. – Proceedings of The Linnean Society of London, 168 Session: 57-60.

Hansen, K. 1957. Les Vases des eaux douces au Danemark. – Revue de l'Institut Francais du Pétrole et Annales des Combustibles Liquides 12: 449-452.

Røen, U. 1957. Contributions to the biology of some Danish free living freshwater copepods. – Det Kongelige Danske Videnskabers Selskabs Biologiske Skrifter 9, 2: 1-101.

Berg, K. (ed.). 1958. Furesøundersøgelser 1950-54. Limnologiske studier over Furesø's kulturpåvirkning. English Summary: Investigations on Fure Lake 1950-54. Limnological studies on cultural influences. – Folia Limnologica Scandinavica 10: 1-189.

Berg, K. 1958. Elektrisk ledningsevne af søvandet (i Furesø). – Folia Limnologica Scandinavica 10: 35.

Berg, K. 1958. Studier over reliktkrebs og over ilt- og temperaturforhold. – Folia Limnologica Scandinavica 10: 148-160.

Berg, K. 1958. Sammenfattende bemærkninger om Furesø. – Folia Limnologica Scandinavica 10: 167-182.

Berg, K. 1958. Professor C. Wesenberg-Lund. – Verhandlungen Internationale Vereinigung für Theoretische und Angewandte Limnologie 13: 975-978.

Berg, K. and Røen, U. 1958. Gennemsigtigheden i Furesø. – Folia Limnologica Scandinavica 10: 34.

Berg, K., Lumbye, J. and Ockelmann, K. W. 1958. Seasonal and experimental variations of the oxygen consumption of the limpet *Ancylus fluviatilis* (O. F. Müller). – Journal of Experimental Biology 35: 43- 73.

Berg, K. and Mathiesen, H. 1958. Opløste stoffer i Furesø ifølge analyser af K. Korsgaard og Gunde Lange. – Folia Limnologica Scandinavica 10: 35-38.

Christensen, T. and Andersen, K. 1958. De større vandplanter i Furesø. – Folia Limnologica Scandinavica 10: 114-128.

Ebent, F. and Lyshede, J. M. 1958. Afstrømningsundersøgelser i Furesøens nedbørsområde 1951-53. – Folia Limnologica Scandinavica 10: 13-17.

Fjerdingstad, E. 1958. Bakteriologiske undersøgelser (i Furesø). – Folia Limnologica Scandinavica 10: 130-135.

Fjerdingstad, E. 1958. Furesøens litorale algevegetation. – Folia Limnologica Scandinavica 10: 135- 147.

Jónasson, P. M. 1958. The mesh factor in sieving techniques. – Verhandlungen Internationale Vereinigung für Theoretische und Angewandte Limnologie 13: 860-866.

Korsgaard, K. and Skadhauge, A. 1958. Tilførsel af næringsstoffer til Furesøen fra vandløb og kloakanlæg. – Folia Limnologica Scandinavica 10: 18-32.

Lumbye, J. 1958. The oxygen consumption of *Theodoxus fluviatilis* (L.) and *Potamopyrgus jenkinsi* (Smith) in brakish and freshwater. – Hydrobiologia 10: 245-262.

Mathiesen, H. 1958. Om Furesøens rørsump. – Folia Lim-

nologica Scandinavica 10: 128-130.

Nygaard, G. 1958. On the productivity of the bottom vegetation in Lake Grane Langsø. – Verhandlungen Internationale Vereinigung für Theoretische und Angewandte Limnologie 13: 144-155.

Nygaard, G. 1958. Produktions- og milieuundersøgelser i Furesø. – Folia Limnologica Scandinavica 10: 97-104.

Nygaard, G. 1958. Furesøens planteplankton. – Folia Limnologica Scandinavica 10:109-113.

Olsen, S. 1958. Phosphate adsorption and isotopic exhange in lake muds. – Verhandlungen Internationale Vereinigung für Theoretische und Angewandte Limnologie 13: 915-922.

Olsen, S. 1958. Fosfatbalancen mellem bund og vand i Furesø. Forsøg med radioaktivt fosfor. – Folia Limnologica Scandinavica 10: 39-94.

Otterstrøm, C. V. 1958. Temperaturmålinger 1932-33 i Furesø. – Folia Limnologica Scandinavica 10: 33.

Røen, U. 1958. Studies on freshwater Entomostraca in Greenland. I. – Meddelelser om Grønland 159, 2: 1-9.

Røen, U. 1958. Tre års arbejde på arktisk station. – Tidsskriftet Grønland: 69-75.

Steemann-Nielsen, E. 1958. Planteplanktonets årlige produktion af organisk stof i Furesøen. – Folia Limnologica Scandinavica 10: 104-109.

Berg, K. and Ockelmann, K. W. 1959. The respiration of freshwater snails. – The Journal of Experimental Biology 36: 690-708.

Hansen, K. 1959. Sediments from Danish lakes. – Journal of Sedimentary Petrology 29: 38-46.

Hansen, K. 1959. The terms gyttja and dy. – Hydrobiologia 13: 309-315.

Jónasson, P. M. and Mathiesen, H. 1959. Measurements of primary production in two Danish eutrophic lakes, Esrom Sø and Furesø. – Oikos 10: 137-167.

Kristiansen, J. 1959. Flagellates from some Danish lakes and ponds. – Dansk Botanisk Arkiv 18, 4: 1- 55.

Olsen, S. 1959. Isskurede træer. – Botanisk Tidsskrift 55: 1-22.

Røen, U. 1959. Lidt østgrønlandsk ferskvandsbiologi. – Tidsskriftet Grønland: 57-64.

Berg, K. and Reiter, H. F. H. 1960. Observations on Schistosome dermatitis in Denmark. – Acta Dermato-Venereoligica 40: 369-380.

Hemmingsen, A. 1960. Energy metabolism as related to body size and respiratory surfaces, and its evolution. – Report Steno Memorial Hospital 9: 1-110.

Kristiansen, J. 1960. Some cases of sexuality in *Kephyriopsis* (Chrysophyceae). – Botanisk Tidsskrift 56: 128-131.

Røen, U. 1960. Iagttagelser over Stor Snegås (*Anser caerulescens atlanticus* (Kennard)) i Thule Distrikt sommeren 1959. – Dansk Ornithologisk Forenings Tidsskrift 54: 127-135.

Publications 1961 – 1970

Berg, K. og Reiter, H. F. H. 1961. Hudsygdom fremkaldt af Cercarier. – Ferskvandsfiskeribladet 59: 58-64.

Berg, K. 1961. Næringsfattige, næringsrige og overernærede ("søsyge") søer. – Danmarks Naturfrednings-forenings Årsskrift 1961: 44-51.

Berg, K. 1961. On the oxygen consumption of some freshwater snails. – Verhandlungen Internationale Vereinigung für Theoretische und Angewandte Limnologie 14: 1019-1022.

Berg, K. 1961. Professor, dr. Frantz Ruttner. – Ferskvandsfiskeribladet 59: 215-216.

Hansen, K. 1961. Lake types and lake sediments. – Verhandlungen Internationale Vereinigung für Theoretische und Angewandte Limnologie 14: 285-290.

Jónasson, P. M. 1961. Population dynamics in *Chironomus anthracinus* Zett. in the profundal zone of Lake Esrom. – Verhandlungen Internationale Vereinigung für Theoretische und Angewandte Limnologie 14: 196-203.

Jónasson, P. M. 1961. Masseforekomst af mosdyr i nogle danske søer. – Ferskvandsfiskeriblandet 59: 48-54.

Røen, U. 1961. Forholdet mellem småkrebsplankton og mængden af produceret organisk stof i Grane Langsø 1950-51. – Flora og Fauna 67: 19-25.

Berg, K., Jónasson, P. M. and Ockelmann, K. W. 1962. The respiration of some animals from the profundal zone of a lake. – Hydrobiologia 19: 1-39.

Berg, K. 1962. Københavns Universitets Ferskvandsbiologiske Laboratorium. – Ferskvandsfiskeribladet 60: 165-167.

Hansen, K. 1962. The dystrophic lake type. – Hydrobiologia 19: 183-191.

Johnsen, P., Mathiesen, H. and Røen, U. 1962. Sorø-Søerne, Lyngby Sø og Bagsværd Sø. Limnologiske studier over fem kulturpåvirkede, sjællandske søer. – Dansk ingeniørforening, Spildevandskomiteen, Skrift 14: 1-135.

Madsen, B. L. 1962. Økologiske undersøgelser i nogle østjyske vandløb. 1. Fysiske og kemiske forhold. – Flora og Fauna 68: 185-195.

Røen, U. 1962. Trues Grønlands ferskvande af forurening? – Tidsskriftet Grønland: 137-143.

Røen, U. 1962. Salt- og brakvandsdyr i grønlandske ferskvande. – Tidsskriftet Grønland: 335-341.

Røen, U. 1962. Studies on freshwater Entomostraca in Greenland II. – Meddelelser om Grønland 170, 2: 1-249.

Jónasson, P. M. 1963. The growth of *Plumatella repens* and *P. fungosa* (Bryozoa Ectoprocta) in relation to external factors in Danish eutrophic lakes. – Oikos 14: 121-137.

Kennedy, C. R. and Chubb, J. C. 1963. Forekomsten af bændelormeslægten *Archigetes* Leuckart 1869 (Cestoda: Caryophyllidea) in Danmark. – Flora og Fauna 69: 9-10.

Madsen, B. L. 1963. Økologiske undersøgelser i nogle østjyske vandløb. 2. Planarier og igler. – Flora og Fauna 69: 113-125.

Mathiesen, H. 1963. Om planteplanktonets produktion af organisk stof i nogle næringsrige søer på Sjælland. – Ferskvandsfiskeribladet 61: 7-9 and 20-25.

Røen, U. 1963. Nogle udbredelsestyper i den grønlandske ferskvandsfauna. – Tidsskriftet Grønland: 361-374.

Thorup, J. 1963. Growth and life-cycle of invertebrates from Danish springs. – Hydrobiologia 22: 55- 84.

Hansen, K. 1964. Lagoon sediments in Greenland. – Developments in Sedimentology 1: 165-169.

Hansen, K. 1964. Oversigt over danske søsedimenter. – Meddelelser fra Dansk Geologisk Forening 15: 428-430.

Jónasson, P. M. 1964. The relationship between primary production and production of profundal bottom invertebrates in a Danish eutrophic lake. – Verhandlungen Internationale Vereinigung für Theoretische und Angewandte Limnologie 15: 471-479.

Olsen, S. 1964. Phosphate equilibrium between reduced sediments and water. – Verhandlungen Internationale Vereinigung für Theoretische und Angewandte Limnologie 15: 333-341.

Olsen, S. 1964. Vegetationsændringer i Lyngby Sø. Bidrag til analyse af kulturpåvirkninger på vand- og sumpplantevegetationen. – Botanisk Tidsskrift 4: 273-300.

Berg, K. and Jónasson, P. M. 1965. Oxygen consumption of profundal lake animals at low oxygen content of the water. – Hydrobiologia 26: 131-143.

Hansen, K. 1965. The post-glacial development of Grane Langsø. – Meddelelser fra Dansk Geologisk Forening 15: 446-458.

Hansen, K. 1965. Vore søers tilfrysning. – Geografisk Tidsskrift 64: 162-173.

Jónasson, P. M. 1965. Esrom Sø og dens fredning. Pages 104-113 *in* A. Nørrevang and T.J. Meyer, editors. Danmarks Natur 4. Politikens Forlag, Copenhagen, Denmark.

Jónasson, P. M. 1965. Factors determining population size of *Chironomus anthracinus* in Lake Esrom. – Verhandlungen Internationale Vereinigung für Theoretische und Angewandte Limnologie 13: 139-162.

Knutzon, L. V. and Lyneborg, L. 1965. Danish Acalypterate flies. 3. Sciomyzidae (Diptera). – Entomologiske Meddelelser 34: 61-101.

Kristiansen, J. and Mathiesen, H. 1965. Phytoplankton of the Tystrup-Bavelse Lakes. Primary production and standing crop. – Oikos 15: 1-43.

Lumbye, J. and Lumbye, L. E. 1965. The oxygen consumption of *Potamopyrgus jenkinsi* (Smith). – Hydrobiologia 25: 489-500

Nygaard, G. 1965. Hydrographic studies, especially on the carbon dioxide system in Grane Langsø. – Det Kongelige Danske Videnskabers Selskabs Biologiske Skrifter 14, 2: 1-110.

Røen, U. and Larsen, A. 1965. Entomostraca from the Skaftafell area, Iceland. – Videnskabelige Meddelelser fra Dansk Naturhistorisk Forening 127: 135-148.

Berg, K. 1966. Sven Ekman. 31-V-1876 – 2-II-1964. – Hydrobiologia 27: 274-279.

Berg, K. 1966. Nekrolog over prof. Ragnar Spärck. – Naturfredningsforeningens årsskrift.

Knutzon, L. V. 1966. Biology and immature stages of Malacophagous flies: *Antichaeta analis, A. atriseta, A. brevipennis,* and *A. obliviosa* (Diptera: Sciomyzidae). – Transactions of the American Entomological Society 92: 67-101.

Madsen, B. L. 1966. Om rytmisk aktivitet hos døgnfluenymfer. – Flora og Fauna 72: 148-154.

Olsen, S. 1966. The phosphate equilibrium between mud and water in Lake Furesø. – Laboratory of Radiation Biology, University of Washington. 27 pp.

Rebsdorf, Aa. 1966. Evaluation of some modifications of the Winkler method for the determination of oxygen in natural waters. – Verhandlungen Internationale Vereinigung für Theoretische und Angewandte Limnologie 16: 459-464.

Thorup, J. 1966. Substrate type and its value as a basis for the delimination of bottom fauna communities in running waters. – Pymatuning Laboratory of Ecology, University of Pittsburg, Special Publication Number 4: 59-74.

Hansen, K. 1967. The general limnology of arctic lakes as illustrated by examples from Greenland. – Meddelelser om Grønland 178, 3: 1-77.

Jónasson, P. M. and Kristiansen, J. 1967. Primary and secondary production in Lake Esrom. Growth of *Chironomus anthracinus* in relation to seasonal cycles of phytoplankton and dissolved oxygen. – Internationale Revue der gesamten Hydrobiologie 52: 163-217.

Madsen, B. L. 1967. Økologiske undersøgelser i nogle østjyske vandløb. – Flora og Fauna 73: 21-36.

Thorup, J. 1967. *Protonemura hrabei* Rauser, ny for Danmark. – Flora og Fauna 73: 7-10.

Foged, N. 1968. Diatomeerne i en post-glacial boreprøve fra bunden af Esrom Sø, Danmark. – Meddelelser fra Dansk Geologisk Forening 18: 161-180.

Hansen, K. 1968. En boring i Esrom Sø. – Meddelelser fra Dansk Geologisk Forening 18: 244-246.

Hemmingsen, A. M. 1968. The role of *Triogma trisulcata* (Schummel) (Diptera, Tipulidae, Cylindrotominae) in the adaptive radiation of the Cylindrotominae. – Folia Limnologica Scandinavica 15: 1-30.

Madsen, B. L. 1968. Vandløbenes organiske drift. – Naturens Verden 1968: 129-136.

Madsen, B. L. 1968. A comparative ecological investigation of two related mayfly nymphs. – Hydrobiologia 31: 337-349.

Madsen, B. L. 1968. The distribution of nymphs of *Brachyptera risi* Mort. and *Nemoura flexuosa* Aub. (Plecoptera) in relation to oxygen. – Oikos 19: 304-310.

Mathiesen, H. 1968. Kulturpåvirkninger af danske søer. – Stads- og Havneingeniøren 12: 1-4.

Nygaard, G. 1968. On the significance of the carrier carbon dioxide in determinations of the primary production in soft-water lakes by the radio-carbon technique. – Verhandlungen Internationale Vereinigung für Theoretische und Angewandte Limnologie 14: 111-121.

Røen, U. 1968. Studies on freshwater Entomostraca in Greenland III. Entomostraca from Peary Land with notes on their biology. – Meddelelser om Grønland 184, 4: 1-59.

Hansen, K. 1969. Saline Lakes in Greenland. – Internationale Revue der gesamten Hydrobiologie 54, 2: 195-200.

Jónasson, P. M. 1969. Dyrelivet på barbunden. Pages 281-302 *in* A. Nørrevang and T. J. Meyer, editors. Danmarks Natur 5. Politikens Forlag, Copenhagen, Denmark.

Jónasson, P. M. 1969. Biologiske undersøgelser af recipienter. – Supplement. Spildevandsrensning, Teknisk Forlag: 28-34.

Jónasson, P. M. 1969. Bottom fauna and eutrophication. – International Symposium on eutrophication. Madison, USA: 274-305.

Steemann Nielsen, E. 1969. Influence of poison on photosynthesis of unicellular algae. – IBP/PP Technical meeting. Productivity of photosynthesis systems. Models and methods. Trebon 1969, 1. supplement.

Steemann Nielsen, E. 1969. General remarks. Mediterranean Productivity Project. NATO Subc. – Oceanographic Research Technical Report 47: 1-16.

Steemann Nielsen, E. 1969. Ecological sciences in Denmark. – Intecol Bulletin 1: 31-33.

Steemann Nielsen, E., Kamp-Nielsen, L. and Wium-Andersen, S. 1969. The effect of deleterious concentrations of copper on the photosynthesis of *Chlorella pyrenoidosa*. – Physiologia Plantarum 22: 1121-1133.

Bjarnov, N. and Thorup, J. 1970. A simple method for rearing running-water insects, with some preliminary results. – Archiv für Hydrobiologie 67: 201-209.

Gargas, E. 1970. Measurements of primary produktion, dark fixation and vertical distribution of the microbenthic algae in the Øresund. – Ophelia 8: 231-253.

Madsen, B. L. and Nygaard, G. 1970. Professor Kaj Berg. – 70 years old. – Hydrobiologia 35: 345-351.

Steemann Nielsen, E. 1970. Produktionsmålinger i havet og vore ferske vande. – Vand 1: 17-21.

Steemann Nielsen, E. and Kamp-Nielsen, L. 1970. Influence of deleterious concentrations of copper on the growth of *Chlorella pyrenoidosa*. – Physiologia Plantarum 23: 828-840.

Steemann Nielsen, E. and Wium-Andersen, S. 1970. Copper ions as poison in the sea and in freshwater. – Marine Biology 6: 93-97.

Steemann Nielsen, E. 1970. Mass and energy exchange between phytoplankton populations and environment. – Discussion Section 6. Proceedings IBP/PP Technical Meeting. Productivity of photosynthesis systems. Models and methods. Trebon 1969. Pudoc, Wageningen 1970: 555-558.

Thorup, J. 1970. The influence of a short-termed flood on

a springbrook community. – Archiv für Hydrobiologie 66: 447-457.

Thorup, J. 1970. Frequency analysis in running waters and its application on a springbrook community. – Archiv für Hydrobiologie 68: 126-142.

Whiteside, M. C. 1970. Danish chydorid Cladocera: Modern ecology and core studies. – Ecological Monographs 40: 79-118.

Publications 1971 – 1980

Hunding, C. 1971. Production of benthic microalgae in the littoral zone of a eutrophic lake. – Oikos 22: 389-397.

Hunding, C. 1971. Om bakterier i ferskvand, I. Bakteriernes rolle, set i økologisk perspektiv. – Vand 2: 63-68.

Iversen, T. M. 1971. *Culex torrentium* Martini (Dipt. Culicidae) ny for Danmark, med en beskrivelse af larvebiotopen. – Entomologiske Meddelelser 39: 235-239.

Iversen, T. M. 1971. The ecology of a mosquito population (*Äedes communis*) in a temporary pool in a Danish beech wood. – Archiv für Hydrobiologie 69: 309-332.

Kamp-Nielsen, L. 1971. The effect of deleterious concentrations of mercury on the photosynthesis and growth of *Chlorella pyrenoidosa*. – Physiologia Plantarum 24: 556-561.

Kamp-Nielsen, L. 1971. Sedimentgeologiske forhold. – Forureningsrådets sekretariat. Publ. 12: 22-23.

Steemann Nielsen, E. 1971. Production in costal areas of the sea. – Thalassia Jugoslavica 7: 383-391.

Steemann Nielsen, E. 1971. Measurement of the fertility of the sea. – Nautilus 11: 2-3.

Steemann Nielsen, E. 1971. Stofproduktionen i havet og i vore ferske vande. – Dansk Natur Dansk Skole, årsskrift 1970: 13-19.

Steemann Nielsen, E. 1971. Recipientgruppens rapport. – Vand 2: 53-55.

Steemann Nielsen, E. 1971. The balance between phytoplankton and zooplankton in the Sea. Pages 116-126 *in* J. Nybakken, editor. Readings in Marine Ecology. Harper & Row, New York 1971.

Steemann Nielen, E. and Wium-Andersen, S. 1971. The influence of Cu on photosynthesis and growth in diatoms. – Physiologia Plantarum 24: 482-484.

Steemann Nielsen, E. and Willemoés, M. 1971. How to measure the illumination rate when investigating the rate of photosynthesis of unicellular algae under various light conditions. – Internationale Revue der gesamten Hydrobiologie 56: 541-556.

Wium-Andersen, S. 1971. Photosynthetic uptake of free CO_2 by the roots of *Lobelia dortmanna*. – Physiologia Plantarum 25: 245-248.

Andersen, J. M. og Hansen, J. L. 1972. Giftstoffers indflydelse på BOD-bestemmelse på ufortyndede spildevandsprøver. – Vatten 2: 115-128.

Bjarnov, N. 1972. Carbohydrases in *Chironomus, Gammarus* and some Trichoptera larvae. – Oikos 23: 261-263.

Hargrave, B. T. 1972. Oxidation-reduction potentials, oxygen concentration and oxygen uptake of profundal sediments in a eutrophic lake. – Oikos 23: 167-177.

Hargrave, B. T. 1972. A comparison of sediment oxygen uptake, hypolimnetic oxygen deficit and primary production in Lake Esrom, Denmark. – Verhandlungen Internationale Vereinigung für Theoretische und Angewandte Limnologie 18: 134-139.

Hargrave, B. T. 1972. Aerobic decomposition of sediment and detritus as a function of particle surface area and organic content. – Limnology and Oceanography 17: 583-596.

Hunding, C. 1972. Om bakterier i Ferskvand, II. Heterotrofe bakterier. – Vand 1972, 1: 2-7.

Jónasson, P. M. 1972. Population studies of *Chironomus anthracinus*. – Proceedings Advanced Institute. Dynamics of numbers in populations (Oosterbeek 1970): 220-231.

Jónasson, P. M. and Thorhauge, F. 1972. Life cycle of *Potamothrix hammoniensis* (Tubificidae) in the profundal of a eutrophic lake. – Oikos 23: 151-158.

Jónasson, P. M. 1972. Ecology and production of the profundal benthos in relation to phytoplankton in Lake Esrom. – Oikos Suppl. 14: 1-148.

Jónasson, P. M. 1972. Kaj Berg 13. maj 1899 – 14. marts 1972. – Universitetets Festskrift: 115-119.

Kamp-Nielsen, L. 1972. Some comments of the determination of copper fractions in natural waters. – Deep-Sea Research 19: 899-902.

Kamp-Nielsen, L. 1972. Eutrofiering. – "FISK OG HAV-72" – Danmarks Fiskeri- og Havundersøgelser 32: 47-50.

Lindegaard-Petersen, C. 1972. An ecological investigation of the Chironomidae (Diptera) from a Danish lowland stream (Linding Å). – Archiv für Hydrobiologie 69: 465-507.

Lindegaard-Petersen, C. 1972. The chironomid fauna in a lowland stream in Denmark compared with other European streams. – Verhandlungen Internationale Vereinigung für Theoretische und Angewandte Limnologie 18: 726-729.

Rebsdorf, Aa. 1972. The Carbon Dioxide System in Freshwater. A set of tables for easy computation of total carbon dioxid and other components of the carbon dioxid system. – Freshwater Biological Laboratory, Hillerød.

Steemann Nielsen, E. and Wium-Andersen, S. 1972. Influence of copper on photosynthesis of diatoms with special reference to an afternoon depression. – Verhandlungen Internationale Vereinigung für Theoretische und Angewandte Limnologie 18: 78-83.

Steemann Nielsen, E. 1972. The rate of primary production and the size of the standing stock of zooplankton in the oceans. – Internationale Revue der gesamten Hydrobiologie 57: 513-516.

Steemann Nielsen, E. 1972. Mekanisk rensning contra biologisk rensning. – Stads- og havneingeniøren 72: 239-241.

Wium-Andersen, S. and Andersen, J. M. 1972. Carbon dioxide content of the interstitial water in the sediment of Grane Langsø, a Danish *Lobelia* lake. – Limnology and Oceanography 17: 943-947.

Wium-Andersen, S. and Andersen, J. M. 1972. The influence of vegetation on the redox profile in the sediment of Grane Langsø, a Danish *Lobelia* lake. – Limnology and Oceanography 17: 948-952.

Arevad, K., Iversen, T. M. and Lodal, J. 1973. Stikmyg (Dipt., Culicidae) i Furesøområdet. – Entomologiske Meddelelser 41: 147-158.

Frey, D. G. 1973. Comparative morphology and biology of three species of *Eurycercus* (Chydoridae, Cladocera) with a description of *Eurycercus macrocanthus* sp. nov. – Internationale Revue der gesamten Hydrobiologie 58: 221-267.

Hunding, C. and Hargrave, B. T. 1973. A comparison of benthic microalgal production measured by C^{14} and oxygen methods. – Journal of the Fisheries Research Board of Canada 30: 309-312.

Hunding, C. 1973. Diel variation in oxygen production and uptake in a microbenthic littoral community of a nutrient-poor lake. – Oikos 24: 352-360.

Iversen, T. M. 1973. Decomposition of autumn-shed beech leaves in a springbrook and its significance for the fauna. – Archiv für Hydrobiologie 72: 305-312.

Iversen, T. M. 1973. Life cycle and growth of *Sericostoma personatum* Spence (Trichoptera, Sericostomatidae) in a Danish spring. – Entomologica Scandinavica 4: 323-327.

Iversen, T. M. 1973. Vores småvande og deres betydning for flora og fauna. – Danmarks Naturfredningsforenings Årsskrift 1973: 51-66.

Steemann Nielsen, E. 1973. Hydrobiologi. En introduktion til belysning af dens forudsætninger. – Polyteknisk Forlag, København, Danmark. 204 pp.

Steemann Nielsen, E. 1973. Kaj Berg 13. april 1899 – 14. marts 1972. – Oversigt over Det Det Kongelige Danske Videnskabernes Selskabs Virksomhed 1972-1973. 6 pp.

Thorup, J. 1973. Interpretation of growth-curves for animals from running waters. – Verhandlungen Internationale Vereinigung für Theoretische und Angewandte Limnologie 18: 1512-1520.

Thorup, J. 1973. Ferske vandes økologi. – P. Haase & Søn, København, Danmark. 92 pp.

Andersen, J. M. 1974. Nitrogen and phosphorus budgets and the role of sediments in six shallow Danish lakes. – Archiv für Hydrobiologie 74: 528-550.

Bosselmann, S. 1974. The crustacean plankton of Lake Esrom. – Archiv für Hydrobiologie 74: 18-31.

Iversen, T. M. 1974. Ingestion and growth in *Sericostoma personatum* (Trichoptera) in relation to the nitrogen content of ingested leaves. – Oikos 25: 278-282.

Jónasson, P. M., Lastein, E. and Rebsdorf, A. 1974. Production, insolation and nutrient budget of eutrophic Lake Esrom. – Oikos 25: 255-277.

Steemann Nielsen, E. 1974. Store P-koncentrationers indflydelse ved varierende pH på encellede algers fotosyntese og vækst. – Aquanalen 1974, 2: 16-20.

Thorup, J. and Iversen, T. M. 1974. Ingestion by *Sericostoma personatum* Spence (Trichoptera, Sericostomatidae). – Archiv für Hydrobiologie 74: 39-47.

Thorup, J. 1974. Occurrence and size-distribution of Simuliidae (Diptera) in a Danish Spring. – Archiv für Hydrobiologie 74: 316-335.

Wium-Andersen, S. 1974. Quantitative changes in the higher vegetation in lake Bastrup Sø caused by eutrophication. – Botanisk Tidsskrift 69: 64-68.

Wium-Andersen, S. 1974. The effect of chromium on the photosynthesis and growth of diatoms and green algae. – Physiologia Plantarum 32: 308-310.

Wium-Andersen, S. 1974. Chroms og kobbers indvirkning på fotosyntesen og væksten hos diatomeer og grønalger. – Nordisk Konference om Miljøgifte 1974.

Wium-Andersen, S. and Franzmann, N. E. 1974. Dør andefugle af at spise blyhagl. – Feltornithologen 1.

Andersen, J. M. 1975. Influence of pH on release of phosphorus from lake sediments. – Archiv für Hydrobiologie 76: 379-383.

Bosselmann, S. 1975. Population dynamics of *Eudiaptomus graciloides* in Lake Esrom. – Archiv für Hydrobiologie 75: 329-346.

Bosselmann, S. 1975. Production of *Eudiaptomus graciloides* in Lake Esrom, 1970. – Archiv für Hydrobiologie 76: 43-64.

Iversen, T. M. 1975. Disappearance of autumn shed beech leaves placed in bags in small streams. – Verhandlungen Internationale Vereinigung für Theoretische und Angewandte Limnologie 19: 1687-1692.

Jacobsen, O. S. and Jørgensen, S. E. 1975. A submodel for nitrogen release from sediments. – Ecological Modelling 1:147-151.

Jacobsen, O. S. 1975. Udtømning af mobilt phosphor fra danske søsedimenter. – Proceedings 4th Nordic Symposium on Sediments. Sweden: 145-159.

Jónasson, P. M. 1975. Kaj Berg. – Archiv für Hydrobiologie 76: 256-264.

Jónasson, P. M. 1975. Population ecology and production of benthic detritivores. – Verhandlungen Internationale Vereinigung für Theoretische und Angewandte Limnologie 19: 1066-1072.

Jørgensen, S. E., Kamp-Nielsen, L. and Jacobsen, O. S. 1975. A submodel for anaerobic mud-water exchange of phosphate. – Ecological Modelling 1: 133-146.

Kamp-Nielsen, L. 1975. A kinetic approach to the aerobic sediment-water exchange of phosphorus in Lake Esrom. – Ecological Modelling 1: 153-160.

Kamp-Nielsen, L. 1975. Seasonal variation in sediment-water exchange of nutrient ions in Lake Esrom. – Verhandlungen Internationale Vereinigung für Theoretische und Angewandte Limnologie 19: 1057-1065.

Lindegaard, C., Thorup, J. and Bahn, M. 1975. The invertebrate fauna of the moss carpet in the Danish spring Ravnkilde and its seasonal, vertical and horizontal distribution. – Archiv für Hydrobiologie 75: 109-139.

Lindegaard, C. and Jónasson, P. M. 1975. Life cycles of *Chironomus hyperboreus* Staeger and *Tanytarsus gracilentus* (Holmgren) (Chironomidae, Diptera) in Lake Mývatn, Northern Iceland. – Verhandlungen Internationale Vereinigung für Theoretische und Angewandte Limnologie 19: 3155-3163.

Sand-Jensen, K. 1975. Biomass, net production and growth dynamics in an eelgrass (*Zostera marina* L.) population in Vellerup Vig, Denmark. – Ophelia 14: 185-201.

Steemann Nielsen, E. 1975. Marine photosynthesis with special emphasis on the ecological aspects. – Elsevier Oceanography Series 13: 1-141, Amsterdam.

Steemann Nielsen, E., Wium-Andersen, S. and Rochon, T. 1975. On problems in G. M. countings in the C^{14}-technique. – Verhandlungen Internationale Vereinigung für Theoretische und Angewandte Limnologie 19: 26-31.

Thorhauge, F. 1975. Reproduction of *Potamothrix hammoniensis* (Tubificidae, Oligochaeta) in Lake Esrom, Denmark. A field and laboratory study. – Archiv für Hydrobiologie 76: 449-474.

Wium-Andersen, S. 1975. The influence of the zooplankton anaesthetising substance physostigmine salicylium on photosynthesis. – Archiv für Hydrobiologie 76: 379-383.

Andersen, J. M. 1976. An ignition method for determination of total phosphorus in lake sediments. – Water Research 10: 329-331.

Hansen, U. K. and Mortensen, E. 1976. Danmarks åer savner træer. – Sportsfiskeren 51, 5: 16-17.

Iversen, T. M. 1976. Biologisk bestemmelse af vandløbsforurening. – Vand 7: 23-26.

Iversen, T. M. 1976. Life cycle and growth of Trichoptera in a Danish spring. – Archiv für Hydrobiologie 78: 482-493.

Jónasson, P. M. and Thorhauge, F. 1976. Population dynamics of *Potamothrix hammoniensis* in the profundal of Lake Esrom with special reference to environmental and competitive factors. – Oikos 27: 193-203.

Jónasson, P. M. and Thorhauge, F. 1976. Production of *Potamothrix hammoniensis* in the profundal of eutrophic Lake Esrom. – Oikos 27: 204-209.

Kamp-Nielsen, L. 1976. Model considerations on sediment-water exchange of phosphorus. – Naturvådsverkets PM-Series. 694: 161-166.

Kamp-Nielsen, L. 1976. Kvælstofomsætning i søer og i havet. – Vand 7: 13-16.

Lastein, E. 1976. Recent sedimentation and resuspension of organic matter in eutrophic Lake Esrom, Denmark. – Oikos 27: 44-49.

Mortensen, E. and Hansen, U. K. 1976. Skånsom grødeskæring gavner både fisk og fiskere mest. – Sportsfiskeren 51, 6: 14-16.

Nygaard, G. 1976. Desmids from an arctic salt lake. – Botanisk Tidsskrift 71: 84-86.

Sand-Jensen, K. 1976. A comparison of chlorophyll *a*

determinations of unstored and stored plankton filters extracted by methanol and acetone. – Vatten 32: 337-341.

Steemann Nielsen, E. and Bruun Laursen, H. 1976. Effect of $CuSO_4$ on the photosynthetic rate of phytoplankton in four Danish lakes. – Oikos 27: 239-242.

Steemann Nielsen, E. 1976. Johannes Krey 1912-1975. – Journal Conseil International pour l'Exploration de la Mer 37: 1-2.

Steemann Nielsen, E. 1976. The Carlsberg Laboratory and the Exploration of the sea. – The Carlsberg Laboratory 1876-1976: 331-345.

Steemann Nielsen, E. 1976. Hvad betyder P-tilførslen fra land til Øresund. – Øresunds Vand-Kvalitet 1971-1974: 111-114.

Steemann Nielsen, E. 1976. The carbon flow in marine ecosystems. – Coastal Pollution Control 1: 193-200.

Steemann Nielsen, E. and Rochon, T. 1976. The influence of extremely high concentrations of inorganic P at varying pH on the growth and photosynthesis of unicellular algae. – Internationale Revue der gesamten Hydrobiologie 61: 407-415.

Thorhauge, F. 1976. Growth and life cycle of *Potamothrix hammoniensis* (Tubificidae, Oligochaeta) in the profundal of eutrophic Lake Esrom. A field and laboratory study. – Archiv für Hydrobiologie 78: 71-85.

Andersen, J. M. 1977. Importance of denitrification in sediments of some Danish lakes. – Proceedings 5th Nordic Symposium on Sediments. Denmark: 15-27.

Andersen, J. M. 1977. Importance of the denitrification process for the rate of degradation of organic matter in lake sediments. – Proceedings International Symposium Amsterdam, 1976: Interactions between sediments and fresh water. The Hague: 357-362.

Andersen, J. M. 1977. Rates of denitrification of undisturbed sediment from six lakes as a function of nitrate concentration, oxygen and temperature. – Archiv für Hydrobiologie 80: 147-159.

Christensen, B. and Wium-Andersen, S. 1977. Seasonal growth of mangrove trees in Southern Thailand. I. The phenology of *Rhizophora apiculata* Bl. – Aquatic Botany 3: 281-286.

Christensen, B. and Wium-Andersen, S. 1977. Mangrove plants, sea grasses and benthic algae at Surin Islands, west coast of Thailand. – Phuket Marine Biological Center, Thailand. Research Bulletin 14: 1-5.

Dahl-Madsen, K. I. and Kamp-Nielsen, L. 1977. Sedimentets rolle ved vurdering og virkninger af forureningsbegrænsende foranstaltninger. – Vand 8: 41-43.

Hargrave, B. T. and Kamp-Nielsen, L. 1977. Accumulation of sedimentary organic matter at the base of steep bottom gradients. – Proceedings International Symposium Amsterdam, 1976: Interactions between sediments and fresh water. The Hague: 168-173.

Iversen, T. M. and Madsen, B. L. 1977. Allochtonous organic matter in streams. – Folia Limnologica Scandinavica 17: 17-20.

Iversen, T. M. and Jessen, J. 1977. Life cycle, drift and production of *Gammarus pulex* L. (Amphipoda) in a Danish spring. – Freshwater Biology 7: 287-296.

Jacobsen, O. S. 1977. The influence of bottom fauna density on the exchange rates of phosphate and inorganic nitrogen in a eutrophic profundal sediment. – Proceedings 5th Nordic Symposium on Sediments. Denmark: 39-47.

Jacobsen, O. S. 1977. Sorption of phosphate by Danish lake sediments. – Vatten 33: 290-298.

Jónasson, P. M. 1977. Lake Esrom Research 1867-1977. – Folia Limnologica Scandinavica 17: 67-89.

Jónasson, P. M., Adalsteinson, H., Hunding, C., Lindegaard, C. and Olafsson, J. 1977. Limnology of Iceland. – Folia Limnologica Scandinavica 17: 11-123.

Jørgensen, S. E., Kamp-Nielsen, L. and Dahl-Madsen, K. I. 1977. Mathematical modelling in Danish limnology. – Folia Limnologica Scandinavica 17: 59-66.

Kamp-Nielsen, L. 1977. Horizontal and temporal variation in sediments of some Danish lakes. – Proceedings 5th Nordic Symposium on Sediments. Denmark: 15-27.

Kamp-Nielsen, L. 1977. Modelling the temporal variation in sedimentary phosphorus fractions. – Proceedings International Symposium Amsterdam, 1976: Interactions between sediments and fresh water. The Hague: 277-285.

Kamp-Nielsen, L. and Andersen, J. M. 1977. A review of the literature on sediment- water exchange of nitrogen compounds. – Progress Water Technology 8: 393-418.

Mortensen, E. 1977. The population dynamics of young trout (*Salmo trutta* L.) in a Danish brook. – Journal of Fish Biology 10: 23-33.

Mortensen, E. 1977. Population survival, growth and production of trout *Salmo trutta* in a small Danish stream. – Oikos 28: 9-15.

Mortensen, E. 1977. Fish production in small Danish streams. – Folia Limnologica Scandinavica 17: 21-26.

Mortensen, E. 1977. Populationsdynamik og bestandsregulering hos ørred i vandløb. – Vand 8: 44-48.

Mortensen, E. 1977. Density dependent mortality of trout fry (*Salmo trutta* L.) and its relationship to the management of small streams. – Journal of Fish Biology 11: 613-617.

Nygaard, G. 1977. New or interesting plankton algae. – Det Kongelige Danske Videnskabers Selskabs Biologiske Skrifter 21,1: 1-107.

Nygaard, G. 1977. Vertical and seasonal distribution of some mobile freshwater plankton algae in relation to some environmental factors. – Archiv für Hydrobiologie Suppl. 51: 67-76.

Nygaard, G. 1977. On making fluid mounts of plankton algae. – Phycologia 16: 351.

Sand-Jensen, K. 1977. Effect of epiphytes on eelgrass photosynthesis. – Aquatic Botany 3: 55-63.

Steemann Nielsen, E. 1977. The carbon-14 technique for measuring organic production by plankton algae. A report on the present knowledge. – Folia Limnologica Scandinavica 17: 45-48.

Thorup, J. and Lindegaard, C. 1977. Studies on Danish springs. – Folia Limnologica Scandinavica 17: 7-15.

Wium-Andersen, S. 1977. Primary production in waters around Surin Islands off the west coast of Thailand. – Phuket Marine Biological Center, Thailand. Research Bulletin 16: 1-4.

Ameen, M. and Iversen, T. M. 1978. Food of *Aëdes* larvae (Diptera, Culicidae) in a temporary forest pool. – Archiv für Hydrobiologie 83: 552-564.

Andersen, J. M. 1978. Plankton primary production and respiration in eutropic Frederiksborg Slotssø, Denmark. – Verhandlungen Internationale Vereinigung für Theoretische und Angewandte Limnologie 20: 702-708.

Iversen, T. M. 1978. Life cycle and growth of three species of Plecoptera in a Danish spring. – Entomologiske Meddelelser 46: 57-62.

Iversen, T. M. og Mortensen, E. 1978. Sammenligning af den danske bedømmelse af vandrecipienters forureningsgrad og det engelske Trent index. – Vand 9: 65-67.

Iversen, T. M., Wiberg-Larsen, P., Hansen, S. B. and Hansen, F. S. 1978. The effect of partial and total drought on the macroinvertebrate communities of three small Danish streams. – Hydrobiologia 60: 235-242.

Jacobsen, O. S. 1978. Nitrogen and phosphorus dynamics in eutrophic Frederiksborg Slotssø. – Verhandlungen Internationale Vereinigung für Theoretische und Angewandte Limnologie 20: 696-701.

Jacobsen, O. S. 1978. A description model for phosphate sorption by lake sediments – Interaction between sediment and water. – Proceedings 6th Nordic Symposium on Sediments. Hurdal, Norway: 127-136.

Jacobsen, O. S. 1978. Sorption, adsorption and chemosorption of phosphate by Danish lake sediments. – Vatten 34: 230-243.

Jónasson, P. M. 1978. Zoobenthos of lakes. – Verhandlungen Internationale Vereinigung für Theoretische und Angewandte Limnologie 20: 13-37.

Kamp-Nielsen, L. and Hargrave, B. T. 1978. Influence of bathymetry on sediment focusing in Lake Esrom. – Verhandlungen Internationale Vereinigung für Theoretische und Angewandte Limnologie 20: 714-719.

Kamp-Nielsen, L. 1978. Modelling the vertical gradients in sedimentary phosphorus fractions. – Verhandlungen Internationale Vereinigung für Theoretische und Angewandte Limnologie 20: 720-727.

Lindegaard, C., Mæhl, P. and Nielsen, B. H. 1978. Zoobenthos of lakes located within and outside the Ilímaussaq intrusion in South Greenland. – Verhandlungen Internationale Vereinigung für Theoretische und Angewandte Limnologie 20: 159-164.

Riemann, B. 1978. Differentiation between heterotrophic and photosynthetic plankton by size fractionation, glucose uptake, ATP and chlorophyll content. – Oikos 31: 358-367.

Riemann, B. 1978. Interference in the quantitative determination of ATP extracted from freshwater microorganisms. Pages 316-332 *in* E. Schram and P. Stanley, editors. Proceedings International Symposium Analythical Applied Bioluminiscence & Chemiluminiscence, Bruxelles.

Sand-Jensen, K. 1978. Metabolic adaptation and vertical zonation of *Litorella uniflora* (L.) Aschers, and *Isoetes lacustris* L. – Aquatic Botany 4: 1-10.

Sand-Jensen, K. and Rasmussen, L. 1978. Macrophytes and chemistry of acidic streams from lignite mining areas. – Botanisk Tidsskrift 72: 105-112.

Sand-Jensen, K. and Søndergaard, M. 1978. Growth and production of isoetids in oligotrophic Lake Kalgaard, Denmark. – Verhandlungen Internationale Vereinigung für Theoretische und Angewandte Limnologie 20: 659-666.

Steemann Nielsen, E. 1978. Growth of plankton algae as a function of N-concentration, measured by means of a batch technique. – Marine Biology 46: 185-189.

Steemann Nielsen, E. 1978. Principal aspects concerning the batch technique in algal assays. – Verhandlungen Internationale Vereinigung für Theoretische und Angewandte Limnologie 20: 81-87.

Steemann Nielsen, E. 1978. Growth of the unicellular alga *Selenastrum capricornutum* as a function of P. With some information also on N. – Verhandlungen Internationale Vereinigung für Theoretische und Angewandte Limnologie 20: 38-42.

Søndergaard, M. and Sand-Jensen, K. 1978. Total autotrophic production in oligotrophic Lake Kalgaard, Denmark. – Verhandlungen Internationale Vereinigung für Theoretische und Angewandte Limnologie 20: 667-673.

Wiberg-Larsen, P. 1978. Species composition, succession of instars and mortality among the immature stages of *Aedes* spp. inhabiting some Danish forest pools. – Archiv für Hydrobiologie 84: 180-198.

Wium-Andersen, S. 1978. Havskildpadder – en truet dyregruppe. – Naturens Verden 1978: 130-136.

Wium-Andersen, S. and Ovesen, C. H. 1978. Light gives life. – Naturopa 31: 26-28.

Wium-Andersen, S. and Christensen, B. 1978. Seasonal growth of mangrove trees in Southern Thailand. II. Phenology of *Bruguiera cylindrica*, *Ceriops tagal*, *Lumnitzera littorea* and *Avicennia marina*. – Aquatic Botany 5: 383-390.

Adalsteinsson, H. 1979. Zooplankton and its relation to available food in Lake Mývatn. – Oikos 32: 162-194.

Adalsteinsson, H. 1979. Seasonal variation and habitat distribution of benthic Crustacea in Lake Mývatn in 1973. – Oikos 32: 195-201.

Adalsteinsson, H. 1979. Size and food of arctic char *Salvelinus alpinus* and stickleback *Gasterosteus aculeatus* in Lake Mývatn. – Oikos 32: 228-231.

Andersen, J. M., Jacobsen, O. S. 1979. Production and decomposition of organic matter in eutrophic Frederiksborg Slotssø, Denmark. With contributions by P. D. Grevy and P. N. Markmann. – Archiv für Hydrobiologie 85: 511-542.

Bosselmann, S. 1979. Population dynamics of *Keratella cochlearis* in Lake Esrom. – Archiv für Hydrobiologie 87: 152-165.

Bosselmann, S. 1979. Production of *Keratella cochlearis* in Lake Esrom. – Archiv für Hydrobiologie 87: 304-313.

Dall, P. C. 1979. Ecology and production of the leeches *Erpobdella octoculata* L. and *Erpobdella testacea* Sav. In Lake Esrom, Denmark. – Archiv für Hydrobiologie Suppl. 57: 188-220.

Dall, P. C. 1979. A sampling technique for littoral stone dwelling organisms. – Oikos 33: 106-112.

Dall, P. C. 1979. Distributional relationship and migration among leeches (Hirudinea) in the exposed littoral of Lake Esrom, Denmark. – Oikos 33: 113-120.

Einarsson, M. Á. 1979. Climatic conditions of the Lake Mývatn area. – Oikos 32: 29-37.

Gardarsson, A. 1979. Waterfowl population of Lake Mývatn and recent changes in numbers and food habits. – Oikos 32: 250-270.

Gudmundsson, F. 1979. The past status and exploitation of the Mývatn waterfowl populations. – Oikos 32: 232-249.

Hansen, S. B. 1979. Livscyklus og vækst hos to arter af *Ptychoptera* (Diptera, Nematocera) i en dansk bæk. – Entomologiske Meddelelser 47: 33-38.

Hunding, C. 1979. The oxygen balance of Lake Mývatn, Iceland. – Oikos 32: 139-150.

Iversen, T. M. 1979. Laboratory energetics of larvae of *Sericostoma personatum* (Trichoptera). – Holarctic Ecology 2: 1-5.

Jeppesen, E. 1979. An evaluation of methods used in measurements of the sediment oxygen uptake in lotic environments. – Proceedings 7th Nordic Symposium on Sediments. Lunds Universitet, Sweden: 5-17.

Jónasson, P. M. (ed.) 1979. Lake Mývatn. – Oikos 32: 1-308.

Jónasson, P. M. 1979. Introduction to the ecology of the eutrophic, subarctic Lake Mývatn and the River Laxá. – Oikos 32: 1-16.

Jónasson, P. M. 1979. The Lake Mývatn ecosystem, Iceland. – Oikos 32: 289-308.

Jónasson, P. M. 1979. The River Laxá ecosystem, Iceland. – Oikos 32: 306-305.

Jónasson, P. M. and Adalsteinsson, H. 1979. Phytoplankton production in shallow eutrophic Lake Mývatn, Iceland. – Oikos 32: 113-138.

Jónasson, P, M. and Lindegaard, C. 1979. Zoobenthos and its contribution to the metabolism of shallow lakes. – Archiv für Hydrobiologie Beiheft Ergebnisse der Limnologie 13: 162-180.

Kamp-Nielsen, L. 1979. Estimation of exchange parameters on some tropical lake sediments. – Proceedings 7th Nordic Symposium on Sediments. Lunds Universitet, Sweden.

Lindegaard, C. 1979. The invertebrate fauna of Lake Mývatn, Iceland. – Oikos 32: 151-161.

Lindegaard, C. 1979. A survey of the macroinvertebrate fauna with special reference to Chironomidae (Diptera) in the rivers Laxá and Kráká, northern Iceland. – Oikos 32: 281-288.

Lindegaard, C. and Jónasson, P. M. 1979. Abundance, population dynamics and production of zoobenthos in Lake Mývatn, Iceland. – Oikos 32: 202-227.

Nygaard, G. 1979. Freshwater phytoplankton from the Narssaq area, South Greenland. – Botanisk Tidsskrift 73: 191-238.

Ólafsson, J. 1979. Physical characteristics of Lake Mývatn and River Laxá. – Oikos 32: 38-66.

Ólafsson, J. 1979. The chemistry of Lake Mývatn and River Laxá. – Oikos 32: 82-112.

Rasmussen, L. and Sand-Jensen, K. 1979. Heavy metals in acid streams from lignite mining areas. – The Science of the Total Environment 12: 61-74.

Riemann, B. 1979. The occurrence and ecological importance of dissolved ATP in fresh water. – Freshwater Biology 9: 481-490.

Rist, S. 1979. Water level fluctuations and ice cover of Lake Mývatn. – Oikos 32: 67-81.

Rist, S. 1979. The hydrology of River Laxá. – Oikos 32: 271-280.

Sand-Jensen, K. and Søndergaard, M. 1979. Distribution and quantitative development of aquatic macrophytes in relation to sediment characteristics in oligotrophic Lake Kalgaard, Denmark. – Freshwater Biology 9: 1-11.

Schierup, H. H. and Riemann, B. 1979. Effects of filtration on concentrations of ammonia and orthophosphate from lake water samples. – Archiv für Hydrobiologie 86: 204-216.

Søndergaard, M. and Sand-Jensen, K. 1979. Carbon uptake by leaves and roots of *Littorella uniflora* (L) Aschers. – Aquatic Botany 6: 1-12.

Søndergaard, M. and Sand-Jensen, K. 1979. The delay in ^{14}C fixation rates by three submerged macrophytes. A source of error in the ^{14}C technique. – Aquatic Botany 6: 111-119.

Søndergaard, M. and Sand-Jensen, K. 1979. Physico-chemical environment, phytoplankton biomass and production in oligotrophic softwater, Lake Kalgaard, Denmark. – Hydrobiologia 63: 241-253.

Thórarinsson, S. 1979. The postglacial history of the Mývatn area. – Oikos 32: 17-28.

Wium-Andersen, S. 1979. Plankton primary production in a tropical mangrove bay at the south-west coast of Thailand. – Ophelia 18: 53-60.

Andersen, J. M. and Sand-Jensen, K. 1980. Discrepancies between O_2 and ^{14}C methods for measuring phytoplankton gross photosynthesis at low light levels. – Oikos 35: 359-364.

Anthoni, U., Christophersen, C., Madsen, J. Ø., Wium-Andersen, S., and Jacobsen, N. 1980. Biologically active sulphur compounds from the green algae *Chara globularis*. – Phytochemistry 19: 1228-1229.

Borum, J. and Wium-Andersen, S. 1980. Biomass and production of epiphytes on eelgrass (*Zostera marina* L.) in the Øresund, Denmark. – Ophelia Suppl. 1: 57-64.

Holopainen, I. J. 1980. Growth of two *Pisidium* (Bivalvia, Sphaeridae) species in the laboratory. – Oecologia (Berl.) 45: 104-108.

Iversen, T. M. 1980. Densities and energetics of two streamliving larval populations of *Sericostoma personatum* (Trichoptera). – Holarctic Ecology 3: 65-73.

Iversen, T. M. and Sand-Jensen, K. 1980. Okker – en trussel mod livet i de vestjyske vandløb. – Natur og Miljø 3: 38-39.

Jeppesen, E. 1980. A comparison between four oxygen balance models for small organic polluted streams with many macrophytes. – Progress in Ecological Engineering and Management by Mathematical Modelling: 183-195.

Jónasson, P. M. 1980. Energy flow in Lake Mývatn. – Nordecol Newsletter 12: 3.

Jónasson, P. M. 1980. Esrom Sø's økosystem. – Miljøværn 6: 11-16.

Kamp-Nielsen, L. 1980. Intensive measurements of sedimentation in Lake Esrom and Lake Glumsø. – Proceedings 8th Nordic Symposium on Sediments. Denmark: 44-56.

Kamp-Nielsen, L. 1980. The influence of sediments on changed phosphorus loading to hypertrophic Lake Glumsø. – Developments in Hydrobiology 2: 29-36.

Kamp-Nielsen, L. 1980. A sediment-water exchange model for lakes in the Upper Nile basin. – Progress in Ecological Engineering and Management by Mathematical Modelling: 557-582.

Kamp-Nielsen, L. 1980. The influence of changed phosphorus loading to hypereutrophic Lake Glumsø. – Proceedings SIL Workshop on Hypertrophic Ecosystems. Växsjö, Sweden.

Lindegaard, C. 1980. Bathymetric distribution of Chironomidae (Diptera) in the oligotrophic Lake Thingvallavatn, Iceland. Pages 225-232 in D. A. Murray , editor. Chironomidae, Ecology, Systematics, Cytology and Physiology. Pergamon Press, Oxford and New York.

Lindegaard, C. 1980. Bundfaunaundersøgelsers anvendelighed til vurdering af søers forureningstilstand. Pages 23-27 in B. L. Madsen, editor. Tilsyn med søer, Supplement. Miljøstyrelsens Ferskvandslaboratorium.

Marker, F. H., Nusch E. A., Rai, H. and Riemann, B. 1980. The measurement of photosynthetic pigments in freshwaters and standardization of methods: Conclusions and recommendations. – Archiv für Hydrobiologie Beiheft Ergebnisse der Limnologie 14: 91-106.

Mejer, H., Jørgensen, S. E. and Kamp-Nielsen, L. 1980. A sediment phosphorus model. Proceedings of the Second Joint MTA/IIASA Task Force on Lake Esrom and Lake Glumsø. – Proceedings 8th Nordic Symposium on Sediments. Denmark: 44-56.

Riemann, B. 1980. A note on the use of methanol as an extraction solvent for chlorophyll *a* determination. –

Archiv für Hydrobiologie Beiheft Ergebnisse der Limnologie 14: 70-78.

Riemann, B. 1980. Diurnal variations in the uptake of glucose by attached and free-living microheterotrophs in lake water (Lake Mossø, Denmark). – Developments in Hydrobiology 3: 153-160.

Sand-Jensen, K. 1980. Balancen mellem autotrofe komponenter i tempererede søer med forskellig næringsbelastning. – Vatten 36: 104-115.

Sand-Jensen, K. 1980. Økologisk balance i danske *Lobelia* søer. – Naturens Verden 1980: 222-227.

Sand-Jensen, K. 1980. Den klarvandede *Lobelia* sø – en truet naturtype med en enestående økologi. – Natur og Miljø 3: 6-9.

Sand-Jensen, K. and Iversen, T. M. 1980. Okkerforurening af vandløb. – Kaskelot 46: 10-15.

Thyssen, N. og Jeppesen, E. 1980. Genluftning i mindre vandløb. – Vatten 36: 231-248.

Whiteside, M. C. and Lindegaard, C. 1980. Complementary procedures for sampling small benthic invertebrates. – Oikos 35: 317-320.

Wium-Andersen, S. and Borum, J. 1980. Biomass and production of eelgrass (*Zostera marina* L.) in the Øresund, Denmark. – Ophelia Suppl. 1: 49-55.

Publications 1981 – 1990

Bosselmann, S. 1981. Population dynamics and production of *Keratella hiemalis* and *K. quadrata* in Lake Esrom. – Archiv für Hydrobiologie 90: 427-447.

Dall, P. C. 1981. A new grab for the sampling of zoobenthos in the upper stony littoral zone. – Archiv für Hydrobiologie 92: 396-405.

Hansen, J. A. and Sand-Jensen, K. 1981. Energihusholdning ved teknisk-hygiejniske anlæg. Recipienter. – Dansk Ingeniørforening, København. 162 pp.

Hansen, S. B. 1981. Bestemmelsesnøgle til larver af danske Ptychopteridae (Diptera, Nematocera), med noter om arternes habitatpræferencer. – Entomologiske Meddelelser 49: 59-64.

Jeppesen, E. 1981. A comparison between four oxygen balance models for small organic polluted streams with many macrophytes. Pages 183-195 *in* D. Dubois, editor. Progress in Ecological Engineering and Management by mathematical Modelling. Proceedings 2nd State of Art in Ecological Modelling. Liege, Belgium.

Jónasson, P. M. 1981. Energy flow in a subarctic, eutrophic lake. – Verhandlungen Internationale Vereinigung für Theoretische und Angewandte Limnologie 21: 389-393.

Jørgensen, S. E., Jørgensen, L. A., Kamp-Nielsen, L. and Mejer, H. F. 1981. Parameter estimation in eutrophication modelling. – Ecological Modelling 13: 111-129.

Kamp-Nielsen, L. 1981. Diurnal variation in phytoplankton oxygen metabolism. – Verhandlungen Internationale Vereinigung für Theoretische und Angewandte Limnologie 21: 431-437.

Kamp-Nielsen, L., Jørgensen, L. A. and Jørgensen, S. E.

1981. A sediment water exchange model for lakes in the Upper Nile Basin. Pages 557-583 *in* D. Dubois, editor. Progress in Ecological Engineering and Management by mathematical Modelling. Proceedings 2nd State of Art in Ecological Modelling. Liege, Belgium.

Nygaard, G. and Sand-Jensen, K. 1981. Light climate and metabolism of *Nitella flexilis* (L.) Ag. in the bottom waters of oligotrophic Lake Grane Langsø, Denmark. – Internationale Revue der gesamten Hydrobiologie 66: 685-699.

Riemann, B. and Wium-Andersen, S. 1981. The ATP and total adenine nucleotide content of four unicellular and colonial green algae. – Oikos 36: 368-373.

Sand-Jensen, K. and Søndergaard, M. 1981. Phytoplankton and epiphyte development and their shading effect on submerged macrophytes in lakes of different nutrient status. – Internationale Revue der gesamten Hydrobiologie 66: 529-552.

Wium-Andersen, S. 1981. Seasonal growth of mangrove trees in southern Thailand. III. Phenology of *Rhizophora mucronata* Lamb. and *Scyphiphora hydrophyllacea* Gaertn. – Aquatic Botany 10: 371- 376.

Dall, P. C. 1982. Diversity in reproduction and general morphology between two *Glossiphonia* species (Hirudinea) in Lake Esrom, Denmark. – Zoologica Scripta 11: 127-133.

Hamburger, K. and Kramhøft, B. 1982. Respiration and fermentation during growth and starvation in the fission yeast, *Schizosaccharomyces pombe*. – Carlsberg Research Cummunications 47: 405-411.

Iversen, T. M., Thorup, J. and Skriver, J. 1982. Inputs and transformation of allochthonous particulate organic matter in a headwater stream. – Holarctic Ecology 5: 10-19.

Jeppesen, E. 1982. Diurnal variation in the oxygen uptake of river sediments in vitro by use of continuous flow-through systems. – Hydrobiologia 91: 189-195.

Jørgensen, S. E., Kamp-Nielsen, L. and Mejer, H. F. 1982. Comparison of a simple and a complex sediment phosphorus model. – Ecological Modelling 16: 99-124.

Jørgensen, S. E., Kamp-Nielsen, L., Jørgensen, L. A. and Mejer, H. F. 1982. An environmental management model of the Upper Nile. – ISEM Journal 4: 5-72.

Kamp-Nielsen, L., Mejer, H. F. and Jørgensen, S. E. 1982. Modelling the influence of bioturbation on the vertical distribution of sedimentary phosphorus in Lake Esrom. – Hydrobiologia 91: 197-206.

Lindegaard, C., Mosegaard, H., Mæhl, P. and Nielsen, B. H. 1982. Limnological investigations in the Narssaq Area. Pages 1-26 (vol 1) and 1-32 (vol 2) *in* J. Rose-Hansen, C. O. Nielsen, and H. Sørensen, editors. The Narssaq Project. Progress Report No. 2, Inst. Petrology, Copenhagen University.

Mæhl, P. 1982. Phytoplankton production in relation to physico-chemical conditions in a small, oligotrophic subarctic lake in South Greenland. – Holarctic Ecology 5: 420-427.

Nygaard, G. 1982. Phytoplankton from lakes and ponds of the Narssaq Area, South Greenland. Pages 27-30 (vol 1) and 33-46 (vol 2) *in* J.Rose-Hansen, C. O. Nielsen, and H. Sørensen, editors. The Narssaq Project. Progress Report No. 2, Inst. Petrology, Copenhagen University.

Riemann, B. 1982. Measurement of chlorophyll a and its degradation products: a comparison of methods. – Archiv für Hydrobiologie Beiheft Ergebnisse der Limnologie 16: 19-24.

Riemann, B., Fuhrman, J. and Azam, F. 1982. Bacterial secondary production in freshwater measured by ^{3}H-thymidine incorporation method. – Microbial Ecology 8: 101-114.

Riemann, B. and Ernst, D. 1982. Extraction of chlorophylls a and b from phytoplankton using standard extraction techniques. – Freshwater Biology 12: 217-223.

Riemann, B., Søndergaard, M. Schierup, H. H., Bosselmann, S., Christensen, G., Hansen, J. and Nielsen, B. 1982. Carbon metabolism during a spring diatom bloom in the eutrophic Lake Mossø. – Internationale Revue der gesamten Hydrobiologie 67: 145-195.

Riemann, B. and Wium-Andersen, S. 1982. Predictive value of adenylate energy charge for metabolic and growth states of planktonic communities in lakes. – Oikos 39: 256-260.

Sand-Jensen, K., Prahl, C. and Stockholm, H. 1982. Oxygen release from roots of submerged aquatic macrophytes. – Oikos 38: 349-354.

Sand-Jensen, K., Prahl, C. and Stockholm, H. 1982. Oxygen exchange with the lacunae and across leaves and roots of the submerged vascular macrophyte, *Lobelia dortmanna* L. – New Phytologist 91: 103-120.

Whiteside, M. C. and Lindegaard, C. 1982. Summer distribution of zoobenthos in Grane Langsø, Denmark. – Freshwater Invertebrate Biology 1: 1-16.

Wium-Andersen, S., Anthoni, U., Christophersen, C. and Houen, G. 1982. Allelopathic effects on phytoplankton by substances isolated from aquatic macrophytes (Charales). – Oikos 39: 187-190.

Borum, J. 1983. The quantitative role of macrophytes, epiphytes and phytoplankton under different nutrient conditions in Roskilde Fjord, Denmark. – Proceedings International Symposium Aquatic Macrophytes, Nijmegen. 1983: 35-40.

Bosselmann, S., Jensen, L. M., Jørgensen, N. O., Riemann, B. and Søndergaard, M. 1983. Omsætning af organisk stof i de frie vandmasser i søer. – Naturens Verden 1983: 377-382.

Dall, P. C., Lindegaard, C. og Kirkegaard, J. 1983. Søernes littoralfauna afspejler eutrofigraden. – Stads- og Havneingeniøren 2: 43-48.

Dall, P. C. 1983. The natural feeding and ressource partitioning of *Erpobdella octoculata* L. and *Erpobdella testacea* Sav. in Lake Esrom, Denmark. – Internationale Revue der gesamten Hydrobiologie 68: 473-500.

Hamburger, K., Møhlenberg, F., Randløv, A. and Riisgaard, H. U. 1983. Size, oxygen consumption and growth in the mussel *Mytilys edulis*. – Marine Biology 75: 303-306.

Holopainen, I. J. and Jónasson, P. M. 1983. Long-term population dynamics and production of *Pisidium* (Bivalvia) in the profundal of Lake Esrom, Denmark. – Oikos 42: 99-117.

Jeppesen, E., Sand-Jensen, K. and Hasholt, B. 1983. Modelling the seasonal variation of organic matter in the sediment of a Danish lowland stream. – Proceedings 11th Nordic Symposium on Sediments. Oslo, Norway: 53-56.

Jørgensen, S. E., Jørgensen, L. A., Mejer, H. F. and Kamp-Nielsen, L. 1983. A water quality model for the Upper Nile Lake system. Pages 631-641 *in* W. K. Lauenroth, G. V. Skogerboe and M. Flug, editors. Analysis of Ecological systems: State-of-the-art in ecological modelling. Amsterdam.

Jørgensen, N. O. G., Søndergaard, M., Hansen, H. J., Bosselmann, S. and Riemann, B. 1983. Diel variation in concentration, assimilation and respiration of dissolved free amino acids in relation to planktonic primary and secondary produktion in two eutrophic lakes. – Hydrobiologia 107: 107-122.

Kamp-Nielsen, L. 1983. Sediment-water exchange models. Pages 387-421 *in* S. E. Jørgensen, editor. Application of ecological modelling in enviromental management, Part A. Amsterdam.

Kamp-Nielsen, L., Jørgensen, L. A. and Jørgensen, S. E. 1983. Modelling the distribution of heavy metals between sediment and water in the Upper Nile Lake system. Pages 623-631 *in* W. K. Lauenroth, G. V. Skogerboe and M. Flug, editors. Analysis of ecological systems: State-of-the-art in ecological modelling. Amsterdam.

Kamp-Nielsen, L. 1983. State-of-the-art in application of ecological models to water resources. Page 989 *in* W. K. Lauenroth, G. V. Skogerboe and M. Flug, editors. Analysis of ecological systems: State-of-the-art in ecological modelling. Amsterdam.

Petersen, F. 1983. Population dynamics and production of *Daphnia galeata* (Crustacea, Cladocera) in Lake Esrom. – Holarctic Ecology 6: 285-294.

Riemann, B. 1983. Biomass and production of phyto- and bacterioplankton in eutrophic Lake Tystrup, Denmark. – Freshwater Biology 13: 389-398.

Sand-Jensen, K. 1983. Photosynthetic carbon sources of stream macrophytes. – Journal of Experimental Botany 3: 198-210.

Sand-Jensen, K. 1983. Physical and chemical parameters regulating growth of periphytic communities. Pages 63-71 *in* R. G. Wetzel, editor. Periphyton of Freshwater Ecosystems. Dr. W. Junk. The Haag.

Sand-Jensen, K. and Borum, J. 1984. Regulation of growth of eelgrass (*Zostera marina* L.) in Danish costal waters. – Marine Technology Society Journal 17: 15-21.

Thyssen, N., Erlandsen, M., Jeppesen, E. and Holm, T. F. 1983. Modelling the rearation capacity of lowland streams dominated by submerged macrophytes. Pages 861-867 *in* W. K. Lauenroth, G. V. Skogerboe and M. Flug, editors. Analysis of Ecological systems: State-of-the-Art in Ecological modelling. Amsterdam.

Windolf, J., Westergaard, B., Christensen, T. and Kamp-Nielsen, L. 1983. Fosfor-massebalance på Glumsø og dens sediment. – Proceedings 11th Nordic Symposium on Sediments. Oslo, Norway: 92-107.

Wium-Andersen, S. 1983. Kemisk krigsførelse hos vandplanter. – Naturens Verden 1983: 217-221.

Wium-Andersen, S., Anthoni, U. and Houen, G. 1983. Elemental sulphur. A possible allelopathic compound from *Ceratophyllum demersum.* – Phytochemistry 22: 2613.

Andersen, M. M., Rigét, F. F. and Sparholt, H. 1984. A modification of the Trent index for use in Denmark. – Water Research 18: 145-151.

Borum, J., Kaas, H. and Wium-Andersen, S. 1984. Biomass variation and autotrophic production of an epiphyte-macrophyte community in a coastal Danish area: II. Epiphyte species composition, biomass and production. – Ophelia 23: 165-179.

Dall, P. C., Heegaard, H. and Fullerton, A. F. 1984. Life-history strategies and production of *Tinodes waeneri* (L) (Trichoptera) in Lake Esrom, Denmark. – Hydrobiologia 112: 93-104.

Dall, P. C., Lindegaard, C., Jónsson, E., Jónsson, G. and Jónasson, P. M. 1984. Invertebrate communities and their environment in the exposed littoral zone of Lake Esrom, Denmark. – Archiv für Hydrobiologie Suppl. 69: 477-524.

Dannisøe, J., Frederiksen, N., Jensen, E., Lindegaard, C. and Nissen, E. 1984. Fødegrundlagets betydning for produktion af ørred (*Salmo trutta* L.) i okkerbelastede vandløb. Bilag nr. 17 til redegørelse om den treårige forsøgsordning til nedbringelse af okkergener i vandløb. – Miljøministeriet. 116 pp.

Hansen, H. J. 1984. Assimilatory sulphate reduction as an approach to determine bacterial growth. – Archiv für Hydrobiologie Beiheft Ergebnisse der Limnologie 19: 43-51.

Iversen, T. M. 1984. Markvandingens betydning for vandløbenes økologiske tilstand. – Vand og Miljø 1, 3: 29-33.

Iversen, T. M., Jeppesen, E., Sand-Jensen, K. and Thorup, J. 1984. Økologiske konsekvenser af reduceret vandføring i Susåen. Bind 1. Den biologiske struktur. – Rapport til Miljøstyrelsen. 178 pp.

Jeppesen, E., Iversen, T. M., Sand-Jensen, K. and Prahl-Jørgensen, C. 1984. Økologiske konsekvenser af reduceret vandføring i Susåen. Bind II. Biologiske proceshastigheder og vandkvalitetsforhold. – Rapport til Miljøstyrelsen. 261 pp.

Jeppesen, E. and Thyssen, N. 1984. Modelling the seasonal variation in structural biological components and oxygen in macrophyte dominated streams: A summary of work in progress. – Water Science and Technology 16: 533-537.

Jónasson, P. M. 1984. The ecosystem of eutrophic Lake Esrom. Pages 177-204 *in* D. G. Goodall and F. B. Taub, editors. Ecosystems of the world. 23. Lakes and Reservoirs. Amsterdam.

Jónasson, P. M. 1984. Oxygen demand and long term changes of profundal zoobenthos. – Hydrobiologia 114: 121-126.

Jónasson, P. M. 1984. Decline of zoobenthos through five decades of eutrophication in Lake Esrom. – Verhandlungen Internationale Vereinigung für Theoretische und Angewandte Limnologie 22: 800-804.

Jónasson, P. M. 1984. Forsuring af ferskvandets økosystemer. – Naturens Verden 1984: 106-115.

Jørgensen, S. E., Kamp-Nielsen, L. and Jørgensen, L. A. 1984. An environmental management model of the Upper Nile lake system. – Journal of the International Society for Ecological Modelling 4: 5-73.

Lindegaard, C. and Jónsson E. 1984. Succession of Chironmidae (Diptera) in Hjarbæk Fjord, Denmark, during a period with change from brackish water to freshwater. – Memoirs of the American Entomological Society 34: 169-185.

Mathiesen, F. D., Langhoff, H., Dahl, J., Jacobsen, B., Stilling, L., Østergaard, K., Nielsen, C. E., Østergaard, H., Lindegaard, C., Christensen, C., Somer, E., Rørdam, E., Korkman, T. E. and Reker, L. 1984. Okker, redegørelse om den treårige forsøgsordning til nedbringelse af okkergener i vandløb. – Miljøministeriet. 182 pp.

Marcussen, B., Nielsen, P. and Jeppesen, M. 1984. Diel changes in bacterial activity determined by means of microautoradiography. – Archiv für Hydrobiologie Beiheft Ergebnisse der Limnologie 19: 141-149.

Mejer, H. M. 1984. A menu driven lake model. – Journal of the International Society for Ecological Modelling 5: 43-51.

Riemann, B. 1984. Determining growth rates of natural assemblages of freshwater bacteria by means of ^{3}H-thymidine incorporation into DNA: comments on methodology. – Archiv für Hydrobiologie Beiheft Ergebnisse der Limnologie 19: 67-80.

Riemann, B. 1984. Alternativ sø-restaurereing. – Vand og Miljø 1, 2: 6-9.

Riemann, B. and Søndergaard, M. 1984. Measurement of diel rates of bacterial secondary production in aquatic environments. – Applied and Environmental Microbiology 47: 632-638.

Riemann, B. and Søndergaard, M. 1984. Bacterial growth in relation to phytoplankton primary production and extracellular release of organic carbon. Pages 233-248 *in* J. E. Hobbie and P. Williams, editors. Heterotrophic activity in the sea. J. Leb. Nato Conference Series, New York.

Riemann, B., Nilsen, P., Jeppesen, M., Marcussen, B. and Fuhrmann, J. A. 1984. Diel changes in bacterial biomass and growth rates in coastal environments, determined by means of thymidine incorporation into DNA, frequency of dividing cells (FDC) and microautoradiography. – Marine Ecology Progress Series 17: 227-235.

Riemann, B. and Bosselmann, S. 1984. *Daphnia* grazing on natural populations of bacteria. – Verhandlungen Internationale Vereinigung für Theoretische und Angewandte Limnologie 22: 795-799.

Sand-Jensen, K. and Gordon, D. M. 1984. Differential ability of marine and freshwater macrophytes to utilize HCO_3 and CO_2. – Marine Biology 80: 247-253.

Sand-Jensen, K. and Borum, J. 1984. Epiphyte shading and its effect on photosynthesis and diel metabolism of *Lobelia dortmanna* L. during the spring bloom in a Danish lake. – Aquatic Botany 20: 109-119.

Wium-Andersen, S. 1984. Landbruget og kvælstofforureningen. – Tolvmandsbladet 4: 131-134.

Wium-Andersen, S. and Borum, J. 1984. Biomass variation and autotrophic production of an epiphyte- macrophyte community in a coastal Danish area. I. Eelgrass (*Zostera marina* L.) biomass and net production. – Ophelia 23: 33-46.

Borum, J. 1985. Development of epiphytic communities on eelgrass (*Zostera marina*) along a nutrient gradient in a Danish estuary. – Marine Biology 87: 211-218.

Hagelskjær, B., Madsen, A., Olsen, M., Sand-Jensen, K. and Iversen, T. M. 1985. Rævsø – en hedesø under forandring. – Vand og Miljø 2: 65-67.

Iversen, T. M., Hansen, T., Thorup, J., Lodal, J. and Olsen, J. 1985. Quantitative estimates and community structure of invertebrates in a macrophyte rich stream. – Archiv für Hydrobiologie 102: 291-301.

Iversen, T. M., Lindegaard, C., Sand-Jensen, K. and Thorup, J. 1985. Vandløbsøkologi. Kompendium til brug ved undervisningen i Akvatisk Økologi ved Københavns Universitet. – Ferskvandsbiologisk Laboratorium, Københavns Universitet. 110 pp.

Jónsson, E. 1985. Population dynamics and production of Chironomidae (Diptera) at 2 m depth in Lake Esrom, Denmark. – Archiv für Hydrobiologie Suppl. 70: 239-278.

Kamp-Nielsen, L. 1985. Katastrofen lurer i vore fjorde og søer. – Have/Park og Miljø: 23.

Kamp-Nielsen, L. 1985. Modelling of eutrophication processes. Pages 5-16 *in* R. de Bernardi et al. Proceedings EWPCA International Congress. Lakes pollution and recovery. Rome, Italy.

Kamp-Nielsen, L., Grevy, P., Rasmussen, E. K. and Krarup, H. 1985. Modelling the recovery and internal loading of Lake Hald. – Proceedings 13th Nordic Symposium on Sediments. Aneboda, Sweden: 74-107.

Kiørboe, T., Møhlenberg, F. and Hamburger, K. 1985. Bioenergetics of the planktonic copepod *Acartia tonsa*: relation between feeding, egg production and respiration, and composition of specific dynamic action. – Marine Ecology Progress Series 26: 85-97.

Lindegaard, C., Dannisøe, J., Frederiksen, N., Nissen, E. and Jensen, E. R. 1985. Nu skal vore okkervandløb gøres rene. – Natur og Miljø 12: 33-35.

Nielsen, L. W., Nielsen, K. and Sand-Jensen, K. 1985. High rates of production and mortality of submerged *Sparganium emersum* Rehman during its short growth season in a eutrophic Danish stream. – Aquatic Botany 22: 325-334.

Riemann, B. 1985. Potential importance of fish predation and zooplankton grazing on natural populations of freshwater bacteria. – Applied and Environmental Microbiology 50: 187-193.

Sand-Jensen, K. 1985. Vandplanternes optagelse af uorganisk kulstof ved fotosyntesen. – Naturens Verden 1985: 175-182.

Sand-Jensen, K. 1985. Forurening fra landbruget skader åer, søer og fjorde. De våde landskaber. Pages 34-39 *in* K. Laursen, éditor. Dansk vildtforskning 1984-85. Kalø, Rønde.

Sand-Jensen, K. 1985. Kvælstof og fosfat virker sammen på plantevækst og sedimentomsætning. – Vand og Miljø 2: 76-78.

Sand-Jensen, K., Revsbech, N. P. and Jørgensen, B. P. 1985. Microprofiles of oxygen in epiphyte communities on submerged macrophytes. – Marine Biology 89: 55-62.

Wium-Andersen, S. 1985. NPO redegørelsens kvalitet og vandkvaliteten. – Tidsskrift for Landøkonomi 2: 124-127.

Wium-Andersen, S. 1985. Botanikkens svar på næbdyret. – Naturens Verden 1985: 152-155.

Wium-Andersen, S. 1985. Gribsøs forsuring accelererer. – Vand og Miljø 2: 267-269.

Wium-Andersen, S. and Gravsholt, S. 1985. Kolonisering af et tømmerdepot i Esrom Sø. – Urt 2: 51- 53.

Bjørnsen, P. K., Larsen, J. B., Geertz-Hansen, O. and Olesen, M. 1986. A field technique for the determination of zooplankton grazing on natural bacterioplankton. – Freshwater Biology 16: 245-253.

Bjørnsen, P. K. 1986. Automatic determination of bacterioplankton biomass by image analysis. – Applied and Environmental Microbiology 51: 1199-1204.

Bjørnsen, P. K. 1986. Bacterioplankton growth yield in continuous seawater cultures. – Marine Ecology Progress Series 30: 191-196.

Borum, J. and Sand-Jensen, K. 1986. En almindelig fejl ved bestemmelse af marine planktonalgers primærproduktion. – Vand og Miljø 3: 74-76.

Bosselmann, S. and Riemann, B. 1986. Zooplankton. Pages 199-236 *in* B. Riemann and M. Søndergaard, editors. Carbon dynamics in eutrophic, temperate lakes. Elsevier, Amsterdam.

Cabecadas, G., Brogueira, M. J. and Windolf, J. 1986. A phytoplankton bloom in shallow Divor reservoir (Portugal). The importance of internal nutrient loading. – Internationale Revue der gesamten Hydrobiologie 71: 795-806.

Christoffersen, K. and Jespersen, A. M. 1986. Gut evacuation rates and ingestion rates of *Eudiaptomus graciloides* measured by means of the gut flourescence method. – Journal of Plankton Research 8: 973- 983.

Hamann, O. and Wium-Andersen, S. 1986. *Scalesia gordilloi* sp. Nov. (Asteraceae) from the Galapagos Islands, Ecuador. – Nordic Journal of Botany 6: 35-38.

Hamburger, K. 1986. Energy flow in the populations of *Eudiaptomus graciloides* and *Daphnia galeata* in Lake Esrom. – Archiv für Hydrobiologie 105: 517-530.

Hoffmann, C. C. 1986. Nitrate reduction in a reedswamp receiving water from a agricultural watershed. – Proceedings 13th Nordic Symposium on Sediments. Aneboda, Sweden: 41-63.

Jørgensen, S. E. and Kamp-Nielsen, L. 1986. Eutrofieringsmodeller som prognoseværktøj. Et case study: Glumsø. – Vand og Miljø 3: 191-195.

Jørgensen, S. E., Kamp-Nielsen, L. and Jørgensen, L. A. 1986. Examination of the generality of eutrophication models. – Ecological Modelling 32: 251-266.

Jørgensen, S. E., Kamp-Nielsen, L., Christensen, T., Windolf-Nielsen, J. and Vestergaard, B. 1986. Validation of a prognosis based upon a eutrophication model. – Ecological Modelling 32: 165-182.

Kamp-Nielsen, L. 1986. Modelling the recovery of hypertrophic Lake Glumsø, Denmark. – Hydrobiological Bulletin 20: 245-255.

Kamp-Nielsen, L. 1986. Pianificazione della qualità delle acque e fonti diffuse di inquinamento. – Annuario Europeo dell'Ambiente. – Istituto de Studi e documenzione pr l'terretorio: 479-483.

Kamp-Nielsen, L., Grevy, P., Rasmussen, E. K. and Karup, H. 1986. Modelling the recovery and internal loading of Lake Hald. – Proceedings 13th Nordic Symposium on Sediments, Aneboda, Sweden: 74-107.

Olesen, M. and Bjørnsen, P. K. 1986. Forudsigeligheden af eutrofieringseffekter i kystvande. – Vand og Miljø 3: 16-18.

Riemann, B. and Søndergaard, M. 1986. Regulation of bacterial secondary production in two eutrophic lakes with experimental enclosures. – Journal of Plankton Research 8: 519-536.

Riemann, B. and Søndergaard, M. (eds.). 1986. Carbon dynamics in eutrophic, temperate lakes. Elsevier, Amsterdam. 284 pp.

Riemann, B., Jørgensen, N. O. G., Lambert, W. and Fuhrman, J. A. 1986. Zooplankton induced changes in dissolved free amino acids and in production rates of freshwater bacteria. – Microbial Ecology 12: 247-258.

Riemann, B., Søndergaard, M., Persson, L. and Johansson, L. 1986. Carbon metabolism and community regulation in eutrophic temperate lakes. Pages 267-280 *in* B. Riemann and M. Søndergaard, editors. Carbon dynamics in eutrophic, temperate lakes. Elsevier, Amsterdam.

Riemann, B., Søndergaard, M. and Jørgensen, N. O. G. 1986. Bacteria carbon dynamics in eutrophic, temper-ate lakes. Pages 137-197 *in* B. Riemann and M. Søndergaard, editors. Carbon dynamics in eutrophic, temperate lakes. Elsevier, Amsterdam.

Sand-Jensen, K. and Gordon, D. M. 1986. Variable HCO_3 affinity of *Elodea canadensis* Michaux in response to different HCO_3 and CO_2 concentrations during growth. – Oecologia 70: 426-432.

Sand-Jensen, K. and Søndergaard, M. 1986. (eds.). Submerged macrophytes: Carbon metabolism, growth regulation and role in macrophyte dominated ecosystems. – Aquatic Botany 26: 201-394.

Søndergaard, M., Riemann, B. and Jørgensen, N. O. G. 1986. Extracellular organic carbon (EOC) released by phytoplankton and bacterial production. – Oikos 45: 323-332.

Wium-Andersen, S. 1986. The development of the acid Lake Gribsø in Denmark after 1950. – Contribution to the COST Workshop on Reversibility of Acidification. Grimstad, Norway, 1986: 1-9.

Wium-Andersen, S. 1986. *Scalesia gordilloi* a new Galapagos plant species. – Noticias de Galapagos 44: 13-14.

Wium-Andersen, S. and Hamann, O. 1986. Manglares de las Islas Galapagos. – Revista Geografica 23: 101-122.

Borum, J. 1987. Dynamics of epiphyton on eelgrass (*Zostera marina* L.) leaves: Relative roles of algal growth, herbivory, and substratum turnover. – Limnology and Oceanography 32: 986-992.

Dall, P. C. 1987. The ecology of the littoral leech fauna (Hirudinea) in Lake Esrom, Denmark. – Archiv für Hydrobiologie Suppl. 76: 256-313.

Dall, P. C. 1987. Larvemigration hos *Sialis lutaria* L. (Megaloptera) i Esrom Sø. – Flora og Fauna 93: 47-50.

Dall, P. C., Iversen, T. M., Kirkegaard, J., Lindegaard, C. and Thorup, J. 1987. En oversigt over danske ferskvandsinvertebrater til brug ved bedømmelse af forureningen i søer og vandløb. – Ferskvandsbiologisk Laboratorium, Københavns Universitet. 237 pp.

Geertz-Hansen, O., Olesen, M., Bjørnsen, P. K., Larsen, J. B. and Riemann, B. 1987. Zooplankton consumption of bacteria in a eutrophic lake and in experimental enclosures. – Archiv für Hydrobiologie 110: 553-563.

Geertz-Hansen, O. 1987. Roskilde Fjord – carbonkredsløb og eutrofieringseffekter. – Nordisk Miljövårdsserien. Nordforsk. Helsingfors.

Hamburger, K. and Boëtius, F. 1987. Ontogeny of growth, respiration and feeding rate of the freshwater calanoid copepod *Eudiaptomus graciloides*. – Journal of Plankton Research 9: 589-606.

Iversen, T. M. and Thorup, J. 1987. Population dynamics and production of *Sialis lutaria* L. (Megaloptera) in the Danish River Suså. – Freshwater Biology 17: 461-469.

Jeppesen, E. and Iversen, T. M. 1987. Two simple models for estimating daily mean water temperatures and diel variations in a Danish low gradient stream. – Oikos 49: 149-155.

Jespersen, A. M. and Christoffersen, K. 1987. Measurements of chlorophyll-a from phytoplankton using ethanol as extraction solvent. – Archiv für Hydrobiologie 109: 445-454.

Jeppesen, E., Thyssen, N., Prahl, C., Sand-Jensen, K. and Iversen, T. M. 1987. Kvælstofakkumulering og omsætning med udgangspunkt i undersøgelser i Suså og Gryde Å. – Vand og Miljø 4: 123-129.

Jónsson, E. 1987. Flight periods of aquatic insects at Lake Esrom, Denmark. – Archiv für Hydrobiologie 110: 259-274.

Jónsson, G. S. 1987. The depth distribution and biomass of epilithic periphyton in Lake Thingvallavatn, Iceland. – Archiv für Hydrobiologie 108: 531-547.

Kairesalo, T., Gunnarsson, K., Jónsson, G. S. and Jónasson, P. M. 1987. The occurence and photosynthetic activity of epiphytes on the tips of *Nitella opaca* Ag. (Charophyceae). – Aquatic Botany 18: 333-340.

Kamp-Nielsen, L. 1987. Modelling of eutrophication processes. Pages 57-68 *in* R. Vismara, R. Marforio, V. Mezzanotte and S. Cernuschi, editors. EWPCA International Congress. Lake Pollution and Recovery.

Lindegaard, C. and Jónsson, E. 1987. Abundance, population dynamics and high production of Chironomidae (Diptera) in Hjarbæk Fjord, Denmark, during a period of eutrophication, – Entomologica Scandinavica Suppl. 29: 293-302.

Madsen, T. V. and Sand-Jensen, K. 1987. Photosynthetic capacity, bicarbonate affinity and growth of *Elodea canadensis* exposed to different concentrations of inorganic carbon. – Oikos 50: 176-182.

Riemann, B. and Bjørnsen, P. K. 1987. Calculation of cell production of coastal marine bacteria based on measured incorporation of (3H) thymidine. – Limnology and Oceanography 32: 471-476.

Riemann, B., Jensen, H. J. and Müller, J. P. 1987. Manipulation af søers fiskebestande. – Vand og Miljø 4: 163-166.

Sand-Jensen, K. 1987. Environmental control of bicarbonate use among freshwater and marine macrophytes. Pages 99-112 *in* R. M. Crawford, editor. Ecology and physiology of intertidal plants. – Blackwell, Oxford.

Sand-Jensen, K. and Nielsen, G. Æ. 1987. Eutrofieringseffekter i danske vandområder: En sammenligning mellem vandløb, søer og kystområder. – Kvælstof og fosfor i vandmiljøet. Rapport fra Konsensuskonference. Planlægningsrådet for Forskningen: 41-48.

Sand-Jensen, K. and Revsbech, N. P. 1987. Photosynthesis and light adaptation in epiphyte- macrophyte associations measured by oxygen microelectrodes. – Limnology and Oceanography 32: 452-457.

Sandlund, O. T., Jonsson, B., Malmquist, H. J., Gydemo, R., Lindem, T., Skúlason, S., Snorrason, S. S. and Jónasson, P. M. 1987. Habitat use of arctic charr *Salvelinus alpinus* in Thingvallavatn, Iceland. – Environmental Biology of Fishes 20: 263-274.

Thorup, J., Iversen, T. M., Absalonsen, N. O., Holm, T., Jessen, J. and Olsen, J. 1987. Life cycles of four species of *Baëtis* (Ephemeroptera) in three Danish streams.

– Archiv für Hydrobiologie 109: 49- 65.

Vermaat, J. E. and Sand-Jensen, K. 1987. Survival, metabolism and growth of *Ulva lactuca* under winter conditions: a laboratory study of bottlenecks in the life cycle. – Marine Biology 95: 55-61.

Wium-Andersen, S. 1987. Allelopathy among aquatic plants. – Archiv für Hydrobiologie Beiheft Ergebnisse der Limnologie 27: 167-172.

Wium-Andersen, S., Jørgensen, K. H., Christophersen, C. and Anthoni, U. 1987. Algal growth inhibitors in *Sium erectum* Huds. – Archiv für Hydrobiologie 111: 317-320.

Ahlgren, I., Frisk, T. and Kamp-Nielsen, L. 1988. Empirical and theoretical models of phosphorus loading, retention and concentration vs. lake trophic state. – Hydrobiologia 170: 285-303.

Borum, J., Bosselmann, S., Iversen, T. M., Lindegaard, C., Riemann, B. and Thorup, J. 1988. Laboratorie- og feltprocedurer ved det limniske semesterkursus (Biologi 1). – Ferskvandsbiologisk Laboratorium, Københavns Universitet. 25 pp.

Bjørnsen, P. K. 1988. Phytoplankton exudation of organic matter: Why do healthy cells do it? – Limnology and Oceanography 33: 151-154.

Bjørnsen, P. K. and Riemann, B. 1988. Towards a quantitative stage in the study of microbial processes in pelagic carbon flows. – Archiv für Hydrobiologie Beiheft Ergebnisse der Limnologie 31: 185-193.

Christoffersen, K. 1988. Effect of food concentration on gut evacuation of *Daphnia pulicaria* and *Daphnia longispina* measured by the fluorescence technique. – Verhandlungen Internationale Vereinigung für Theoretische und Angewandte Limnologie 23: 2050-2055.

Dall, P. C., Lindegaard, C. and Jónasson, P. M. 1988. Water Pollution in Denmark. A review with special reference to freshwater resources. – Vatten 44: 96-103.

Dall, P. C. 1988. The morphological differences and the occurrence of *Oulimnius tuberculatus* (P. W. J. Müller, 1806) and *Oulimnius troglodytes* (Gyllenhal, 1827) (Coleoptera: Elminthidae) in Lake Esrom, Denmark. – Entomologiske Meddelelser 56: 113-122.

Dall, P. C., Slot, H. and Kirkegaard, J. 1988. Vandløbsforurening på Lolland og Falster. – Vand og Miljø 5: 199-205.

Holtan, H., Kamp-Nielsen, L. and Stuanes, A. O. 1988. Phosphorus in soil, water and sediment: an overview. – Hydrobiologia 170: 19-34.

Horsted, S. J., Nielsen, T. G., Riemann, B., Pock-Steen, J. and Bjørnsen, P. K. 1988. Regulation of zooplankton by suspension-feeding bivalves and fish in estuarine enclosures. – Marine Ecology Progress Series 48: 217-224.

Iversen, T. M. 1988. Secondary production and trophic relationships in a spring invertebrate community. – Limnology and Oceanography 33: 582-592.

Iversen, T. M. and Lindegaard, C. 1988. Biologisk bedømmelse af vandløb forurenet med organisk stof.

Kompendium til brug ved undervisningen i kursusmodulet "Biologisk bedømmelse af vandløbsforurening" ved Københavns Universitet. – Ferskvandsbiologisk Laboratorium, Københavns Universitet. 56 pp.

Iversen, T. M. and Thorup, J. 1988. A three years' study of life cycle, population dynamics and production of *Asellus aquaticus* L. in a macrophyte rich stream. – Internationale Revue der gesamten Hydrobiologie 73: 73-94.

Jespersen, A. M., Christoffersen, K. and Riemann, B. 1988. Annual carbon fluxes between phyto-, zoo-, and bacterioplankton in eutrophic, Lake Frederiksborg Slotssø, Denmark. – Verhandlungen Internationale Vereinigung für Theoretische und Angewandte Limnologie 23: 440-444.

Jónasson, P. M. and Lindegaard, C. 1988. Ecosystem studies of North Atlantic Ridge Lakes. – Verhandlungen Internationale Vereinigung für Theoretische und Angewandte Limnologie 23: 394-402.

Jonsson, B., Skúlason, S. Snorrason, S. S., Sandlund, O. T., Malmquist, H.J., Jónasson, P. M., Gydemo, R. and Lindem, T. 1988. Life history variation of polymorphic arctic charr (*Salvelinus alpinus*) in Thingvallavatn, Iceland. – Canadian Journal of Fisheries and Aquatic Sciences 45: 1537-1547.

Jørgensen, N. O. G. and Bosselmann, S. 1988. Concentrations of free amino acids and their bacterial assimilation rates in vertical profiles of two Danish lakes: Relations to diel changes in zooplankton grazing activity. – Archiv für Hydrobiologie Beiheft Ergebnisse der Limnologie 31: 289-300.

Kemp, W. M., Murray, L., Borum, J. and Sand-Jensen, K. 1988. Diel growth in eelgrass, *Zostera marina*. – Marine Ecology Progress Series 41: 79-86.

Lindegaard, C. and Dall, P. C. 1988. Abundance and distribution of Oligochaeta in the exposed littoral zone of Lake Esrom, Denmark. – Archiv für Hydrobiologie Suppl. 81: 533-562.

Lindegaard, C. and Dall, P. C. 1988. Eutrofiering af og recipient-kvalitetsplanlægning for søer. Kompendium til brug ved undervisningen i kursusmodulet "Biologisk recipientbedømmelse af søer" ved Københavns Universitet. – Ferskvandsbiologisk Laboratorium, Københavns Universitet. 69 pp.

Lindegaard, C. and Mortensen, E. 1988. Abundance, life history and production of Chironomidae (Diptera) in a Danish lowland stream. – Archiv für Hydrobiologie Suppl. 81: 563-587.

Rasmussen, K. and Lindegaard, C. 1988. Effects of iron compounds on macroinvertebrate communities in a Danish lowland river system. – Water Research 22: 1101-1108.

Riemann, B. 1988. Struktur og omsætning af kulstof i de frie vandmasser i søer. Kompendium til brug ved undervisningen i akvatisk økologi ved Københavns Universitet. – Ferskvandsbiologisk Laboratorium, Københavns Universitet. 22 pp.

Riemann, B., Nielsen, T. G., Horsted, S. J., Bjørnsen, P. K. and Pock-Steen, J. 1988. Regulation of phytoplankton biomass in estuarine enclosures. – Marine Ecology Progress Series 48: 205-215.

Sand-Jensen, K. 1988. Minimum light requirements for growth in *Ulva lactuca*. – Marine Ecology Progress Series 50: 187-193.

Sand-Jensen, K. 1988. Photosynthetic responses of *Ulva lactuca* at very low light. – Marine Ecology Progress Series 50: 195-201.

Sand-Jensen, K., Møller, J. and Olesen, B. H. 1988. Biomass regulation of microbenthic algae in Danish lowland streams. – Oikos 53: 332-340.

Sandlund, O. T., Malmquist, H. J., Jonsson, B. Skúlason, S., Snorrason, S. S., Jónasson, P. M., Gydemo, R. and Lindem, T. 1988. Density, length distribution, and diet of age-0 arctic charr *Salvelinus alpinus* in the surf zone of Thingvallavatn, Iceland. – Environmental Biology of Fishes 23: 183-195.

Schwaerter, S., Søndergaard, M., Riemann, B. and Jensen, L. M. 1988. Respiration in eutrophic lakes: the contribution of bacterioplankton and bacterial growth yield. – Journal of Plankton Research 10: 515-531.

Smits, J. D. and Riemann, B. 1988. Calculation of cell production from (^{3}H) thymidine incorporation with freshwater bacteria. – Applied and Environmental Microbiology 54: 2213-2219.

Søndergaard, M., Riemann, B., Jensen, L. M., Jørgensen, N. O. G., Bjørnsen, P. K., Olesen, M., Larsen, J. B., Geertz-Hansen, O. and Hansen, J. 1988. Pelagic food web processes in an oligotrophic lake. – Hydrobiologia 164: 271-286.

Bijl, L. van der, Sand-Jensen, K. and Hjermind, A. L. 1989. Regulation of photosynthetic rates of submerged rooted macrophytes. – Oecologia 81: 364-368.

Bijl, L. van der, Sand-Jensen, K. and Hjermind, A. L. 1989. Photosynthesis and canopy structure of a submerged plant, *Potamogeton pectinatus* in a Danish lowland stream. – Journal of Ecology 77: 947- 962.

Bjørnsen, P. K., Riemann, B., Pock-Steen, J., Nielsen, T. G. and Horsted, S. J. 1989. Regulation of bacterioplankton production and cell volume in a eutrophic estuary. – Applied and Environmental Microbiology 55: 1512-1518.

Borum, J., Murray, L. and Kemp, W. M. 1989. Aspects of nitrogen acquisition and conservation in eelgrass plants. – Aquatic Botany 35: 289-300.

Dall, P. C. 1989. The life history and occurrence of *Sialis lutaria* L. (Megaloptera) in the littoral zone of Lake Esrom, Denmark. – Internationale Revue der gesamten Hydrobiologie 74: 273-281.

Frandsen, F., Malmquist, H. J. and Snorrason, S. S. 1989. Ecological parasitology of polymorphic arctic charr, *Salvelinus alpinus* (L.), in Thingvallavatn, Iceland. – Journal of Fish Biology 34: 281-297.

Hamburger, K. 1989. Plante- og dyrelivet i Esrom Sø. – Carlsbergfondet, Frederiksborgmuseet, Ny Carlsbergfondet. Årskrift 1989: 27-34.

Holopainen, I. J. and Jónasson, P. M. 1989. Bathymetric distribution and abundance of *Pisidium* (Bivalvia: Sphaeriidae) in Lake Esrom, Denmark, from 1954 to 1988. – Oikos 55: 324-334.

Holopainen, I. J. and Jónasson, P. M. 1989. Reproduction of *Pisidium* (Bivalvia, Sphaeriidae) at different depths in Lake Esrom, Denmark. – Archiv für Hydrobiologie 116: 85-95.

Iversen, T. M. and Dall, P. C. 1989. The effect of growth pattern, sampling interval and number of size classes on benthic invertebrate production estimated by the size-frequency method. – Freshwater Biology 22: 323-331.

Jensen, L. M., Sand-Jensen, K., Marcher, S. and Hansen, M. 1989. Plankton community respiration along a nutrient gradient in a shallow Danish estuary. – Marine Ecology Progress Series 61: 75-85.

Kairesalo, T., Jónsson, G. S., Gunnarson, K. and Jónasson, P. M. 1989. Macro- and microalgal production within a. *Nitella opaca* bed in Lake Thingvallavatn, Iceland. – Journal of Ecology 77: 332-342.

Kamp-Nielsen, L. 1989. The relation between external P-load and inlake-P concentration (subsystem 1). Pages 61-101 *in* H. Sas, editor. Lake restoration by reduction of nutrient loading: Expectations, Experiences, Extrapolations.

Kamp-Nielsen, L. 1989. Sediment-water exchange models. Pages 371-398 *in* S. E. Jørgensen, and M. J. Gromiec, editors. Developments in Environmental Modelling 14. Mathematical Submodels in Water Quality Systems.

Lindegaard, C. 1989. A review of secondary production of zoobenthos in freshwater ecosystems with special reference to Chironomidae (Diptera). – Acta Biologica Debrecina Supplementum Oecologica Hungarica 3: 231-240.

Markager, S. and Sand-Jensen, K. 1989. Patterns of night-time respiration in a dense phytoplankton community under a natural ligth regime. – Journal of Ecology 77: 49-61.

Nielsen, S. L. and Sand-Jensen, K. 1989. Regulation of photosynthetic rates of submerged rooted macrophytes. – Oecologia 81: 364-368.

Nielsen, S. L., Borum, J., Geertz-Hansen, O. and Sand-Jensen, K. 1989. Marine bundplanters dybdegrænse. – Vand og Miljø 6: 217-220.

Nygaard, G. 1989. Some observations on the irradiance and carbon fixation in Grane Langsø. – Internationale Revue der gesamten Hydrobiologie 74: 293-319.

Riemann, B., Simonsen, P. and Stensgaard, L. 1989. The carbon and clorophyll content of phytoplankton from various nutrient regimes. – Journal of Plankton Research 11: 1037-1045.

Sand-Jensen, K. 1989. Environmental variables and their effect on photosynthesis of aquatic plant communities. – Aquatic Botany 34: 5-25.

Sand-Jensen, K. 1989. Næringssalt- og predatorkontrol i søernes frie vandmasser. – Vand og Miljø 6: 15-19.

Sand-Jensen, K. and Madsen, T. V. 1989. Invertebrates graze submerged rooted macrophytes in lowland streams. – Oikos 55: 420-423.

Sand-Jensen, K., Borg, D. and Jeppesen, E. 1989. Biomass and oxygen dynamics of the epiphyte community in a Danish lowland stream. – Freshwater Biology 22: 431-443.

Sand-Jensen, K., Jeppesen, E., Nielsen, K., Bijl, L. van der, Hjermind, L., Nielsen, L. W. and Iversen, T. M. 1989. Growth of macrophytes and ecosystem consequences in a Danish lowland stream. – Freshwater Biology 22: 15-32.

Borum, J., Geertz-Hansen, O., Sand-Jensen, K. and Wium-Andersen, S. 1990. Eutrofiering – effekter på marine primærproducenter. – NPO-forskning fra Miljøstyrelsen, C3: 52 pp.

Christoffersen, K., Riemann, B., Hansen, L. R., Klysner, A. and Sørensen, H. B. 1990. Qualitative importance of the microbioal loop and plankton community structure in a eutrophic lake during a bloom of cyanobacteria. – Microbial Ecology 20: 253-272.

Christoffersen, K. 1990. Evaluation of *Chaoborus* predation on natural populations of herbivorous zooplankton in a eutrophic lake. – Hydrobiologia 200/201: 459-466.

Dall, P. C. 1990. Det våde univers. – Forskning & Samfund 3/4: 36-38.

Dall, P. C., Lindegaard, C. and Jónasson, P. M. 1990. In-lake variations in the compositions of zoobenthos in the littoral of Lake Esrom, Denmark. – Verhandlungen Internationale Vereinigung für Theoretische und Angewandte Limnologie 24: 193-620.

Duarte, C. M. and Sand-Jensen, K. 1990. Seagrass colonization: patch formation and patch growth in *Cymodocea nodosa*. – Marine Ecology Progress Series 65: 193-200.

Duarte, C. M. and Sand-Jensen, K. 1990. Seagrass colonization: biomass development and shoot demography in *Cymodocea nodosa* patches. – Marine Ecology Progress Series 67: 97-103.

Frost-Christensen, H. and Sand-Jensen, K. 1990. Growth rate and carbon affinity of *Ulva lactuca* under controlled levels of carbon, pH and oxygen. – Marine Biology 104: 497-501.

Gordon, D. M. and Sand-Jensen, K. 1990. Effects of O_2, pH and DIC on photosynthetic net-O_2 evolution by marine macroalgae. – Marine Biology 106: 445-451.

Hamburger, K. and Dall, P. C. 1990. The respiration of common benthic invertebrate species from the shallow littoral zone of Lake Esrom, Denmark. – Hydrobiologia 199: 117-130.

Hamburger, K., Dall, P. C. and Jónasson, P. M. 1990. The role of *Dreissena polymorpha* Pallas (Mollusca) in the energy budget of Lake Esrom, Denmark. – Verhandlungen Internationale Vereinigung für Theoretische und Angewandte Limnologie 24: 621-625.

Jensen, L. M., Sand-Jensen, K., Marcher, S. and Hansen,

M. 1990. Plankton community respiration along a nutrient gradient in a shallow Danish estuary. – Marine Ecology Progress Series 61: 75-85.

Jeppesen, E., Søndergaard, M., Mortensen, E., Kristensen, P., Riemann, B., Jensen, H. J., Müller, J. P., Sortkjær, O., Jensen, J. P., Christoffersen, K., Bosselmann, S. and Dall, E. 1990. Fish manipulation as a lake restoration tool in shallow, eutrophic temperate lakes. 1: cross-analysis of three Danish case-studies. – Hydrobiologia 200/201: 205-218.

Jónasson, P. M., Lindegaard, C. and Hamburger, K. 1990. Energy budget of Lake Esrom, Denmark. – Verhandlungen Internationale Vereinigung für Theoretische und Angewandte Limnologie 24: 632-640.

Jónasson, P. M., Lindegaard, C., Dall, P. C., Hamburger, K. and Adalsteinsson, H. 1990. Ecosystem studies on temperate Lake Esrom and the subarctic lakes Mývatn and Thingvallavatn. – Limnologica 20: 259-266.

Lindegaard, C., Hamburger, K. and Dall, P. C. 1990. Population dynamics and energy budget of *Stylodrilus heringianus* Clap. (Lumbriculidae, Oligochaeta) in the shallow littoral of Lake Esrom, Denmark. – Verhandlungen Internationale Vereinigung für Theoretische und Angewandte Limnologie 24: 626-631.

Markager, S. and Sand-Jensen, K. 1990. Heterotrophic growth of *Ulva lactuca* (Chlorophyceae). – Journal of Phycology 26: 670-673.

Nielsen, S. L. and Sand-Jensen, K. 1990. Allometric scaling of maximal photosynthetic growth rate to surface/volume ratio. – Limnology and Oceanography 35: 177-181.

Rasmussen, L., Henriksen, D. and Sand-Jensen, K. 1990. Härjedalen i Jämtland: En naturpark ødelagt af menneskelig påvirkning. – Naturens Verden 1990: 188-193.

Riemann, B., Christoffersen, K., Jensen, H. J., Müller, J. P., Lindegaard, C. and Bosselmann, S. 1990. Ecological consequences of a manual reduction of roach and bream in a eutrophic, temperate lake. – Hydrobiologia 200/201: 241-250.

Riemann, B., Sørensen, H. M., Bjørnsen, P. K., Horsted, S. J., Jensen, L. M., Nielsen, T. G. and Søndergaard, M. 1990. Carbon budgets of the microbial food web in estuarine enclosures. – Marine Ecology Progress Series 65: 159-170.

Sand-Jensen, K., Jensen, L., Marcher, S. and Hansen, M. 1990. Pelagic metabolism in eutrophic coastal waters during a late summer period. – Marine Ecology Progress Series 65: 63-72.

Sand-Jensen, K. 1990. Epiphyte shading: Its role in resulting depth distribution of submerged aquatic macrophytes. – Folia Geobotanica et Phytotaxonomica 25: 315-320.

Sandlund, O. T., Jonsson, B. and Jónasson, P. M. 1990. Reproductive investment patterns in the polymorphic arctic charr of Thingvallavatn, Iceland. – Physiological Ecology, Japan 1: 383-392.

Søndergaard, M. 1990. The effect of environmental variables on release of extracellular organic carbon by freshwater macrophytes. – Folia Geobotanica et Phytotaxonomica 25: 321-332.

Søndergaard, M. 1990. Picophytoplankton in Danish lakes. – Verhandlungen Internationale Vereinigung für Theoretische und Angewandte Limnologie 24: 609-612.

Publications 1991 – 1996

Adams, M. S. and Sand-Jensen, K. 1991. Ecology of submersed aquatic macrophytes. – Aquatic Botany 41: 1-4.

Beer, S., Sand-Jensen, K., Madsen, T. V. and Nielsen, S. L. 1991. The carboxylase activity of *Rubisco* and the photosynthetic performance in aquatic plants. – Oecologia 87: 429-434.

Borum, J., Geertz-Hansen, O., Sand-Jensen, K. and Wium-Andersen, S. 1991. Eutrophication: Effects on marine plant communities. Nitrogen and phosphorus in fresh and marine waters. The N, P and organic matter research programme: 41-54.

Borum, J., Lomstein, B. and Riemann, B. 1991. Effekter af ændringer i kvælstof- og fosforbelastningen på fjorde og estuarier. Rapport fra konsensuskonference. Undervisningsministeriets forskningsafdeling. 15 pp.

Iversen, T. M., Thorup, J., Kjeldsen, K. and Thyssen, N. 1991. Spring bloom development of microbenthic algae and associated invertebrates in two reaches of a small lowland stream with contrasting sediment stability. – Freshwater Biology 26: 189-198.

Jacobsen, D. og Sand-Jensen, K. 1991. Vandplanter som føde. – Vand og Miljø 8: 407-409.

Jónasson, P. M. 1991. Einer Steemann Nielsen, 13. juni 1907 – 17. april 1989. – Det Kongelige Danske Videnskabernes Selskab. Oversigt over Selskabets virksomhed 1990-91: 135-141.

Kamp-Nielsen, L. 1991. I hvor høj grad tilbageholdes fosfor i de forskellige typer våd- og vandområder. – Rapport fra konsensuskonference. Undervisningsministeriets forskningsafdeling: 23 pp.

Madsen, T. V. and Sand-Jensen, K. 1991. Photosynthetic carbon assimilation in aquatic macrophytes. – Aquatic Botany 41: 5-40.

Nielsen, S. L., Gacia, E. and Sand-Jensen, K. 1991. Land plants of amphibious *Littorella uniflora* (L.) Aschers maintain utilization of CO_2 from the sediment. – Oecologia 88: 258-262.

Nielsen, S. L. and Sand-Jensen, K. 1991. Variation in growth rates of submerged rooted macrophytes. – Aquatic Botany 39: 109-120.

Perez, M., Romero, J., Duarte, C. M. and Sand-Jensen, K. 1991. Phosphorus limitation of *Cymodocea nodosa* growth. – Marine Biology 109: 129-133.

Prahl, C., Jeppesen, E., Sand-Jensen, K. and Iversen, T. M. 1991. A continuous-flow system for measuring *in vitro* oxygen and nitrogen metabolism in separated stream communities. – Freshwater Biology 26: 495-506.

Sand-Jensen, K. and Borum, J. 1991. Interactions among phytoplankton, periphyton and macrophytes in temperate freshwaters and estuaries. – Aquatic Botany 41: 137-175.

Sand-Jensen, K. and Madsen, T. V. 1991. Minimum light requirements of submerged freshwater macrophytes in laboratory growth experiments. – Journal of Ecology 79: 749-764.

Sand-Jensen, K., Geertz-Hansen, O. and Pedersen, O. 1991. Et encellet dyr med egen urtehave. – Naturens Verden 1991: 432-439.

Sand-Jensen, K., Iversen, T. M. and Lindegaard, C. 1991. Basisbog i ferskvandsøkologi. – Gads Forlag, København, Danmark. 63 pp.

Søndergaard, M. 1991. Phototrophic picoplankton in temperate lakes: Seasonal abundance and importance along a trophic gradient. – Internationale Revue der gesamten Hydrobiologie 76: 505-522.

Søndergaard, M., Jensen, L. M. and Ærtebjerg, G. 1991. Picoalgae in Danish coastal waters during summer stratification. – Marine Ecology Progress Series 79: 139-149.

Adalsteinsson, H., Jónasson, P. M. and Rist, S. 1992. Physical characteristics of Thingvallavatn, Iceland. – Oikos 64: 121-135.

Andersen, T. H., Hansen, H. O., Iversen, T. M., Jacobsen, D., Krøjgaard, L. and Poulsen, N. 1992. Growth and feeding of 0 + brown trout (*Salmo trutta* L.) introduced to two small Danish streams. – Archiv für Hydrobiologie 125: 339-346.

Antonsson, Ú. 1992. The structure and function of the zooplankton community in Thingvallavatn. – Oikos 64: 188-221.

Einarsson, E. 1992. Vascular plants of the Thingvallavatn area. – Oikos 64: 117-120.

Einarsson, M. Á. 1992. Climatic conditions of the Thingvallavatn area. – Oikos 64: 96-104.

Flindt, M. R. and Nielsen, J. B. 1992. Heterotrophic bacterial activity in Roskilde Fjord sediment during an autumn sedimentation peak. – Hydrobiologia 235/236: 283-293.

Flindt, M., Madsen, M. and Sørensen, P. S. 1992. Modelsystem for fjorde og bugter. – Havforskning fra Miljøstyrelsen nr. 9. 79 pp.

Frost-Christensen, H. and Sand-Jensen, K. 1992. The quantum efficiency of photosynthesis in macroalgae and submerged angiosperms. – Oecologia 91: 377-384.

Geertz-Hansen, O. and Sand-Jensen, K. 1992. Growth rates and photon yield of growth in natural populations of a marine macroalga *Ulva lactuca*. – Marine Ecology Progress Series 81: 179-183.

Geertz-Hansen, O. and Sand-Jensen, K. 1992. Søsalat er reguleret af græsning. – Vand og Miljø 9: 109- 113.

Haflidason, H. 1992. The recent sedimentation history of Thingvallavatn, Iceland. – Oikos 64: 80-95.

Hersteinsson, P. 1992. Mammals of the Thingvallavatn area. – Oikos 64: 396-404.

Jacobsen, D. and Sand-Jensen, K. 1992. Herbivory of invertebrates on submerged macrophytes from Danish freshwaters. – Freshwater Biology 28: 301-308.

Jespersen, A. M., Nielsen, J., Riemann, B. and Søndergaard, M. 1992. Carbon-specific phytoplankton growth rates: a comparison of methods. – Journal of Plankton Research 14: 637-648.

Jónasson, P. M. (ed.) 1992. Ecology of oligotrophic, subarctic Thingvallavatn. – Oikos 64: 1-439.

Jónasson, P. M. 1992. Iceland – an island astride the Mid-Atlantic Ridge. – Oikos 64: 9-13.

Jónasson, P.M. 1992. Thingvallavatn research history. – Oikos 64: 15-31.

Jónasson, P. M. 1992. Exploitation and conservation of the Thingvallavatn catchment area. – Oikos 64: 32-39.

Jónasson, P. M., Adalsteinsson, H. and Jónsson, G. S. 1992. Production and nutrient supply of phytoplankton in subarctic, dimictic Thingvallavatn, Iceland. – Oikos 64: 162-187.

Jónasson, P. M. 1992. The ecosystem of Thingvallavatn: a synthesis. – Oikos 64: 405-434.

Jónasson, P. M. 1992. Epilogue. Why study Thingvallavatn? – Oikos 64: 436.

Jónsson, G. S. 1992. Photosynthesis and production of epilithic algal communities in Thingvallavatn. – Oikos 64: 222-240.

Kairesalo, T., Jónsson, G. S., Gunnarsson, K., Lindegaard, C. and Jónasson, P. M. 1992. Metabolism and community dynamics within *Nitella opaca* (Charaphyceae) beds in Thingvallavatn. – Oikos 64: 241-256.

Kamp-Nielsen, L. 1992. Benthic-pelagic coupling of nutrient metabolism along an estuarine eutrophication gradient. – Hydrobiologia 235/236: 457-470.

Letarte, Y., Hansen, H. J., Søndergaard, M. and Pinel-Alloul, B. 1992. Production and abundance of different bacterial size-classes: relationships with primary production and chlorophyll concentration. – Archiv für Hydrobiologie 126: 15-26.

Lindegaard, C. 1992. The role of zoobenthos in energy flow in deep, oligotrophic Lake Thingvallavatn, Iceland. – Hydrobiologia 243/244: 185-195.

Lindegaard, C. 1992. Zoobenthos ecology of Thingvallavatn: vertical distribution, abundance, population dynamics and production. – Oikos 64: 257-304.

Magnússon, K. G. 1992. Birds of the Thingvallavatn area. – Oikos 64: 381-395.

Malmquist, H. J., Snorrason, S. S., Skúlason, S., Jonsson, B., Sandlund, O. T. and Jónasson, P. M. 1992. Diet differentiation in polymorphic arctic charr in Thingvallavatn, Iceland. – Journal of Animal Ecology 61: 21-35.

Markager, S., Jespersen, A. M., Madsen, T. V., Berdal, E. and Weisburd, R. 1992. Diel changes in dark respiration in a plankton community. – Hydrobiologia 238: 119-130.

Markager, S. and Sand-Jensen, K. 1992. Light requirements and depth zonation of marine macroalgae. – Marine Ecology Progress Series 88: 83-92.

Middelboe, M., Nielsen, B. and Søndergaard, M. 1992.

Bacterial utilization of dissolved organic carbon (DOC) in coastal waters – determination of growth yield. – Archiv für Hydrobiologie Beiheft Ergebnisse der Limnologie 37: 51-61.

Middelboe, M., Nielsen, B. and Søndergaard, M. 1992. Nye studier af bakterieplanktonets økologi. – Vand og Miljø 9: 115-118.

Nybroe, O., Christoffersen, K. and Riemann, B. 1992. Survival of *Bacillus licheniformis* in seawater model ecosystems. – Applied Environment Microbiology 58: 252-259.

Ólafsson, J. 1992. Chemical characteristics and trace elements of Thingvallavatn. – Oikos 64: 151-161.

Pedersen, M. F. and Borum, J. 1992. Nitrogen dynamics of eelgrass *Zostera marina* during a late summer period of high growth and low nutrient availability. – Marine Ecology Progress Series 80: 65-73.

Pedersen, O. and Sand-Jensen, K. 1992. Adaptions of submerged *Lobelia dortmanna* to aerial life form: morphology, carbon sources and oxygen dynamics. – Oikos 65: 89-96.

Sand-Jensen, K. and Madsen, T. V. 1992. Patch dynamics of the stream macrophyte, *Callitriche cophocarpa*. – Freshwater Biology 27: 277-282.

Sand-Jensen, K., Pedersen, M. F. and Nielsen, S. L. 1992. Photosynthetic use of inorganic carbon among primary and secondary water plants in streams. – Freshwater Biology 27: 283-293.

Sandlund, O. T., Gunnarsson, K., Jónasson, P. M., Jonsson, B., Lindem, T., Magnússon, K. P., Malmquist, H. J., Sigurjónsdottír, H., Skúlason, S., and Snorrason, S. S. 1992. The arctic charr *Salvelinus alpinus* in Thingvallavatn. – Oikos 64: 305-351.

Sandlund, O. T., Jónasson, P. M., Jonsson, B., Malmquist, H. J., Skúlason, S. and Snorrason, S. S. 1992. Three-spine stickleback *Gasterosteus aculeatus* in Thingvallavatn: habitat and food in a lake dominated by arctic charr *Salvelinus alpinus*. – Oikos 64: 365-370.

Snorrason, S. S., Jónasson, P. M., Jonsson, B., Lindem, T., Malmquist, H. J., Sandlund, O. T. and Skúlason, S. 1992. Population dynamics of the planktivorous arctic charr *Salvelinus alpinus* ("murta") in Thingvallavatn. – Oikos 64: 352-364.

Snorrason, S. S., Sandlund, O. T. and Jonsson, B. 1992. Production of fish stocks in Thingvallavatn, Iceland. – Oikos 64: 371-380.

Sveinbjörnsdóttír, Á. E. and Johnsen, S. J. 1992. Production of fish stocks in Thingvallavatn area. Groundwatter origin, age and evaporation models. – Oikos 64: 136-150.

Sædmundsson, K. 1992. Geology of the Thingvallavatn area. – Oikos 64: 40-68.

Søndergaard, M. and Borch, N. H. 1992. Decomposition of dissolved organic carbon (DOC) in lakes. – Archiv für Hydrobiologie Beiheft Ergebnisse der Limnologie 37: 9-20.

Thors, K. 1992. Bedrock, sediments, and faults in Thing-vallavatn. – Oikos 64: 69-79.

Thorsteinsson, I. and Arnalds, Ó. 1992. The vegetation and soils of the Thingvallavatn area. – Oikos 64: 105-116.

Andersen, T. H., Friberg, N., Hansen, H. O., Iversen, T. M., Jacobsen, D. and Krøjgaard, L. 1993. The effects of introduction of brown trout (*Salmo trutta* L.) on *Gammarus pulex* L. drift and density in two fishless Danish streams. – Archiv für Hydrobiologie 126: 361-371.

Christoffersen, K., Riemann, B., Klysner, A. and Søndergaard, M. 1993. Potential role of fish predation and natural populations of zooplankton in structuring a plankton community in eutrophic lake water. – Limnology and Oceanography 38: 561-573.

Dahl, A. and Winther, L. B. 1993. Life-history and growth of the prosobranch snail *Potamopyrgus jenkinsi* in Lake Erom, Denmark. – Verhandlungen Internationale Vereinigung für Theoretische und Angewandte Limnologie 25: 582-586.

Dall, P. C., Hamburger, K. and Lindegaard, C. 1993. Short cutting energy budgets of littoral zoobenthos communities. – Verhandlungen Internationale Vereinigung für Theoretische und Angewandte Limnologie 25: 567-573.

Enríques, S., Duarte, C. M. and Sand-Jensen, K. 1993. Patterns in decomposition rates among photosynthetic organisms: the importance of detritus C:N:P content. – Oecologia 94: 457-471.

Geertz-Hansen, O., Sand-Jensen, K., Hansen, D. F. and Christiansen, A. 1993. Growth and grazing control of abundance of the marine macroalga *Ulva lactuca* L. in a eutrophic Danish estuary. – Aquatic Botany 46: 101-109.

Hamburger, K., Dall, P. C. and Lindegaard, C. 1993. Studies on the energy metabolism of *Chironomus anthracinus* from the profundal zone of Lake Esrom, Denmark. – Verhandlungen Internationale Vereinigung für Theoretische und Angewandte Limnologie 25: 574-575.

Jacobsen, D. 1993. Trichopteran larvae as consumers of submerged angiosperms in running waters. – Oikos 67: 379-383.

Jónasson, P. M. 1993. Continental rifting. Energy pathways of two contrasting rift lakes. – Verhandlungen Internationale Vereinigung für Theoretische und Angewandte Limnologie 25: 1-14.

Jónasson, P. M. 1993. Lakes as a basic ressource for development: the role of limnology. Pages 9-26 *in* R. Pagnotta and A. Pugnetti, editors. Strategies for lake ecosystems beyond 2000. – Memorie dell' Istituto Italiano di Idrobiologia 52.

Kamp-Nielsen, L. and Flindt, M. R. 1993. On-line recording of porewater profiles from in situ dialysis. – Verhandlungen Internationale Vereinigung für Theoretische und Angewandte Limnologie 25: 151-156.

Kandel, A., Christoffersen, K. and Nybro, O. 1993. Filtration rates of *Daphnia cucullata* on *Alcaligenes eutrophus* JMP 134 estimated by a fluorescent antibody method. – FEMS Microbial Ecology 12: 1-8.

Lindegaard, C., Dall, P. C. and Hansen, S. B. 1993. Natural and imposed variability in the profundal fauna of Lake Esrom, Denmark. – Verhandlungen Internationale Vereinigung für Theoretische und Angewandte Limnologie 25: 576-581.

Lindegaard, C. and Mæhl, P. 1993. Abundance, population dynamics and production of Chironomidae (Diptera) in an ultraoligtrophic lake in South Greenland. – Netherlands Journal of Aquatic Ecology 26: 297-308.

Lindeneg, S. 1993. Application of the gut fluorescence method on the littoral herbivore, *Tinodes waeneri* (Trichoptera), in Lake Esrom, Denmark. – Verhandlungen Internationale Vereinigung für Theoretische und Angewandte Limnologie 25: 587-592.

Madsen, T. V., Sand-Jensen, K. and Beer, S. 1993. Comparison of photosynthetic performance and carboxylation capacity in a range of aquatic macrophytes of different growth forms. – Aquatic Botany 44: 373-384.

Middelboe, M. and Søndergaard, M. 1993. Bacterioplankton growth yield: Seasonal variations and coupling to substrate lability and ß-glucosidase activity. – Applied Environmental Microbiology 59: 3916-3921.

Nielsen, S. L. 1993. A comparison of aerial and submerged photosynthesis in some Danish amphibious plants. – Aquatic Botany 45: 27-40.

Nielsen, S. L. and Sand-Jensen, K. 1993. Photosynthetic implications of heterophylly in *Batrachium peltatum* (Schrank) Presl. – Aquatic Botany 44: 361-371.

Oertli, B. and Dall, P. C. 1993. Population dynamics and energy budget of *Nemoura avicularis* (Plecoptera) in Lake Esrom, Denmark. – Limnologica 23: 115-122.

Olesen, B. and Sand-Jensen, K. 1993. Seasonal acclimatization of eelgrass *Zostera marina* growth to light. – Marine Ecology Progress Series 94: 91-99.

Pedersen, M. F. and Borum, J. 1993. An annual nitrogen budget for a seagrass *Zostera marina* population. – Marine Ecology Progress Series 101: 169-177.

Pedersen, O. 1993. Long-distance water transport in aquatic plants. – Plant Physiology 103: 1369-1375.

Pedersen, O. and Sand-Jensen, K. 1993. Water transport in submerged macrophytes. – Aquatic Botany 44: 385-406.

Pedersen, O. and Sand-Jensen, K. 1993. Luftåndedræt hos vandplanter og vandinsekter. – Naturens Verden 1993: 260-267.

Riemann, B. and Christoffersen, K. 1993. Microbial trophodynamics in temperate lakes. – Marine Microbial Food Webs 7: 69-100.

Søndergaard, M. and Middelboe, M. 1993. Measurements of particulate organic carbon: a note on the use of glass fiber (GF/F) and anodisc filters. – Archiv für Hydrobiologie 127: 73-85.

Søndergaard, M. 1993. Organic carbon pools in two Danish lakes: Flow of carbon to bacterioplankton. – Verhandlungen Internationale Vereinigung für Theoretische und Angewandte Limnologie 25: 593-598.

Søndergaard, M. 1993. Næringssalte og regulering af vækst og biomasse hos akvatiske primærproducenter. – Tidsskrift for Landøkonomi 180: 118-125.

Agustí, S., Enríquez, S., Frost-Christensen, H., Sand-Jensen, K. and Duarte, C. M. 1994. Light harvesting among photosynthetic organisms. – Functional Ecology 8: 273-279.

Borum, J., Pedersen, M. F., Kær, L. and Pedersen, P. M. 1994. Vækst og næringsstofdynamik hos marine planter. – Havforskning fra Miljøstyrelsen 41. 59 pp.

Brodersen, K. 1994. Subfossile dansemyg i sø-sedimenter. – Miljøforskning 12: 12-15.

Christensen, P. B., Revsbech, N. P. and Sand-Jensen, K. 1994. Microsensor analysis of oxygen in the rhizosphere of the aquatic macrophyte *Littorella uniflora* (L.) Ascherson. – Plant Physiology 105: 847-852.

Christoffersen, K. 1994. Variations of feeding activities of heterotrophic nanoflagellates on picoplankton. – Marine Microbial Food Webs 8: 111-123.

Christoffersen, K. 1994. Picoplankton: biomasse, produktion og græsning af heterotrofe nanoflagellater. Pages 63-73 *in* B. Riemann, editor. Struktur og funktion af kystnære marine økosystemer. Vandkvalitetsinstituttet, ATV.

Dall, P. C. and Stańczykowska, A. 1994. Viborg jest miastem w Jutlandii. – Aura 11: 12.

Duarte, C. M., Marbá, N., Agawin, N., Cebrián, J., Enríquez, S., Fortes, M. D., Gallegos, M. E., Merino, M., Olesen, B., Sand-Jensen, K., Uri, J. and Vermaat, J. 1994. Reconstruction of seagrass dynamics: age determinations and associated tools for the seagrass ecologist. – Marine Ecology Progress Series 107: 195-209.

Flindt, M. R. 1994. Measurements of nutrient fluxes and mass balances by on-line situ dialysis in *Zostera marina* bed culture. – Verhandlungen Internationale Vereinigung für Theoretische und Angewandte Limnologie 25: 2259-2264.

Friberg, N. and Jacobsen, D. 1994. Feeding plasticity of two detritivore-shredders. – Freshwater Biology 32: 133-142.

Geertz-Hansen, O., Enríquez, S., Duarte, C. M., Agustí, S., Vaque, D. and Vidondo, B. 1994. Functional implications of the form of *Codium bursa*, a balloon-like Mediterranean macroalga. – Marine Ecology Progress Series 108: 153-160.

Hambuger, K., Dall, P. C. and Lindegaard, C. 1994. Energy metabolism of *Chironomus anthracinus* (Diptera: Chironomidae) from the profundal zone of Lake Esrom, Denmark, as a function of body size, temperature and oxygen concentration. – Hydrobiologia 294: 43-50.

Jacobsen, D. and Sand-Jensen, K. 1994. Invertebrate herbivory on the submerged macrophyte *Potamogeton perfoliatus* in a Danish stream. – Freshwater Biology 31: 43-52.

Jacobsen, D. and Sand-Jensen, K. 1994. Growth and energetics of a trichopteran larva feeding on fresh submerged and terrestrial plants. – Oecologia 97: 412-418.

Jacobsen, D. 1994. Food preference of the caddis larva *Anabolia nervosa* feeding on aquatic macrophytes. – Verhandlungen Internationale Vereinigung für Theoretische und Angewandte Limnologie 25: 2478-2481.

Jeppesen, E., Søndergaard, M., Lauridsen, T., Berg, S., Jacobsen, L. and Christoffersen, K. 1994. Mod klarvandede søer. – Miljøforskning 12: 8-11.

Jespersen, A. M. 1994. Comparison of $^{14}CO_2$ and $^{12}CO_2$ uptake and release rates in laboratory cultures of phytoplankton. – Oikos 69: 460-468.

Jónasson, P. M. 1994. In memoriam: Einer Steemann-Nielsen. – Archiv für Hydrobiologie 131: 253-254.

Lindegaard, C., Dall, P. C. and Jacobsen, D. 1994. Biologisk bedømmelse af vandløb forurenet med organisk stof. Kompendium til brug ved undervisningen i kursusmodulet "Biologisk bedømmelse af vandløbsforurening" ved Københavns Universitet. – Ferskvandsbiologisk Laboratorium, Københavns Universitet. 60 pp.

Lindegaard, C., Hamburger, K. and Dall, P. C. 1994. Population dynamics and energy budgets of *Marionina southerni* (Cernisvitov) (Enchytraeidae, Oligochaeta) in the shallow littoral of Lake Esrom, Denmark. – Hydrobiologia 278: 291-301.

Lindegaard, C. 1994. The role of zoobenthos in energy flow in two shallow lakes. – Hydrobiologia 275/276: 313-322.

Madsen, T. V. and Sand-Jensen, K. 1994. The interactive effects of light and inorganic carbon on aquatic plant growth. – Plant, Cell and Environment 17: 955-962.

Markager, S. 1994. Open-water measurement of areal photosynthesis in a dense phytoplankton community. – Archiv für Hydrobiologie 129: 405-424.

Markager, S. and Sand-Jensen, K. 1994. The physiology and ecology of ligth-growth relationship in macroalgae. – Progress in Phycological Research 10: 209-298.

Markager, S., Hansen, B. and Søndergaard, M. 1994. Pelagic carbon metabolism in a eutrophic lake during a clear-water phase. – Journal of Plankton Research 16: 1247-1267.

Mebus, J. R., Sand-Jensen, K. and Jespersen, T. S. 1994. Strømningsmønstre omkring planter i vandløb. – Vand og Jord 1: 172-176.

Michelsen, K., Pedersen, J., Christoffersen, K. and Jensen, F. 1994. Ecological consequences of food partitioning for the fish population structure in a eutrophic lake. – Hydrobiologia 291: 35-45.

Mortensen, E., Jeppesen, E., Søndergaard, M. and Kamp-Nielsen, L. (eds.) 1994. Nutrient dynamics and biological structure in shallow freshwater and brachish lakes. – Developments in Hydrobiology 94: 1-507.

Nygaard, G. 1994. A community of epiphytic diatoms living on low irradiances. – Nordic Journal of Botany. Section of phycology: 345-360.

Olesen, B. and Sand-Jensen, K. 1994. Patch dynamics of eelgrass *Zostera marina*. – Marine Ecology Progress Series 106: 147-156.

Olesen, B. and Sand-Jensen, K. 1994. Demography of shallow eelgrass (*Zostera marina*) populations – shoot dynamics and biomass development. – Journal of Ecology 82: 379-390.

Olesen, B. and Sand-Jensen, K. 1994. Biomass-density patterns in the temperate seagrass *Zostera marina*. – Marine Ecology Progress Series 109: 283-291.

Pedersen, M. F. 1994. Transient ammonium uptake in the macroalga *Ulva lactuca* (Chlorophyta): Nature, regulation and the consequences for choice of measuring technique. – Journal of Phycology 30: 980-986.

Pedersen, O. 1994. Acropetal water transport in submerged plants. – Botanica Acta 107: 61-65.

Perez, M., Duarte, C. M., Romero, J., Sand-Jensen, K. and Alcoverro, T. 1994. Growth plasticity in *Cymodocea nodosa* stands: the importance of nutrient supply. – Aquatic Botany 47: 249-264.

Ravn, H., Pedersen, M. F., Borum, J., Andary, C., Anthoni, U., Christoffersen, C. and Nielsen, P. H. 1994. Seasonal variation and distribution of two phenolic compounds, rosmarinic acid and caffeic acid, in leaves and roots-rhizomes of eelgrass (*Zostera marina* L.). – Ophelia 40: 51-61.

Sand-Jensen, K. 1994. Vandløbene og det fysiske miljø. – Miljøforskning 13: 23-25.

Sand-Jensen, K. 1994. Pumping iron in the Pacific. – Vand og Jord 1: 105.

Sand-Jensen, K. 1994. Synspunkt: Sø og å er andet end N og P. – Vand og Jord 1: 194.

Sand-Jensen, K., Nielsen, S. L., Borum, J. and Geertz-Hansen, O. 1994. Fytoplankton- og makrofytudvikling i danske kystområder. – Havforskning fra Miljøstyrelsen 30. 43 pp.

Sand-Jensen, K., Borum, J. and Geertz-Hansen, O. 1994. Resuspension of stofomsætning i Roskilde Fjord. – Havforskning fra Miljøstyrelsen 51. 69 pp.

Sand-Jensen, K., Borum, J. and Duarte, C. M. 1994. Havgræsserne i verdenshavene. – Naturens Verden 1994: 27-40.

Sand-Jensen, K., Jacobsen, D. and Duarte, C. M. 1994. Herbivory and resulting plant damage. – Oikos 69: 545-549.

Sand-Jensen, K., Geertz-Hansen, O., Pedersen, O. and Nielsen, H. S. 1994. Size dependence of composition, photosynthesis and growth in the colony-forming freshwater ciliate, *Ophrydium versatile*. – Freshwater Biology 31: 121-130.

Sand-Jensen, K. and Pedersen, M. F. 1994. Photosynthesis by symbiotic algae in the freshwater sponge, *Spongilla lacustris*. – Limnology and Oceanography 39: 551-561.

Schlütter, L. and Riemann, B. 1994. Anvendelse af algepigmenter til overvågning. – Vand og Jord 1: 257-260.

Snorrason, S. S., Skúlason, S., Jonsson, B., Malmquist, H. J., Jónasson, P. M., Sandlund, O. T. and Lindem, T. 1994. Trophic specialization in arctic charr *Salvelinus alpinus* (Pisces Salmoidae): morphological divergence and ontogenetic niche shifts. – Biological Journal of the Linnean Society of London 52: 1-18.

Snorrason, S. S., Malmquist, H. J., Jonsson, B., Jónasson, P. M., Sandlund, O. T. and Skúlason, S. 1994. Modification in life history characteristics of planktivorous arctic charr *Salvelinus alpinus* in Thingvallavatn, Iceland. – Verhandlungen Internationale Vereinigung für Theoretische und Angewandte Limnologie 25: 2108-2112.

Vaque, D., Agustí, S., Duarte, C. M., Enríquez, S. and Geertz-Hansen, O. 1994. Microbial heterotrophs within *Codium bursa*: a naturally isolated microbial food web. – Marine Ecology Progress Series 109: 275-282.

Søndergaard, M. 1994. Picoalger, de mindste primærproducenter. – Vand og Jord 1: 199-203.

Søndergaard, M. and Nielsen, J. T. 1994. Makrofyter – en kulstofkilde for mikrobiel omsætning. – Miljøforskning 13: 12-17.

Ahl, T., Christoffersen, K., Nybroe, O. and Riemann, B. 1995. A combined microcosm and mesocosm approach to examine factors affecting survival and mortality of *Pseudomonas fluorescens* Ag1 in seawater. – Microbiology 17: 107-116.

Brodersen, K. P. 1995. The effect of wind exposure and filamentous algae on the distribution of surf zone macroinvertebrates in Lake Esrom, Denmark. – Hydrobiologia 297: 131-148.

Christoffersen, K., Ahl, T. and Nybroe, O. 1995. Grazing of nonindigenous bacteria by nano-sized protozoa in a natural coastal system. – Microbial Ecology 30: 67-68.

Dall, P. C. 1995. Commonly used methods for assessment of water quality. Pages 49-70 *in* M. J. Toman and F. Steinman, editors. Biological assessment of stream water quality (Theory, application and comparison of methods). – Proceedings of TEMPUS workshop, University of Ljubljana, Slovenien.

Dall, P. C. and Lindegaard, C. (eds.) 1995. En oversigt over danske ferskvandsinvertebrater til brug ved bedømmelse af forureningen i søer og vandløb. – Ferskvandsbiologisk Laboratorium, Københavns Universitet. 240 pp.

Dall, P. C., Friberg, N., Lindegaard, C. and Toman, M. J. 1995. A practical guide for biological assessment of stream water quality. Pages 97-126 *in* M. J. Toman and F. Steinman, editors. Biological assessment of stream water quality (Theory, application and comparison of methods). – Proceedings of TEMPUS workshop, University of Ljubljana, Slovenien.

Dall, P. C., Lindegaard, C. and Toman, M. J. 1995. Biological assessment of stream water quality. Result section and summary of discussion. Pages 127-145 *in* M. J. Toman and F. Steinman, editors. Biological assessment of stream water quality (Theory, application and comparison of methods). – Proceedings of TEMPUS workshop, University of Ljubljana, Slovenien.

Duarte, C., Sand-Jensen, K., Nielsen, S. L., Enríquez, S. and Agustí, S. 1995. Comparative functional plant ecol-

ogy: rationale and potentials. – TREE, Trends in Ecology and Evolution 10: 253-257.

Enríques, S., Duarte, C. M. and Sand-Jensen, K. 1995. Patterns in the photosynthetic metabolism of Mediterranean macrophytes. – Marine Ecology Progress Series 119: 243-252.

Frost-Christensen, H. and Sand-Jensen, K. 1995. Comparative kinetics of photosynthesis in floating and submerged *Potamogeton* leaves. – Aquatic Botany 51: 121-134.

Hamburger, K., Dall, P. C. and Lindegaard, C. 1995. Effects of oxygen deficiency on survival and glycogen content of *Chironomus anthracinus* (Diptera, Chironomidae) under laboratory and field conditions. – Hydrobiologia 297: 187-200.

Hansen, B. and Christoffersen, K. 1995. Specific growth rates of heterotrophic plankton organisms in a eutrophic lake during a spring bloom. – Journal of Plankton Research 17: 413-430.

Hein, M., Pedersen, M. F. and Sand-Jensen, K. 1995. Size-dependent nitrogen uptake in micro- and macroalgae. – Marine Ecology Progress Series 118: 247-253.

Hein, M. and Riemann, B. 1995. Nutrient limitation of phytoplankton biomass or growth rate: an experimental approach using marine enclosures. – Journal of Experimental Marine Biology and Ecology 188: 167-180.

Jacobsen, D. and Sand-Jensen, K. 1995. Variability of invertebrate herbivory on the submerged macrophyte *Potamogeton perfoliatus*. – Freshwater Biology 34: 357-365.

Jacobsen, D. and Friberg, N. 1995. Food preference of the trichopteran larva *Anabolia nervosa* from two streams with different food availability. – Hydrobiologia 308: 139-144.

Jespersen, A., Søndergaard, M., Richardson, K. and Riemann, B. 1995. Måling af primærproduktion hos marint fytoplankton. – Havforskning fra Miljøstyrelsen 55. 53 pp.

Jónasson, P. M. 1995. G. S. Hutchinson, 1993. A treatise on limnology. V. 4. The zoobenthos. – Limnology and Oceanography 40: 839-841.

Jørgensen, S. E., Kamp-Nielsen, L., Ipsen, L. G. S. and Nicolaisen, P. 1995. Lake restoration using a reed swamp to remove nutrients from non-point sources. – Wetlands Ecology and Management 3: 87-95.

Lindegaard, C. 1995. Classification of water bodies and pollution. Pages 385-404 *in* P. Armitage, P. Cranston, and L. C. V. Pinder, editors. Chironomidae, Biology and Ecology of non-biting Midges. Chapman and Hall, UK.

Lindegaard, C. 1995. Chironomidae (Diptera) of European cold springs and factors influencing their distribution. – Journal of Kansas Entomological Society 68/2 Suppl: 108-131.

Lindegaard, C. 1995. The faunas response on human impacts in running waters with special reference to lowland conditions. Pages 11-48 *in* M. J. Toman and F.

Steinman, editors. Biological assessment of stream water quality (Theory, application and comparison of methods). – Proceedings of TEMPUS workshop, University of Ljubljana, Slovenien.

Lindegaard, C. and Brodersen, K. P. 1995. Distribution of Chironomidae (Diptera) in the river continuum. Pages 257-271 *in* P. Cranston, editor. Chironomids: From genes to ecosystems. CSIRO Publications, Melbourne, Australia.

Middelboe, M., Borch, N. H. and Kirchman, D. L. 1995. Bacterial utilization of dissolved free amino acids, dissolved combined amino acids and ammonium in the Delaware Bay estuary: effects of carbon and nitrogen limitation. – Marine Ecology Progress Series 128: 109-120.

Middelboe, M. and Søndergaard, M. 1995. Concentration and bacterial utilization of submicron particles and dissolved organic carbon in lakes and a coastal area. – Archiv für Hydrobiologie 133: 129-147.

Middelboe, M., Søndergaard, M., Letarte, Y. and Borch, N. H. 1995. Attached and free-living bacteria: Production and polymer hydrolysis during a diatom bloom. – Microbial Ecology 29: 231-248.

Pedersen, M. F. 1995. Nitrogen limitation of photosynthesis and growth: comparison across aquatic plant communities in a Danish estuary (Roskilde Fjord). – Ophelia 41: 261-272.

Pedersen, M. F. 1995. Næringshusholdning hos marine planter. – Vand og Jord 2: 95-98.

Pedersen, O., Sand-Jensen, K. and Revsbeck, N.P. 1995. Diel pulses of O_2 and CO_2 in sandy lake sediments inhabited by *Lobelia dortmanna*. – Ecology 76: 1536-1545.

Sand-Jensen, K. 1995. Biologisk Orientering. Nøgleorganismer og biologiske ingeniører. – Vand og Jord 2: 249-250.

Sand-Jensen, K. 1995. Nye bøger om vandløbsøkologi. – Vand og Jord 2: 89.

Sand-Jensen, K. 1995. Furesøen gennem 100 år. – Naturens Verden 1995: 176-188.

Sand-Jensen, K. and Pedersen, O. 1995. *Lobelia* – en vedvarende fascination. – Vand og Jord 2: 114, 129.

Sand-Jensen, K., Brodersen, P., Madsen, T. V., Jespersen, T. S. and Kjøller, B. 1995. Planter og CO_2 overmætning i vandløb. – Vand og Jord 2: 72-77.

Søndergaard, M., Hansen, B. and Markager, S. 1995. Dynamics of dissolved organic carbon lability in a eutrophic lake. – Limnology and Oceanography 40: 46-54.

Søndergaard, M. and Middelboe, M. 1995. A cross-system analysis of labile dissolved organic carbon. – Marine Ecology Progress Series 118: 283-294.

Vermeulen, A. C. 1995. Elaborating chironomid deformities as bioindicators of toxic sediment stress: The potential application of mixture toxicity concepts. – Annales Zoologica Fennici 32: 265-285.

Borum, J. and Sand-Jensen, K. 1996. Is total primary production in shallow coastal marine waters stimulated by nitrogen loading ? – Oikos 76: 406-410.

Borum, J. 1996. Shallow waters and land/sea boundaries. – Coastal and Estuarine Studies 52: 179-203.

Christensen, P. B., Møhlenberg, F., Lund-Hansen, L. C., Borum, J., Christiansen, C., Larsen, S. E., Hansen, M. E., Andersen, J. and Kirkegaard, J. 1996. Havmiljøet under forandring. – Havforskning fra Miljøstyrelsen 61. 120 pp.

Christoffersen, K. 1996. Effect of microcystin on growth of single species and on mixed natural populations of heterotrophic nanoflagellates. – Natural Toxins 4: 215-220.

Christoffersen, K. 1996. Ecological implications of cyanobacterial toxins in aquatic food webs. – Phycologia 35: 42-50.

Christoffersen, K. and Olrik, K. 1996. Giftige alger i Esrum Sø og Å. – Vand og Jord 3: 21-24.

Christoffersen, K., Bernard, C. and Ekebom, J. 1996. A comparison of the ability of different heterotrophic nanoflagellates to incorporate dissolved macromolecules. – Archiv für Hydrobiologie, Advances in Limnology 48: 73-84.

Dall, P. C. and Hamburger, K. 1996. Recruitment and growth of *Dreissena polymorpha* in Lake Esrom, Denmark. – Limnologica 26: 27-37.

Dall, P. C. 1996. Bevar os vel. – Vand og Jord 3: 176.

Dall, P. C. and Gørtz, P. 1996. Forureningsbedømmelse med raflebæger. – Vand og Jord 3: 204-206.

Duarte, C. M. and Sand-Jensen, K. 1996. Nutrient constraints on establishment from seed and on vegetative expansion of the Mediterranean seagrass *Cymodocea nodosa*. – Aquatic Botany 54: 279-286.

Enríques, S., Duarte, C. M., Sand-Jensen, K. and Nielsen, S. L. 1996. Broad-scale comparison of photosynthetic rates across phototrophic organisms. – Oecologia 108: 197-206.

Hamburger, K., Lindegaard, C. and Dall, P. C. 1996. The role of glycogen during the ontogenesis of *Chironomus anthracinus* (Chironomidae, Diptera). – Hydrobiologia 318: 51-59.

Jónasson, P. M. 1996. Presidential Address. Limits for life in the lake ecosystem. – Verhandlungen Internationale Vereinigung für Theoretische und Angewandte Limnologie 26: 1-33.

Krause-Jensen, D. and Sand-Jensen, K. 1996. Primærproduktion i vand. – Vand og Jord 3: 260-264.

Lyche, A., Andersen, T., Christoffersen, K., Hessen, D. O., Hansen, P. H. B. and Klysner, A. 1996. Zooplankton as sources and sinks for phosphorus in mesocosms in a P-limited lake. – Limnology and Oceanography 41: 460-474.

Lyche, A., Andersen, T., Christoffersen, K., Hessen, D. O., Hansen, P. H. B. and Klysner, A. 1996. The fate of primary production and the role of consumers in carbon cycling in a mesotrophic lake. – Limnology and Oceanography 41: 475-489.

Marbà, N., Duarte, C. M., Cebrián, J., Gallegos, M. E., Olesen, B. and Sand-Jensen, K. 1996. Growth and population dynamics of *Posidonia oceanica* on the Spanish Mediterranean coast: elucidating seagrass decline. – Marine Ecology Progress Series 137: 203-213.

Markager, S. and Sand-Jensen, K. 1996. Implications of thallus thickness for growth-irradiance relationships of marine macroalgae. – European Journal of Phycology 31: 79-87.

Nielsen, S. L., Enríquez, S., Duarte, C. M. and Sand-Jensen, K. 1996. Scaling maximum growth rates across photosynthetic organisms. – Functional Ecology 10: 167-175.

Nygaard, G. 1996. Temporal and spatial development of individual species of plankton algae from European lakes. – Hydrobiologia 332: 71-91.

Pedersen, M. F., Borum, J. and McGlathery, K. J. 1996. Changes in intracellular nitrogen pools and feedback controls on nitrogen uptake in *Chaetomorpha linum* (Chlorophyta). – Journal of Phycology 32: 393-401.

Pedersen, M. F. and Borum, J. 1996. Phosphorus recycling in the seagrass *Zostera marina* L. – Biology and Ecology of Shallow Coastal Waters. 28 EMBS Symposium: 45-50.

Pedersen, M. F. and Borum, J. 1996. Nutrient control of algal growth in estuarine waters. Nutrient limitation and the importance of nitrogen requirements and nitrogen storage among phytoplankton and species of macroalgae. – Marine Ecology Progress Series 142: 261-272.

Pedersen, O. 1996. De højere planters tilpasninger til livet i vand. Pages 10-15 *in* G. Kyhn, editor. Carlsbergfondets Årskrift – Carlsbergfondet, København, Denmark.

Pedersen, O. 1996. Das Geheimnis des Transportes von Wasser und Gas in Wasserpflanzen. – Aquarium Heute 14: 374-377.

Pedersen, O. 1996. Il mistero del trasporto di acqua e gas nelle piante acquatiche. – Aquarium Oggi 3. 42-45.

Riis, T., Sand-Jensen, K. and Jørgensen, T. B. 1996. Danmarks reneste sø gennem 40 år. – Vand og Jord 3: 199-203.

Sand-Jensen, K. 1996. Fra mikroalger til bøgetræer – overordnede mønstre i planteriget. – Naturens Verden 1: 28-40.

Sand-Jensen, K. and Mebus, J. R. 1996. Fine-scale patterns of water velocity within macrophyte patches in streams. – Oikos 76: 169-180.

Sand-Jensen, K. and Lindegaard, C. 1996. Økologi i søer og vandløb. – G. E. C. Gads Forlag, København, Danmark. 188 pp.

Sand-Jensen, K. 1996. Biologisk Orientering. Love for biologisk variation. – Vand og Jord 3: 174-175.

Sand-Jensen, K., Mebus, J. and Madsen, T. V. 1996. Vandplanter påvirker strøm og sediment. – Vand og Jord 3: 191-195.

Søndergaard, M. 1996. Lake Veluwe. – Aquatic Botany 55: 145-147.

Sørensen, O. and Christoffersen, K. 1996. Algetoksin i søer. – Vand og Jord 3: 17-20.

Sørensen, O. 1996. *Artemia* bioassay og algetoksiner. – Vand og Jord 3: 212-217.

Vermulen, A., Dall, P. C. and Hansen, F. G. 1996. Screening for sære tænder. – Vand og Jord 3: 83-86.

Williams, P. J. le, Robinson, C., Søndergaard, M., Jespersen, A. M., Bentley, T. L., Lefevre, D., Richardson, K. and Riemann, B. 1996. Algal ^{14}C and total carbon metabolisms. 2. Experimental observations with the diatom *Skeletonema costatum*. – Journal of Plankton Research 18: 1961-1974.